Contents

Preface   *ix*

I. INTRODUCTION

1. The Importance of Sedimentary Rocks   **3**

The History of the Earth   3
Utility to Man   6
Selected References   8

2. The Present as the Key to the Past   **9**

A Rational Approach to Sandstone, Shale, and Limestone   9
Sedimentation Rates and the Stratigraphic Record   13
Selected References   14

II. SEDIMENTS, TIME, AND THE STRATIGRAPHIC RECORD

3. The Properties of Sediments and Sedimentary Rocks   **17**

The Basic Elements of Classification
of Sedimentary Rocks   17
Size Distribution   28
Stratification   30
Primary Structures   30
Sequential Relationships in Sedimentary Rocks   39
Electrical Properties of Sedimentary Rocks   43
Citations and Selected References   45

4. Dynamics in the Stratigraphic Record   48

Isostasy and the Stratigraphic Record   49
Global Dynamics of the Lithosphere   60
Eustatic Sea Level Fluctuations   69
Transgression and Regression in the Stratigraphic Record   71
Citations and Selected References   72

5. Geologic Framework of Sediment Accumulation   74

The Stable Craton   74
The Mobile Belts   76
The Coastal Plain and Continental Margin   82
Citations and Selected References   88

6. Physical Stratigraphy   90

The Basic Rationale of Physical Stratigraphy   90
The Utility of Physical Stratigraphy   92
Citations and Selected References   95

7. Biostratigraphy: Introduction to Temporal Correlation   96

Rock Units, Time-Rock Units, and Time Units   97
Fossils and Time-Rock Units   98
A Biostratigraphic Overview of Evolution   98
The Complications of Niche and Biogeography   105
Biostratigraphic Nomenclature   110
Citations and Selected References   113

8. Time-Stratigraphic Correlation based on Physical Events of Short Duration   114

Magnetic Stratigraphy   114
Time-Stratigraphic Correlation based on Climatic Fluctuations   119
Time-Stratigraphic Correlation based on Transgressive-Regressive Events   121
Rapid Depositional Events   123
Citations and Selected References   123

9. Absolute Time in the Stratigraphic Record   125

Radioactive Isotopes: The Clocks of Geologic Time   125
Radiometric Dating of the Stratigraphic Record   127

Dynamic Stratigraphy

Dynamic Stratigraphy

AN INTRODUCTION TO SEDIMENTATION AND STRATIGRAPHY

ROBLEY K. MATTHEWS

Department of Geological Sciences
Brown University

PRENTICE-HALL, INC., *Englewood Cliffs, New Jersey*

Library of Congress Cataloging in Publication Data

MATTHEWS, ROBLEY K.
 Dynamic stratigraphy.

  Contains bibliographies
  1. Geology, Stratigraphic. 2. Sedimentation
and deposition. I. Title.
QE651.M354   551.7   73-11050
ISBN 0-13-222273-6

PRENTICE-HALL INTERNATIONAL, INC., *London*
PRENTICE-HALL OF AUSTRALIA, PTY, LTD., *Sydney*
PRENTICE-HALL OF CANADA, LTD., *Toronto*
PRENTICE-HALL OF INDIA PRIVATE LIMITED, *New Delhi*
PRENTICE-HALL OF JAPAN, INC., *Tokyo*

The Geologic Time Scale   132
Citations and Selected References   132

III. SEDIMENTARY ENVIRONMENTS
FROM THE MOUNTAINS TO THE DEEP SEA

10. *Clastic Sedimentation in Stream Environments*   137

Recent Clastic Sedimentation in Stream Environments   137
A Basic Model
for Clastic Sedimentation in Stream Environments   155
An Ancient Example: The Shawangunk Conglomerate
and Tuscarora Quartzite, Silurian of the Pennsylvania,
New Jersey and New York   162
An Ancient Example: Interaction between Meandering
Stream Cycles and Regional Submergence,
the Catskill "Delta," Devonian of New York State   168
Citations and Selected References   171

11. *Clastic Sedimentation in Coastal Environments*   173

Recent Clastic Sedimentation in Coastal Environments   174
A Basic Model for Clastic Sedimentation
in Coastal Environments   194
An Ancient Example: Bottomset to Meandering
Stream Transition. Reedsville Shale—Oswego Sandstone,
Upper Ordovician of Central Pennsylvania   204
An Ancient Example: Coastal Clastic Sediments
in the Subsurface Lower Oligocene
of Southeast Texas   208
An Ancient Example: A Stratigraphic Oil Trap
in Coastal Clastic Sands of the Subsurface
Upper Cretaceous of New Mexico   214
Citations and Selected References   220

12. *Carbonates of the Shelf-Margin*
and Subtidal Shelf Interior   224

Recent Carbonate Sedimentation   226
A Basic Model for Shelf-Margin
and Subtidal Shelf-Interior Sedimentation   251
An Ancient Example: Pleistocene Coral Reefs
of Barbados, the West Indies   259
Citations and Selected References   264

13. Carbonates and Evaporites of the Intertidal and Supratidal Shelf Interior   *268*

Recent Intertidal and Supratidal Environments   268
Continuing Development of a Basic Model
for Carbonate and Evaporite Sedimentation   276
An Ancient Example:
The Helderberg Group, Devonian of New York   277
An Ancient Example:
The Permian of West Texas   285
Citations and Selected References   289

14. Shelf-to-Basin Transitions at Continental Margins   *292*

Recent Sedimentation along Continental Margins   292
A Basic Model for Shelf-to-Basin Transition
at the Continental Margin   306
An Ancient Example: The Taconic Sequence,
Cambro-Ordovician, Eastern New York State   309
Citations and Selected References   314

IV. CYCLICITY IN THE STRATIGRAPHIC RECORD

15. The Quaternary as the Key to the Past   *321*

Early Interest in Pleistocene Geology   322
Pleistocene History of the Ocean Basins   322
The Milankovitch Hypothesis
Concerning Pleistocene Climatic Fluctuations   329
Citations and Selected References   333

16. Cyclic Sedimentation in Paleozoic Epeiric Seas of Central North America   *334*

Cyclothems of Eastern Kansas   334
A Model for Clear-Water Sedimentation in Epeiric Seas   337
An Alternate View of the Kansas Cyclothem   342
Citations and Selected References   342

17. The Multimodel Approach to Cyclic Sedimentation 344

Stratigraphic Relationships of the Shelf-to-Basin Transition,
Guadalupian (Middle Permian) of West Texas 344
A Multimodel Approach to the Middle Permian of West Texas 347
The Scale of Cyclicity in the Middle Permian of West Texas 347
Citations and Selected References 352

Glossary 355

Index 361

Preface

There are commonly three stages to the development of a field of science. First, the identification and cataloging of the fundamental building blocks that are to be the subject of this field. Second, the identification and cataloging of the major configurations that exist among the fundamental materials. Finally, the consideration of the kinetics by which these configurations were achieved. Upon reaching the kinetics level, all aspects of the science can be considered simultaneously for the first time.

The field of Global Tectonics has recently passed through one of these development cycles. Continental crust, oceanic crust, mid-ocean ridges, deep sea trenches, and velocity structure of the upper mantle were the building blocks. Global seismicity and magnetic patterns along mid-ocean ridges revealed the major configuration of lithosphere plates. Dating by radiometric methods, magnetic stratigraphy and biostratigraphy revealed kinetics of plate motions, both present and past. Thus, the years since the mid-sixties have given us a new, dynamic framework within which to consider the stratigraphic record.

Stratigraphy itself has been moving systematically toward its own stage three in a somewhat slower fashion. We have cataloged the sediment types, the formations, and the biostratigraphic time framework. We have seen that map configuration of Recent sedimentary facies can often be related to facies configurations in Ancient lithologic units. We are beginning to think of lithologic sequences in terms of sedimentation rates, rates of eustatic sea level fluctuation, and rates of tectonic deformation.

If we are to fully appreciate the stratigraphic record, all aspects of the science must be considered within the dynamic context. This book attempts to pull together the essentials of (a) dynamics of Recent sedimentation, (b) dynamics of tectonism on the present earth's surface, and (c) dynamics of Quaternary eustatic sea-level fluctuations. With these basic input parameters, we begin to devise models which generate various stratigraphic relationships from the interplay of the various rates involved. Examples from the stratigraphic record tend to confirm the potential utility of such models.

This book is intended as an introduction suitable for all earth science majors, not just for those anticipating a career in stratigraphic geology. While it is important that future stratigraphers get a start in the right directions, it is equally important that future geophysicists, meteorologists, etc. acquire confidence that stratigraphers really can make accurate statements about the history of the earth. The book is designed to instill such a confidence; *not* that the student knows the stratigraphic record, but rather that *much can be known* if stratigraphers approach the record properly.

This book was developed for a one-semester course intended for sophomores and juniors. The format of Sections III and IV is easily expandable by addition of more complicated Ancient examples. In this manner, the book could serve as a focal point for a two-semester undergraduate course or Sections III and IV could serve as the starting point for a graduate-level course.

Finally, some important topics in stratigraphy are omitted. Lacustrine deposits, deep-water evaporites, flysch-molasse, glacial sediments, and desert sediments simply do not fit the format of Section III. It is suggested that these topics be deferred to a more advanced stratigraphy course.

Acknowledgements. To a large extent, this book is the product of Brown University; it would not have come out the same if I had written it anywhere else. I am particularly grateful for the interaction with W. M. Chapple, B. J. Giletti, J. Imbrie, L. F. Laporte, and T. A. Mutch.

Preliminary drafts of the manuscript, or portions of it, were reviewed by L. V. Benson, W. M. Chapple, R. S. Harrison, S. I. Husseini, G. deV. Klein, K. J. Mesolella, R. C. Murray, T. A. Mutch, N. D. Smith, and R. P. Steinen, to whom I express sincere thanks. Nevertheless, I accept full responsibility for any errors or misconceptions the book may contain.

The manuscript was typed primarily by Margaret T. Cummings. J. A. Creaser and Paul R. Jones contributed significantly to the preparation and organization of illustrative materials.

Finally, I acknowledge with thanks the sufferings of Nancy, Betty, Gretchen, Sandra, and Charles throughout the writing process.

R. K. Matthews

Dynamic Stratigraphy

Section *I*

Introduction

Chapter 1 examines the fundamental motivations for man's interest in sedimentation and stratigraphy. Chapter 2 attempts to demonstrate that the stratigraphic record can be approached on a rational basis. We shall note, however, that the situation becomes rather complicated. These complications must be brought under our intellectual control if we are to explore the interesting questions posed in Chapter 1. Section II is intended to provide some necessary background. Sections III and IV go on to investigate Recent sedimentary environments and to apply this knowledge, along with sound stratigraphic principals, to specific sequences of Ancient rocks.

The Importance of Sedimentary Rocks 1

Man's interest in sedimentary rocks can be conveniently divided into two parts: a curiosity about the history of the earth and a desire to utilize the natural resources contained within the sediments.

The History of the Earth

The written record of man's experience with the earth is extremely limited. In the New World, it is measured in 10^2 years, in the eastern Mediterranean, 10^4 years. Yet evidences in the stratigraphic record, which we are about to discuss, lead us to recognize that man's history on earth goes back as far as 10^6 years. Indeed, complex forms of life have existed on the face of the earth for 10^8 years. Thus, if we should for any reason wish to consider the history of the planet, Earth, we must look beyond the written record of man's personal experience with the planet, for this written record covers only a trivial amount of time.

The stratigraphic record, the archives of earth history. Sediments have been deposited in various environments throughout the history of the earth. These sediments, today preserved as sequences of sedimentary rocks, provide a spectacular wealth of information concerning earth history. Consider Figure 1.1, for example. Limestones of Mesozoic age overlie an angular unconformity. Alternating layers of sandstone and shale of Paleozoic age underlie the unconformity. If we know how to recognize sedimentary environments in the stratigraphic record, we have a glimpse here of the various environments that have occupied this spot on the face of the earth at varying times in its history. Furthermore, the rocks contain fossils.

Figure 1.1 Photograph of an outcrop, Sahara desert, Libya, North Africa. Paleozoic sandstones and shales are unconformably overlain by Mesozoic carbonate rocks. The area is today a desert. Thus, the rocks provide a record of changing conditions on the face of the earth.

Thus, we have also a record of the kinds of life that existed in these various environments at the various times in earth history. Finally, the existence of the angular unconformity records structural deformation of this portion of the earth's crust at some time in the past.

Figure 1.1 gives us one small glimpse at the history of one small spot on the face of the earth. How does that history fit together with the history of other areas? How do the life-forms preserved in these rocks fit together with life-forms preserved in other rocks? These are the broad questions that first led man to intellectual consideration of the stratigraphic record and then to the accumulation of a data base sufficiently large that he could begin to ask important questions concerning the history of earth.

Recognition of large-scale earth processes. Life-forms occupying the various environments on the face of the earth have evolved with time. The orientation of the earth's magnetic field has from time to time become reversed. Both of these interesting facts are the subject of considerable scholarly activity, and both sub-jects have a voluminous literature. We shall discuss them in more detail later. For the time being, let us accept them as fact and note how stratigraphy can be used to document a large-scale earth process.

Ocean-floor spreading is one of the most gigantic and all-encompassing earth processes yet documented by man. For the present, let us consider the process only as an example of how stratigraphic data can aid in our understanding of the dynamics of the earth. Figure 1.2 schematically depicts two types of stratigraphic data that allow documentation of the process of ocean-floor spreading.

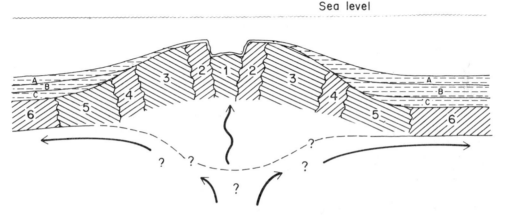

Figure 1.2 Schematic cross section depicting stratigraphic evidence concerning the generation of new oceanic crust along mid-ocean ridges. The crust itself, labeled (1–6), acquired its magnetic properties at the time of formation. Thus, the various segments of crust possess unique magnetic properties indicative of the time when that particular segment of crust was formed. Magnetic surveys, such as indicated in Figure 4.10, may therefore be interpreted in the light of the earth's magnetic history and the sea-floor spreading hypothesis. Biostrati-graphic determination of the relative age of deep-ocean sediments overlying the crust provides an independent check on the sea-floor spreading hypothesis. In this schematic diagram, the age of sediments overlying the basalt crust becomes younger toward the center of the ridge.

The late Cenozoic magnetic history of the earth was first worked out in vol-canic stratigraphic sequences on mid-ocean islands. Then it was recognized that magnetic patterns in the sea floor could be explained by continuous generation of new oceanic crust beneath mid-ocean ridges and rises. As new crust cooled, it took on its magnetic properties within the field existing at the time of cooling. Thus, as the magnetic field reversed, each new generation of oceanic crust took on magnetic properties that were distinct from the crust preceding it.

A further test of the concept of ocean-floor spreading is provided by the bio-stratigraphic estimation of the age of the sediment overlying oceanic crust. As indicated in Figure 1.2, it can be shown that oceanic crust closer to the mid-ocean ridge is overlain by the younger sediment, whereas oceanic crust farther removed from the ridge is overlain by older sediments.

In this manner, sea-floor spreading rates of centimeters to tens of centimeters per year are estimated. Such data, for example, lead us to understand earthquakes in the context of lithosphere dynamics on a global scale. Earthquakes are not "freaks of nature" or "acts of God"; they are simply the inevitable and recurring consequences of a very large-scale earth process. Clearly, an understanding of such large-scale earth processes is important and relevant to man.

Dynamics of "the environment." To many people "the environment" seems both important and relevant, but what can they do about it? Once again, we must note that man's experience on earth is limited to an extremely short time. Indeed, we face numerous environmental hazards, the likes of which man has never experienced. Consider for example the Antarctic Ice Cap.

The volume of ice perched above sea level on the Antarctic continent is sufficient to raise the sea level by some 50 meters if it should melt or slide into the sea. Throughout the world, many large cities are built at or near sea level. A 50-meter sea-level rise would be an environmental problem of the first magnitude! Is this rise likely to happen? That is essentially a very large-scale engineering problem, and the data are not conclusive. Has such a rise ever happened in the past? Now there is a stratigraphic question that we might be able to answer! The deep-sea sediments adjacent to the Antarctic continent would contain the record of such an event, had it occurred. Piston cores and rotary drill cores allow us access to the stratigraphic record contained in these and other deep-sea sediments. Work continues on this fascinating question.

Utility to Man

Natural processes, acting over large periods of geologic time, have concentrated many of the raw materials utilized by man. Consider for example common table salt. Sodium chloride is a major component of seawater. It would be a simple matter to design a processing plant to evaporate seawater and recover the sodium chloride for man's use. However, nature has already constructed such "facilities" numerous times in the past. The stratigraphic record contains enough available rock salt to serve our needs for some time to come.

Fossil fuels. The industrialized Western world runs on fossil fuels: coal, oil, and natural gas. Petroleum exploration is particularly demanding upon the sedimentologist and the stratigrapher. Modern drilling techniques allow us to probe 5 to 8 kilometers into the earth in search of concentrations of hydrocarbons. Earth processes and earth history must be unraveled and shown to have occurred in a sequence that experience tells us is favorable for petroleum accumulation (see

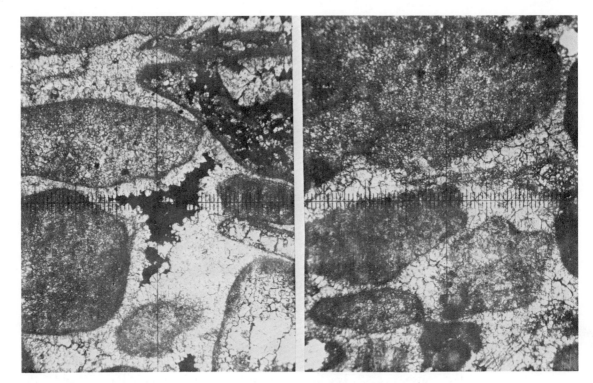

Figure 1.3 Photomicrographs of a core of carbonate rocks in an oil field. Both sediments were deposited as relatively clean, washed skeletal sands. The rock on the left has its pore space filled with petroleum, whereas the rock on the right has its pore space completely filled with calcite cement. The former rock type will produce oil; the latter will not. The petroleum geologist must attempt to understand the time and the mechanism of emplacement of both petroleum and calcite cement in order to predict where to find oil.

Figure 1.3). In the past, petroleum exploration activities have provided major employment for sedimentologists and stratigraphers. These activities will undoubtedly remain important for some years to come.

Other uses for holes in rocks. Exploitation of groundwater resources also requires considerable knowledge of sedimentology and stratigraphy. The well must encounter rock with good porosity and permeability if water is to be extracted in large volumes. Likewise, aquifer recharge and potential pollution of aquifers are questions requiring considerable stratigraphic knowledge.

Although we have long been interested in taking things out of holes in rocks, we are becoming more and more interested in putting things into holes in rocks; that is, in subsurface waste disposal. Where will the waste go once it is pumped down into the earth? Once again, the answer will require considerable sedimentologic and stratigraphic knowledge. If we do not want these wastes turning up in municipal water supplies, we had best pay close attention to what we are doing from the beginning.

Earth processes, finite resources, and the future of mankind. Sedimentologists and stratigraphers think on a different time scale from most people. This ability is perhaps the greatest gift that such scientists have to offer mankind. The problem of finite resources provides the most clear-cut example of the divergence between popular opinion and the thought processes of the stratigrapher.

Economists talk of an equilibrium between supply and demand. The Western world will consume copious quantities of petroleum products, for example, if the price is right. In general, the price is right, because past petroleum exploration activities discovered abundant quantities of petroleum at relatively small cost. To the economist's mind, there is an equilibrium here. To the stratigrapher's mind, this idea is absurd. True equilibrium can never exist, simply because new petroleum concentrations form much more slowly than the rate at which existing supplies are being exploited.

Thus, even as the geologist explores for new reserves of oil and gas, the geologist is also the man in the best position to point out to the world's nations that eventually there will be no supply at any price. How and when we phase out fossil fuels is a fundamental problem facing modern man.

Sedimentologists and stratigraphers can potentially offer similar insight into innumerable problems concerning earth processes. A perfect short-term engineering solution to an environmental problem may, for example, run afoul of some long-term earth process. The long-term earth process is likely to remain beyond the range of everyday engineering thought. It will likely fall to the sedimentologist and stratigrapher to keep society advised concerning the hazards of long-term disequilibrium between man and the natural processes of his environment.

Selected References

DENTON, G. H., and S. C. PORTER. 1970. Neoglaciation. Sci. American 222 (6.): 100–111.
   Stratigraphic data record the history of glacial advance and retreat beyond the time scale of man's observation, yet within a time range that clearly has implication concerning future climates.

HALBOUTY, M. T. 1967. Our profession's challenge and responsibility. Bull. Amer. Assoc. Petrol. Geol. 51: 124–125.
   Then-president of the American Association of Petroleum Geologists gives some views concerning the role of geologists in a changing world.

——— (ed.). 1971. Geology of giant petroleum fields. Amer. Assoc. Petrol. Geol. Mem. 14. 575 p.

KUKLA, G. J., R. K. MATTHEWS, and J. M. MITCHELL (eds.). 1972. The end of the present interglacial. Quaternary Res. 2: 261–269.
   Whole issue devoted to the application of stratigraphic data to the problem of global climatic forecasting.

LANDES, K. K. 1970. Petroleum geology of the United States. Wiley-Interscience, New York. 568 p.

The Present

as the Key to the Past

The major tenet of this book is that late Cenozoic processes and history provide a model that will help us to understand the details of earth history recorded by the stratigraphic record. Sediments are being deposited today in a wide variety of environments. We must study them in detail and learn to infer the depositional environment of ancient sediments by analogy with Recent sediments. With recognition of depositional environment comes a whole series of insights concerning the processes that must have been active and the sequence of events required to produce a certain stratigraphic sequence.

As often happens with such statements, "the present as the key to the past" sounds deceptively simple. In fact, this approach requires sufficient sophistication that we shall have to cover a large amount of background material (Section II) before we can come to grips with the major subject matter of the book (Sections III and IV). In the meantime, the following discussion is offered as a simplified overview of how the environmental approach can make the stratigraphic record come alive in our minds.

A Rational Approach to Sandstone, Shale, and Limestone

Figure 2.1 depicts a stratigraphic sequence that commonly occurs, with varying dimensions, throughout the stratigraphic record. Figure 2.2 portrays lithologic correlation of three measured sections containing this sequence of lithologies. In early stratigraphic work, there was a tendency to equate lithology with time. With this simple view of the stratigraphic record, a geologist of the old school might have looked at Figures 2.1 and 2.2 and written a scenario of earth history that would read as follows: "A time" of deformation and peneplanation was followed

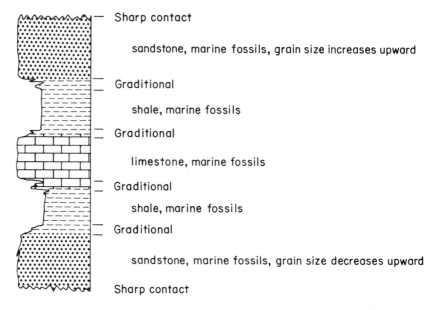

Sharp contact

sandstone, marine fossils, grain size increases upward

Gradational

shale, marine fossils

Gradational

limestone, marine fossils

Gradational

shale, marine fossils

Gradational

sandstone, marine fossils, grain size decreases upward

Sharp contact

Figure 2.1 Common sequential relationships among sandstone, shale, and lime-stone.

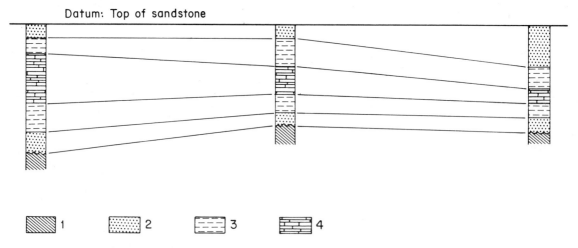

Datum: Top of sandstone

1   2   3   4

Figure 2.2 Lithologic correlation of three measured sections, displaying variations on the general sedimentary sequence depicted in Figure 2.1. Pattern (1) represents basement rock; (2), sandstone; (3), shale; and (4), limestone. By the *ad hoc* model discussed in the text, each lithology may be taken to represent "a time" in earth history: "a time" of sandstone sedimentation, "a time" of shale sedimentation, and so on.

by "a time" of marine sandstone deposition. (The quotation marks are added for special significance in later discussions.) The sandy nature of this basal marine unit indicates that nearby mountains stood high at "this time." As the mountains became worn down, "the time" of sandstone deposition gave way to "a time" of marine shale deposition. As the sources of clastic sediment supply became completely peneplained, there came "a time" of marine carbonate deposition. Subsequent to the deposition of the marine limestone, there was "a time" of tectonic rejuvenation of the source area, leading once again to deposition of marine shale and finally to deposition of marine sandstone.

The preceding outline of earth history is probably a well-reasoned *ad hoc* explanation of the data. This is the so-called "layer cake" approach to stratigraphy. Things may have happened just that way. We cannot argue conclusively against it on the basis of the limited amount of data presented. On the other hand, this explanation does not fit our study of Recent sediments. Thus, we are led to ask if the data contained in Figures 2.1 and 2.2 could be equally well explained in terms of processes and products with which we are familiar from studying Recent sediments.

Figure 2.3 presents an exceedingly simplified and generalized schematic cross

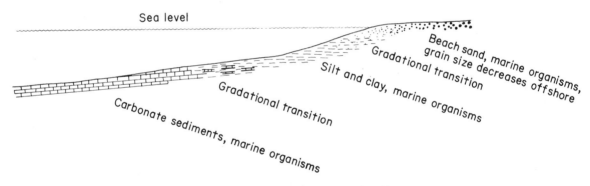

Figure 2.3 Hypothetical cross section depicting a common relationship among sand, clay, and carbonate sediments in the Recent epoch.

section of a situation common in Recent sedimentation. Scale of the model and depth relationship among the sediment types may vary widely. On the one hand, Figure 2.3 may generally describe the transition from intertidal deltaic sediments to globigerina ooze accumulating in oceanic depths. On the other hand, in some Recent environments, the transition from clastics to carbonates may involve little or no change in water depth. For the moment, let us accept the following discussion as a reasonable and moderate generalization useful only to convey an initial feeling of security. As we move into Sections III and IV, we shall develop more specific models based on specific examples of Recent sedimentation.

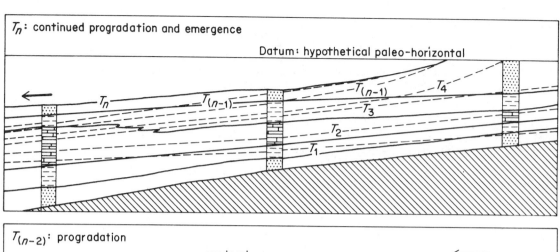

T_n: continued progradation and emergence

Datum: hypothetical paleo-horizontal

T_n $T_{(n-1)}$ $T_{(n-1)}$ T_4 T_3 T_2 T_1

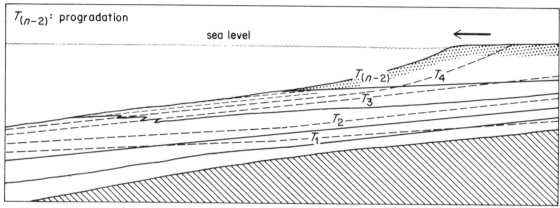

$T_{(n-2)}$: progradation

sea level

$T_{(n-2)}$ T_4 T_3 T_2 T_1

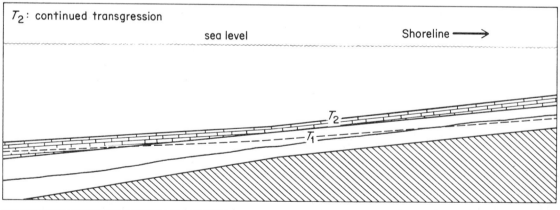

T_2: continued transgression

sea level   Shoreline ⟶

T_2 T_1

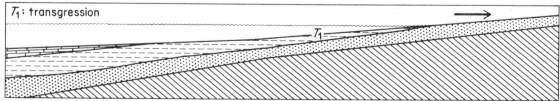

T_1: transgression

T_1

Figure 2.4 (Left) Dynamic model for the generation ·of the stratigraphic sections depicted in Figure 2.2. Transgression and regression superimposed upon the general model of Figure 2.3 generates an alternate hypothesis concerning the earth history recorded by the sections. The sequence of events begins with transgression at time T_1. Note that time lines T_1 through T_{n-1} cut across sediment types. Throughout the deposition of the entire sequence, sand, shale, and limestone are all being deposited somewhere within the model.

Marine sands are commonly high-energy nearshore deposits: deltas, beaches, and the like. Marine shales usually occur in deeper waters seaward of the high-energy nearshore sand deposits. Still farther seaward, beyond the influence of clay input from the land area, clear-water carbonate sedimentation occurs. Here, in the absence of a high influx of terrigenous clay, carbonate-secreting organisms such as foraminifers, molluscs, corals, bryozoa, and calcareous algae produce biogenic accumulations of calcium carbonate that are quite similar to limestones of the stratigraphic record.

To the modern sedimentologist, therefore, Figure 2.2 might suggest transgression followed by regression. As the sea level rises, the site of high-energy sand accumulation might be expected to move landward. Likewise, the environments of shale and limestone deposition would migrate landward. Thus, marine sandstone becomes overlain by marine shale, which in turn becomes overlain by marine limestone. With subsequent regression, shale comes to overlie limestone, and the sandstone in turn comes to overlie shale. Such a reinterpretation of the data in Figures 2.1 and 2.2 are given in Figure 2.4.

Note that the two interpretations of each set of data are quite different. For the present time, suffice it to say that the interpretation given in Figures 2.3 and 2.4 "makes sedimentologic sense" in that all sediment types exist at the same time, whereas the previous interpretation of "a time" of sand, "a time" of clay sedimentation, and so forth seems rather foreign to the sedimentologist familiar with the Recent epoch. In subsequent chapters, we shall discuss additional sedimentological criteria that might indicate more clearly how we should interpret the earth's history as recorded by stratigraphic sequences.

Sedimentation Rates and the Stratigraphic Record

Although the preceding examination of a hypothetical stratigraphic example encourages our trying to understand stratigraphy in terms of Recent sedimentation, comparison of Recent sedimentation rates with the thickness of the total stratigraphic record complicates the problem. Consider, for example, sedimentation rates in Recent calcium carbonate environments. Reasonable rates for Recent shallow-water calcium carbonate vertical accumulation rates are from .10 meter/ 1000 years to 1 meter/1000 years. Similar deposits occur in Mississippian through Permian strata over much of the central United States. Mississippian through the Permian periods represent approximately 10^8 years. Thus, application of Recent sedimentation rates would suggest that some 10^4 to 10^5 meters of sediment should have accumulated within that time. In reality, these sediments seldom exceed 10^3

meters in thickness. Thus, we have a problem. During the Recent epoch, there could have been ten to a hundred times as much sediment accumulation as is actually recorded.

Another way of fitting Recent sedimentation data to stratigraphic record is by considering the lengths of time represented by the Recent and by the classical stratigraphic units of Ancient deposits. Recent shallow-marine sedimentation began some 5000 years ago as the post-Wisconsin transgression brought sea level up to approximately where it now stands. From our knowledge of sedimentation dynamics within 5000 years, we must attempt to build models that will apply to the stratigraphic record. In contrast, biostratigraphic zonation of Ancient rock sequences usually provides us with working units of geologic time that are on the order of 1 to 10 million years long. If the processes involved in Recent sedimentation are indeed responsible for the sedimentation of stratigraphic units representing 1 to 10 million years, then we must suspect (1) that our Recent sedimentation model has barely begun to run its course, or (2) that our Recent sedimentation model has been repeated over and over again within single biostratigraphic interval, or (3) that the record is missing for large portions of many biostratigraphic intervals, or (4) some combination of (1), (2), and (3).

Thus, understanding the stratigraphic record in terms of Recent sedimentation will not be as simple as we might have originally anticipated. To begin with, we must study the Recent sediments as we see them today. Next, we must seek to understand the dynamics of Recent sedimentation over the short time span for which it has been operating. Then we must construct a dynamic model that will extend Recent sediment models to a time scale appropriate to the stratigraphic record. Finally, we must apply these models to the stratigraphic record in an iterative fashion; that is, crude models leading to an improved understanding of the stratigraphic record, which in turn leads to an improved model, which in turn leads to still a better understanding of the stratigraphic record, and so on. These four activities are treated in Sections III and IV. But first we must organize the materials and dynamics with which we will be dealing. This organization is the subject of Section II.

Selected References

IMBRIE, J., and N. NEWELL (eds.). 1964. Approaches to paleoecology. John Wiley & Sons, New York. 432 p.
Collection of topical papers. Fairly advanced level.

LAPORTE, L. F. 1968. Ancient environments. Prentice-Hall, Inc., Englewood Cliffs, N. J. 115 p.
Introductory treatment of analogies between Recent sediments and Ancient sedimentary rocks.

Sediments, Time,
and the Stratigraphic Record

In the following chapters, we shall begin to think carefully of sediments, sedimentary rocks, and sequences of sedimentary rocks. What are their properties and how should we go about studying them? What are some reasonable conclusions that we can draw from various observations? What are the pitfalls?

At this stage, our approach must be somewhat traditional. Many generations of stratigraphers and sedimentologists have gone before us. They have done some things well, some things poorly, and a great many things that are somewhere in between. Yet, if each new generation is to build upon the structure left it by previous generations, then each new generation must acquaint itself with what has gone before. Thus, in Section II, we gather some basic facts and ideas upon which to build. Then, in Sections III and IV, we shall begin to go beyond the traditional and seek our own understanding of the stratigraphic record.

3

The Properties

of Sediments and Sedimentary Rocks

The sedimentary cycle begins with the mechanical and chemical weathering of pre-existing rocks. Mechanical processes produce boulders, sand, and even clay-sized particles. Chemical processes make new minerals, predominantly clays, out of old minerals; take soluble salts into aqueous solution; and leave behind a chemically inert residue that is predominantly sand-sized quartz. The solid products of mechanical and chemical weathering become the terrigenous clastic sediments of the stratigraphic record. Soluble salts may ultimately become the chemical rocks, such as carbonates and evaporites, or the cement within the terrigenous clastic sediments.

The following discussion concerns what happens after the weathering process. Sediments continually accumulate in a variety of sedimentary environments. Which of their properties can be used to decipher the earth history recorded in the accumulated layers?

The Basic Elements of Classification of Sedimentary Rocks

Rocks with certain attributes recur over and over again in the stratigraphic record. We shall consider the following classifications to be equally applicable to sediments and sedimentary rocks. A sediment is just a rock that has not been cemented; a sedimentary rock is just a sediment that has been cemented—whichever way you choose to look at it.

Grains, matrix, and cement. In the broadest sense, sediments consist of these three components in varying proportions. By grains, we generally mean particles of sand size or larger. Matrix refers to detrital sediment of silt or clay size. Cement is the mineral matter that is a chemical precipitate within the interstices of a sediment composed of larger particles. Where grains are abundant, the sediment is referred to as a conglomerate (large grains) or a *sandstone* (sand-sized grains). Where large particles are not present, the terrigenous clastic rock will be a *siltstone* or a *shale*. Finally, if terrigenous sediment forms an insignificant portion of the rock, we speak of chemical sediments; *limestones, dolomites,* and *evaporites* are the most common.

Further classification of sediment types is strongly dependent upon our study methods. In particular, conglomerates, sandstones, and limestones can be studied quite effectively in hand specimens and in thin sections with a relatively low-power optical microscope. These sediments are traditionally classified on the basis of hand-specimen and thin-section observations. On the other hand, shales, dolomites, and evaporites commonly require X-ray diffraction mineralogy data and chemical data to be meaningfully classified. These studies are traditionally beyond the scope of the stratigrapher-sedimentologist, who instead concentrates on studying outcrops, hand specimens, and thin sections. Thus, our discussions concerning the classification of sediments will center around the classification of sandstones and limestones.

Composition of sandstones. The composition of sand grains provides important information concerning the provenance and predepositional history of the sediments. Rock fragments and mineralogical suites commonly allow the distinction of granitic, metamorphic, volcanic, and sedimentary source areas. Recognition of these general provenances is one factor in the classification of sandstones.

Compositional maturity of the sand is another major concern underlying most classifications of sandstones. Rigorous weathering and transportation tend to reduce many igneous and metamorphic minerals to clays. However, quartz, zircon, tourmaline, and rutile generally resist chemical alteration and thus remain as the ultrastable sand-sized component even after the sediment has undergone weathering and transporting.

Textural maturity of sandstones. The concept of the maturity of a sandstone may also be applied to the overall texture of the sediment. For example, prolonged transportation of sediment should ultimately result in separation of the various-sized fractions. Clay minerals, for example, should become separated from sand-sized particles, because clay is transported in true suspension (*suspended load*), whereas sand tends to be bounced or dragged along the bottom (*bed load*). If a sand contains a relatively large amount of clay, it is considered texturally immature. Conversely, clean sands are considered to be texturally mature. In this fashion, the presence or absence of fine-grained matrix is regarded as a major element in most classifications of sandstones.

Textural maturity may be further refined by consideration of the rounding of sand grains. Angular sand particles are the result of mechanical and chemical weathering. Either rigorous or prolonged transportation tends to wear down the sharp edges and produces rounded sand grains.

In general, compositional maturity and textural maturity go together. We can expect, however, complex deviation from this generality. Compositional maturity and textural maturity, as observed in any hand specimen of sedimentary rock, are the end products of chemical and mechanical processes acting in the source area, during transportation, and during and following the ultimate deposition in the sedimentary environment.

Classification of sandstones. Guided by a general concern for the provenance of the sediment and for its general state of maturity, geologists have formulated numerous classifications of sandstones. Four of these schemes are indicated in Figures 3.1 through 3.4. Although memorization of these classifications is not necessarily desirable, you will find the various rock names frequently used throughout the literature. Unfortunately, no single classification suffices, for the simple reason that older literature may use any of these classifications or even some other classification.

Figure 3.1 Sandstone classification according to Krynine (1948). [After **P. D.** Krynine, "The Megascopic Study and Field Classification of Sedimentary Rocks," *Jour. Geol.,* **56**, 130–165 (1948).]

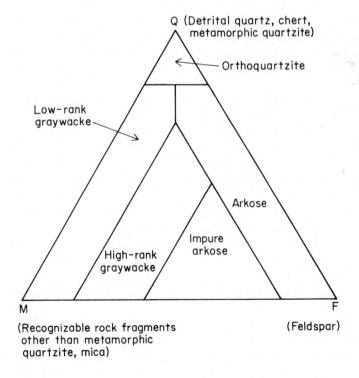

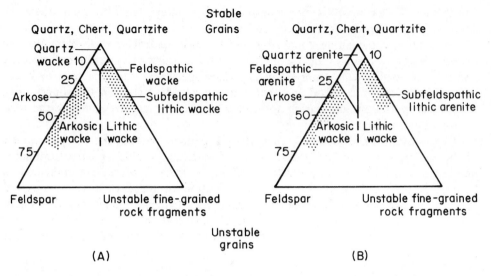

Figure 3.2 Sandstone classification according to Gilbert (1954). Rocks that are named wacke contain significant mud matrix; rocks named arenite contain essentially clean sands.

All of these classifications are essentially arbitrary. Each geologist has spent a certain amount of time working on the petrology of sandstones and has found it useful to make certain distinctions. Modern statistical techniques may provide a more rigorous basis for classification. We discuss this subject later under natural classification.

Figure 3.1 presents the classification of sandstones according to Krynine (1948). Triangular diagrams are quite popular among those workers who classify sandstones. Although tripartite division is not necessarily a fundamental property of sandstones, three components are all we can handle conveniently in a single graphic illustration. In the Krynine classification, quartz and chert constitute the supermature subdivision of the classification. *F* (feldspar) may be viewed as shorthand for a granitic source area. *M* (rock fragments and mica) may generally be read as a volcanic or low-rank metamorphic source area. Orthoquartzite, arkose, and graywacke are the major sandstone types recognized in this classification. Note that texture *per se* does not enter into this classification. One orthoquartzite might contain as much as 20% clay minerals. Alternatively, another orthoquartzite might contain as much as 20% well-rounded feldspar sand grains. These two orthoquartzites would be genetically quite different, but the classification would place them together.

Figure 3.2 presents the Gilbert (1954) classification of sandstone. This classification uses two words to describe the sandstone. The first word, based on the triangular diagram, is an adjective describing the general composition of the sand grains. The second word describes the presence or absence of fine-grained detrital

Figure 3.3 Sandstone classification according to Pettijohn (1954, 1957). [After F. J. Pettijohn, *Jour. Geol.,* **56,** 130–165 (1948).]

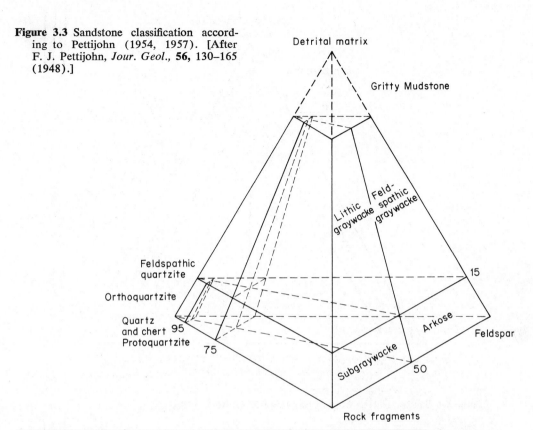

Cement or matrix	Detrital matrix exceeds 15% Chemical cement absent	Detrital matrix less than 15%. Voids empty or filled with Chemical cement		
Sand or detrital fraction — Feldspar exceeds rock fragments — Graywackes	Feldspathic graywacke	**Arkosic** Arkose	**Sandstones** Subarkose or feldspathic Ss	Orthoquartzites	Chert <5%
Rock fragments exceed felds — Graywackes	Lithic graywacke	**Lithic** Subgraywacke	**Sandstones** Protoquartzites		Chert >5%
Quartz content	Variable; generally <75%	<75%	>75% <95%	>95%

21

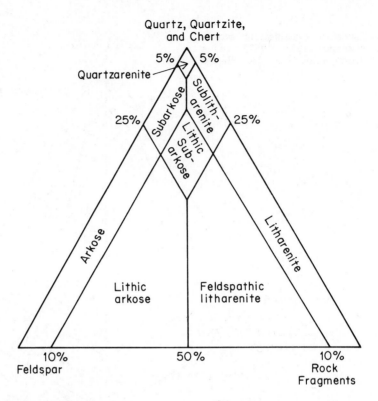

Figure 3.4 Sandstone classification according to McBride (1963).

matrix. Once again: quartz and chert denote the supermature end of the classification spectrum; feldspar denotes granitic source area; and unstable rock fragments denote volcanic or metamorphic source area. In comparison to the Krynine classification (Figure 3.1), note especially the appearance of a number of special words in the impure regions around the quartz-chert corner of the diagram. A quartz arenite of the Gilbert classification must contain greater than 90% quartz, whereas the orthoquartzite of the Krynine classification need contain only 80% quartz. Krynine's graywacke may be either a lithic wacke or a lithic arenite, depending on the presence or absence of a fine-grained detrital matrix.

Figure 3.3 summarizes the classification of sandstones after Pettijohn (1954, 1957). By and large, the input concepts of this classification are similar to those of Gilbert (1954). Only the names and style of presentation have changed. The use of the word "orthoquartzite" is now restricted to rocks containing greater than 95% quartz.

Figure 3.4 presents a sandstone classification of McBride (1962). Once again, the major elements of the classification are those of Gilbert (1954), yet we still find more names added to the classification on the basis of arbitrary subdivision within the triangular diagram.

Table 3.1 PARTICLE-SIZE NOMENCLATURE

Name	Millimeters (mm)	Size in Microns (μ)	Phi φ	U. S. Standard Sieve Mesh #
Boulder				
	256.0	——	−8.0	——
Cobble				
	64.0	——	−6.0	——
Pebble				
	4.0	——	−2.0	5
Granule				
	2.0	——	−1.0	10
Very coarse sand				
	1.0	——	0.0	18
Coarse sand				
	0.5	500.0	1.0	35
Medium sand				
	——	250.0	2.0	60
Fine sand				
	——	125.0	3.0	120
Very fine sand				
	——	62.5	4.0	230
Coarse silt				
	——	31.0	5.0	——
Medium silt				
Fine silt				
Very fine silt				
	——	3.9	8.0	——
Clay				

GRAVEL

SAND

MUD

Classification of fine-grained clastic rocks. Because of small particle size, these rocks are difficult to classify on the basis of petrographic examination such as we use to classify sandstones and carbonates. Useful subdivision often requires mineralogical studies by X-ray techniques. In the following discussion, we shall speak only in general terms of clays, silts, shales, and siltstones. Clays and silts are sediments, their names denoting primarily particle size (see Table 3.1). Shales and siltstones are the rocks formed from the respective sediments.

Classification of limestones. The classification of limestones runs somewhat parallel to the sandstone classification of Gilbert. The classifications of Folk (1962) and of Dunham (1962) are the most generally accepted ones. Both schemes name the sediment on the basis of grain types and the amount of mud matrix.

The grain types of carbonate rocks are commonly the chemical or biochemical products of the sedimentary environment. Whereas terrigenous clastic grains were derived from preexisting rock and transported to depositional environment, the grains within carbonate sediments are commonly precipitated from the water by chemical or biochemical processes active in the depositional environment. Folk (1962) recognizes four basic categories of grain types. These are bioclastic debris, oolites, intraclasts, and pellets. Bioclastic debris is the skeletal remains of carbonate-secreting organisms that live in or near the depositional environment. Oolites are sand-sized grains constructed by concentric laminations of calcium carbonate deposited under agitated water conditions. Intraclasts are small chips of semi-indurated calcium carbonate sediment commonly formed by alternate wetting and desiccation in intratidal and supratidal carbonate environments. Pellets are the semi-indurated to indurated excreta of benthonic organisms.

In the Folk classification, abbreviations identifying the various grain types constitute the prefix of a shorthand word that describes the rock type. If the rock contains no mud, it is a sparite (taken from "spar," which is the cement for the sediment). If the sediment contains considerable mud matrix, it is a micrite ("micrite" being a contraction for microcrystalline calcite). Thus, a pelsparite is a clean sand composed of pellets, and a oopelmicrite is a muddy sand containing pellets and oolites. If we wish to emphasize that the sand grains are larger than 2 millimeters, the rock may be named a rudite. Pelsparrudites and oopelmicrudites, therefore, contain larger grains than pelsparites and oopelmicrites. Adjectives may also be added to name rocks like mollusc-bearing oopelmicrudite, and so on.

The basic elements of the Folk classification of carbonate rocks is given in Figure 3.5. As with the sandstone classification discussed earlier, the boundaries among the various sediment names are placed at arbitrary percentages.

Dunham's classification hinges around the concept of "grain support" versus "mud support" of the depositional fabric. Note that this emphasis on the ability of the grains to support the depositional fabric allows wide variation in such parameters as the percentage of mud. Dunham contends that the percentage of

Volumetric Allochem Composition	$>10\%$ Allochems *Allochemical Rocks*		$<10\%$ Allochems *Microcrystalline Rocks*		
	Sparry Calcite Cement $>$ Microcrystalline Ooze Matrix	Microcrystalline Ooze Matrix $>$ Sparry Calcite Cement	1–10% Allochems	$<1\%$ Allochems	Undisturbed Bioherms Rocks
	Sparry Allochemical Rocks	Microcrystalline Allochemical Rocks	Most Abundant Allochem		
$>25\%$ Intraclasts	Intrasparrudite Intrasparite	Intramicrudite Intramicrite	Intraclasts: Intraclast-bearing Micrite	Micrite	
$<25\%$ Intraclasts, $>25\%$ Oölites	Oösparrudite Oösparite	Oömicrudite Oömicrite	Oölites: Oölite-bearing Micrite		Biolithite
$<25\%$ Intraclasts, $<25\%$ Oölites; Volume Ratio of Fossils to Pellets $>3{:}1$	Biosparrudite Biosparite	Biomicrudite Biomicrite	Fossils: Fossiliferous		
$<25\%$ Intraclasts, $<25\%$ Oölites; Volume Ratio of Fossils to Pellets 3:1–1:3	Biopelsparite	Biopelmicrite	Pellets: Pelletiferous Micrite		
$<25\%$ Intraclasts, $<25\%$ Oölites; Volume Ratio of Fossils to Pellets $<3{:}1$	Pelsparite	Pelmicrite			

Figure 3.5 A simplified version of the classification of limestones according to Folk (1962).

DUNHAM CLASSIFICATION OF CARBONATE ROCKS ACCORDING TO DEPOSITIONAL TEXTURE

Depositional texture recognizable				Depositional texture not recognizable
Original Components Not Bound Together During Deposition			*Original components were bound together during deposition . . . as shown by intergrown skeletal matter, lamination contrary to gravity, or sediment-floored cavities that are roofed over by organic or questionably organic matter and are too large to be interstices.*	*Crystalline Carbonate*
Contains mud (particles of clay and fine silt size)		*Lacks mud and is grain-supported*		
Mud-supported	*Grain-supported*			
Less than 10 percent grains	*More than 10 percent grains*			
Mudstone	Wackestone	Packstone	Grainstone	Boundstone
				(Subdivide according to classifications designed to bear on physical texture or diagenesis.)

Figure 3.6 Classification of carbonate rocks according to Durham (1962).

mud is not the fact that should demand our attention. This approach carries with it some very important implications. For example, if a rock is truly mud-supported, then the mud and the grains must have been deposited at the same time and in an environment that was quiet enough to allow mud to accumulate. If, on the other hand, the sediment is grain-supported, then we may entertain the suggestion that the grains were deposited under high-energy conditions and that the mud filtered down in among the grains at some later time. Thus, our interpretation of the conditions of sedimentation may vary considerably, depending on whether we think the depositional fabric to be grain-supported or mud-supported. The classification of Dunham (1962) is given in Figure 3.6. The names mudstone, wackestone, packstone, and grainstone are widely used.

Natural classification. With the exception of the Dunham limestone classification, we have noted considerable usage of arbitrary boundaries. If the only purpose of a classification is communication among people who have memorized it, then there is nothing wrong with arbitrary boundaries. If, on the other hand,

Figure 3.7 Hypothetical example indicating the utility of natural classification as opposed to arbitrary classification. Where two groups of data exist in nature, to rely on an arbitrary classification to subdivide the data would be unnecessarily complicated. (After Imbrie and Purdy, 1962, with modification.)

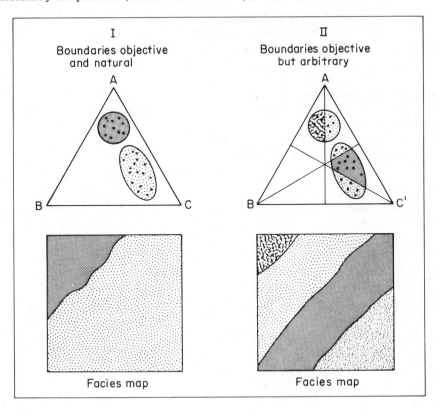

we think these various names convey important sedimentological distinctions, we may be making our studies unnecessarily complicated. Imbrie and Purdy (1962) pointed this fact out in their classification of modern Bahamian carbonate sediments.

Consider a group of samples for which there is quantitative data concerning the abundance of *A, B,* and *C*. A brief glance at the triangle diagrams in Figure 3.7 confirms that the samples fall into two naturally occurring populations. Observe, however, that arbitrary classification of these sediments might require that we name four lithologies. Clearly, we are complicating the problem when we employ an arbitrary classification in a situation like this one.

Instead, there are available multivariate statistical techniques, which allow the recognition of naturally occurring clusters in *n*-dimensional space. *N*-dimensional space is simply a mathematical abstraction beyond the three-dimensional array portrayed in Figure 3.7. (See Imbrie, 1964, for further discussion.)

Size Distribution

Naming the sediment according to its composition is just the first small step toward learning what the sediment can tell us about its history. Size and size distribution within sediments is also of significance for studying the paleogeography and depositional environments. Standard terminology concerning description of mean grain size in sediments is given in Table 3.1. Size data are usually reported in millimeters or in phi units, where

$$\phi = -\log_2 X, \tag{3.1}$$

X being the size measure in millimeters.

The standard for description and comparison of size distribution data is the normal distribution function:

$$f(z) = \frac{\exp(-z^2/2)}{\sqrt{2\pi}}. \tag{3.2}$$

In this equation, $z = (x - \mu)/\sigma$, where μ is the population mean and σ is the population standard deviation. Where size data are in phi units, size distribution is expected to be normally distributed; where the data are in millimeters, the data are expected to be lognormally distributed.

As indicated in Figure 3.8, the function in Eq. (3.2) plots as a straight line on a cumulative probability graph. The mean is the 50th percentile, and the standard deviation is the difference in size between any spread of 34 percentiles. In common usage, deviation is reported as $(\phi_{84} - \phi_{16})/2$. The mean is used to name the size of the sediment as indicated in Table 3.1; the standard deviation, or sorting, is used as a first estimate of textural maturity. Sediments that have a standard deviation of .35 ϕ are considered well sorted; .50 ϕ, moderately well

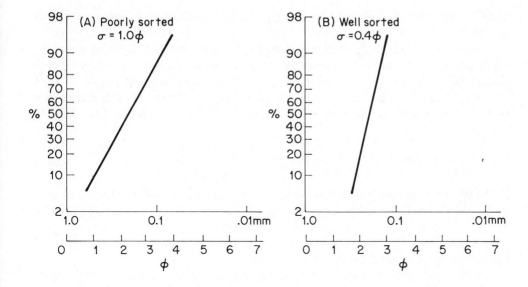

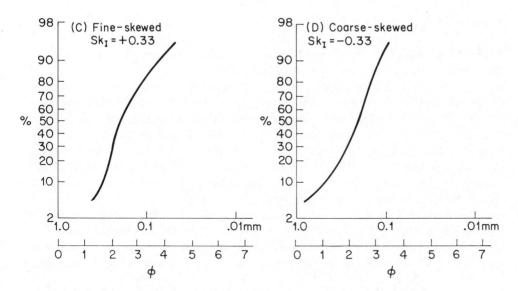

Figure 3.8 Various examples of size distribution plotted with a cumulative probability function as the *Y*-axis. Rather than attempt to construct the traditional bell-shaped frequency distribution curve, we find it convenient to plot size data by using a cumulative probability function as the *Y*-axis. On such a plot, a normal distribution of the form in Eq. (3.2) is a straight line. Thus, variations from normality are easily detectable upon visual examination.

sorted; .70 ϕ, moderately sorted; 1.0 ϕ, poorly sorted; and 2.0 ϕ, very poorly sorted.

Skewness is a measure of the departure of a size distribution from normality or lognormality. If a sediment sample contains excess fine-grained material, it is said to be $(+)$ skewed; if it contains excess coarse-grained material, it is $(-)$ skewed. In theory, skewness should compare the symmetry of size distribution about the mean for each pair of percentiles of the curve. In practice, Folk (1968) suggests:

$$\text{Sk}_I = \frac{\phi_{16} + \phi_{84} - 2\,\phi_{50}}{2\,(\phi_{84} - \phi_{16})} + \frac{\phi_5 + \phi_{95} - 2\,\phi_{50}}{2\,(\phi_{95} - \phi_5)}, \qquad \textbf{(3.3)}$$

which shows inclusive graphic skewness as an adequate measure for most purposes. The comparison of a large number of size distribution curves suggest that Sk_I of $+.5$ is strongly fine-skewed; $+.2$, fine-skewed; $+.1$ to $-.1$, more or less symmetrical; $-.2$ coarse-skewed; and $-.5$ strongly coarse-skewed. Figure 3.8 plots some curves having these attributes.

In this book, discussions of size distribution in sediments will not get beyond the "well-sorted fine-grained sand" level. Consult Blatt, Middleton, and Murray (1972) for further size-distribution studies.

Stratification

Sedimentary rocks are commonly arranged in layers. Indeed, the very name of this subject, *stratigraphy,* comes from this attribute.

In the field description of sedimentary sequences, it is customary to make note of the thickness of stratification and its degree of development. McKee and Weir (1953) provide a simple and convenient discussion of terminology. Strata less than 1 centimeter thick are referred to as *laminae.* Strata thicker than 1 centimeter are *beds,* ranging from very thin to very thick: 1 to 5 centimeters is very thin; 5 to 60 centimeters, thin; 60 to 120 centimeters, thick; and greater than 120 centimeters, very thick. To these words are added such modifiers as "well-developed," "irregular," and so on, as appropriate.

If a single bed exhibits decreasing average grain size from bottom to top, it may be referred to as a *graded* bed. Similarly, if grain size increases upward in a bed, it is referred to as a *reverse-graded* bed. Beds involving other evidences of stratification are said to contain primary structures and require further discussion.

Primary Structures

Sediments deposited by moving air or water commonly show a depositional fabric indicative of the conditions under which sedimentation occurred. Furthermore, activities of burrowing organisms may impart recognizable characteristics to the sediment even though the organism is not preserved. These and similar features are referred to as *primary sedimentary structures;* that is, rock fabrics indicating the conditions of deposition.

Primary sedimentary structures are especially valuable to the sedimentologist-stratigrapher because important interpretations can often be made by visually examining the outcrop or core. Our discussions of sedimentary environments will rely heavily on primary structures as a quick and easy way to distinguish several important environments.

Laminar and turbulent flow. When a liquid or a gas moves in such a fashion that the particles remain in parallel flow lines, the flow is said to be laminar. When the net forward movement of the liquid or gas is accomplished by irregular, non-parallel motion of the particles, the flow is said to be turbulent. All of the features that we shall discuss are the result of turbulent flow. Turbulent flow is further divided into tranquil (sometimes called streaming, lower, or subcritical) and rapid (shooting, upper, or supercritical) flow regime.

Flume studies. A flume provides a convenient method for investigating bed form and primary structures as a function of increasing flow velocity with flow depth held constant. Figure 3.9 depicts the results of a series of these experiments carried out by Simons *et al.* (1961). With low current velocity, small ripples were formed. As velocity increases, the height of the ripples increases. With still greater velocity, the shape of the large ripples begins to flatten down. With even greater velocity, a planar bed form is achieved and no sediment is deposited. This picture is the transition from tranquil to rapid flow regime. Finally, with still greater stream velocity, antidunes are formed. Sediment accumulation on anti-dunes occurs on the updip side of the bed form; the antidune may actually migrate upstream with continuing sedimentation.

Simons *et al.* (1961) discuss only the evolution of bed form with increasing velocity. Let us now translate these discussions into equivalent primary structures.

Consider the small changes in flow velocity associated with the passage of water over a tranquil-ripple bed form. Inasmuch as the depth of the flow is finite and the top of the flow is more or less a smooth surface, then the velocity of flow must increase as the water passes from the trough up onto the crest of the ripple. Similarly, the velocity must decrease as the water passes from the crest of the ripple on toward the next trough. Figure 3.10 depicts these relationships. With increasing velocity, the bottom sediment will be mobilized by the flow. With decreasing velocity, some of this sediment load will be dropped back to the stream bed. Consequently, material from the upstream side of a ripple will be transferred (by *erosion*) to the downstream side of the same ripple (causing *sedimentation*). The net result is the formation of high-angle cross-stratification (*lower flow regime*).

With increasing velocity the sediment is transported somewhat further beyond the crest of the ripple. This process results in lowering the angle of cross-stratifica-

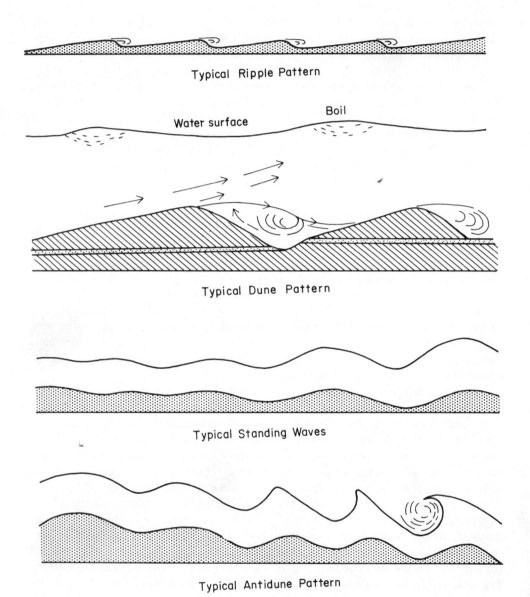

Water surface

Typical Ripple Pattern

Water surface

Boil

Typical Dune Pattern

Typical Standing Waves

Typical Antidune Pattern

Figure 3.9 Schematic summary of flume studies relating bed form in medium-sized sand to flow regime. The upper two diagrams represent tranquil flow, and the lower diagram represents rapid turbulent flow. (After Simons *et al.*, 1961.)

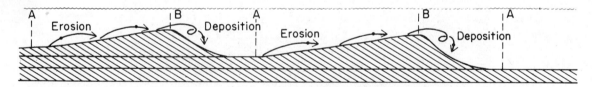

Figure 3.10 Illustration of erosion and deposition associated with ripple migration and the resultant development of planar cross-stratification. As the tranquil flow passes from point *A* to point *B*, velocity will be increasing; as the flow passes from *B* to *A*, velocity is decreasing. Increasing velocity results in net erosion of the sediment, whereas decreasing velocity results in net deposition.

tion to the point where the bed form becomes planar as critical flow (*upper flow regime*) is approached.

Most of the primary sedimentary structures of interest to us form in the lower flow regime. Supercritical flow is sometimes encountered in turbidity currents (where high velocities are attained by density currents) and in beach and alluvial fan deposits (where upper flow regime results from extremely shallow water depth).

Cross-stratification. Flume studies are carried out in long straight channels in the laboratory under very specific conditions. In sharp contrast, natural channels are seldom straight, seldom of constant depth, and seldom operate under specified conditions. Thus, we must supplement knowledge gained from flume studies with empirical observations on cross strata as they are actually observed under field conditions.

McKee and Weir (1953) suggest a convenient nomenclature and classification for cross-stratification in sandstones (see Figure 3.11). Cross strata that owe their origin to the migration of a single ripple form are referred to as a *set*. Similar sets of cross strata are referred to as a *coset*. Where cosets of differing morphology are usually arranged sequentially in the stratigraphic section, the sequence of cosets may be referred to as a *composite set*.

Individual sets of cross strata are classified as simple, planar, or trough. Simple cross strata do not have erosional boundaries. Planar cross strata have erosional set boundaries that are more or less flat planes. Trough cross strata have lower set boundaries that are curved surfaces of erosion. This scheme is further modified by the designation of small, medium, and large sets. Small-scale sets of cross strata are less than 30 centimeters thick. Medium-scale sets are 30 centimeters to 6 meters thick. Large-scale sets are thicker than 6 meters.

Figures 3.12, 3.13, and 3.14 provide examples of naturally occurring cross-stratification associated with alluvial and marginal marine sedimentation.

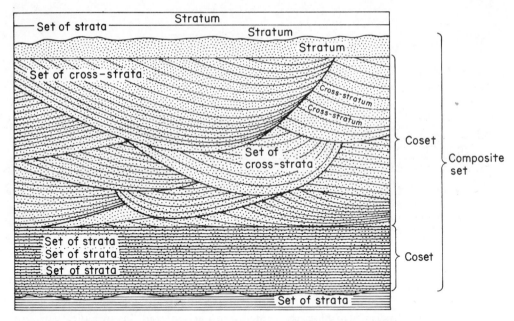

Figure 3.11 Terminology for stratified and cross-stratified units as proposed by McKee and Weir (1953).

Figure 3.12 Cross-stratification shown in dip section through sand deposits of a modern meandering stream. Cosets of medium-scale high-angle, planar cross strata are overlain by cosets of low-angle simple cross strata. Blade width on the large knife is approximately 8 centimeters. (Photo by L. V. Benson.)

Figure 3.13 Strike section revealing cross-stratification in sand deposits of a modern meandering stream. Scale at the bottom of the photograph is 15 centimeters total length. Simple low-angle cross strata, such as seen in the upper portions of Figure 3.12, are overlain by medium-scale trough cross strata, which are in turn overlain by cosets of small-scale simple cross strata. In the upper portion of the photograph, note that burrows and plant roots have reworked the sediment to the extent that primary cross-stratification has been almost obliterated. The burrows, root structures, and general obliteration of the cross-stratified fabric are themselves the sedimentological record of subaerial exposure of this river-sand deposit. (Photo by L. V. Benson.)

Figure 3.14 An example of simple low-angle cross-stratification in beach sands of the high intertidal environment. Extremely shallow water and the relative rapidity with which swash comes on to and off the beach places this environment in the transitional to upper flow regime. (Photo by R. P. Steinen.)

Vertical sequences of cross strata. Cross strata can be studied set by set; each set records the hydrodynamic conditions under which it was sedimented. On the other hand, certain vertical sequences of cross strata commonly repeat themselves in sedimentary rocks. By recognizing the broad general significance of these vertical sequences, we can rapidly identify in the field major features of the paleogeography in which sedimentation occurred. Geologists have developed two models that particularly relate vertical sequences of primary structures to flow regime in natural environments. These are the point-bar model of fluvial sedimentation and the turbidite model for deep-water sedimentation from turbidity currents.

The point-bar model and related interpretations concerning flow regime are given in Figure 3.15. The deep portion of the channel carries the most water and is floored with the coarsest-grained sediment within the system. High-current velocity combined with relatively deep water commonly places this environment in transitional to lower flow regime. Closer to the point bar, stream velocity is slightly lower, but water depth is much less. In this position, decreased water depth more than compensates for slightly lower stream velocities, and the combination usually places this environment in transitional to upper flow regime. Simple low-angle cross-stratification, such as depicted in Figures 3.12 and 3.13, are typical of this condition. On top of the bar, water depth is extremely shallow

Figure 3.15 Schematic representation of the vertical distribution of sedimentary structures and grain size on a point bar. [After G. S. Visher, "Fluvial Processes as Interpreted from Ancient and Recent Fluvial Deposits," *Primary Structures and Hydrodynamic Interpretation,* (G. V. Middleton [ed.]). Soc. Econ. Paleontologists and Mineralogists Spec. Pub. **12**, 116–132 (1965).]

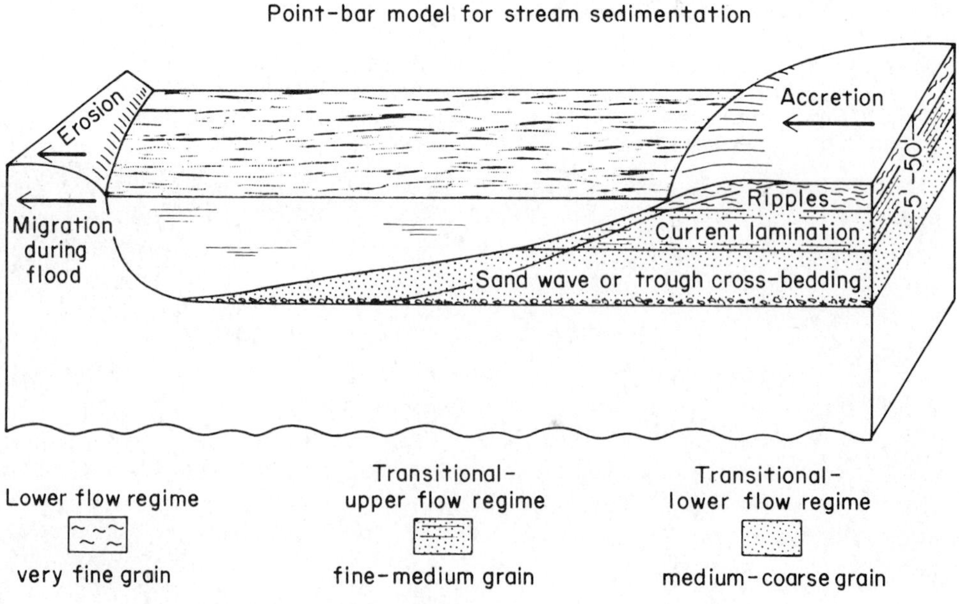

Point-bar model for stream sedimentation

Lower flow regime	Transitional– upper flow regime	Transitional– lower flow regime
very fine grain	fine–medium grain	medium–coarse grain

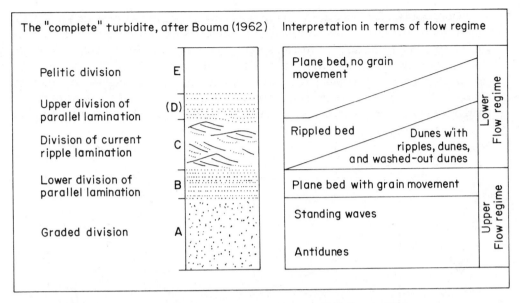

The "complete" turbidite, after Bouma (1962)   Interpretation in terms of flow regime

Pelitic division	E		Plane bed, no grain movement	Lower Flow regime
Upper division of parallel lamination	(D)		Rippled bed	
Division of current ripple lamination	C		Dunes with ripples, dunes, and washed-out dunes	
Lower division of parallel lamination	B		Plane bed with grain movement	Upper Flow regime
Graded division	A		Standing waves	
			Antidunes	

Figure 3.16 Schematic representation of the "complete turbidite" according to Bouma (1962), with hydrodynamic interpretation according to Walker (1967). [After R. G. Walker, "Turbidite Sedimentary Structures and their Relationship to Proximal and Distal Depositional Environments," *Jour. Geol.* **37**, 25–43 (1967).]

and current velocity is minimal. Such a combination places this environment well down into lower flow regime. Small-scale, high-angle cross-stratification is typical of this environment. (See the middle portions of Figure 3.13.) Stream sedimentation is discussed in more detail in Chapter 10.

The hydrodynamic interpretation of a typical turbidite bed is presented in Figure 3.16. Turbidites are the sedimentary record of turbidity currents. Presumably, sediments occupying the outer portions of the continental shelf are: (1) gravitationally unstable and (2) very poorly consolidated. After an initial perturbation of this unstable condition, by an earthquake perhaps, large masses of unconsolidated sediment and their contained fluid may flow downslope into adjacent deep water. Because this turbid suspension of sediment is more dense than the surrounding water, the turbidity current gains speed as it flows down the slope. As the turbidity current reaches the bottom of the slope, its velocity begins to decrease and sedimentation is initiated. Initial sedimentation may occur within the upper flow regime and generally involves very coarse material. Upper flow regime then gives way to transitional and lower flow regimes accompanied by decrease in grain size. Finally, normal processes of deep-water clay sedimentation again dominate the area, and the upper fine-grained portion of the typical sequence is deposited. The erosion of the upper portions of the sequence precedes the deposition of the next *A* layer.

Tool marks. The cohesive properties of muddy sediments provide the opportunity for the formation of current-associated markings on bedding planes. Clay mineral particles are difficult to resuspend once they have been deposited on the sediment-water interface. Thus, relatively strong currents may flow across sedimented mud without inducing large-scale erosion of the sediment-water interface (Figure 3.17). Such currents may drag or roll larger objects, like plant debris or perhaps even pebbles, across a mud bottom and plow a small furrow into the mud.

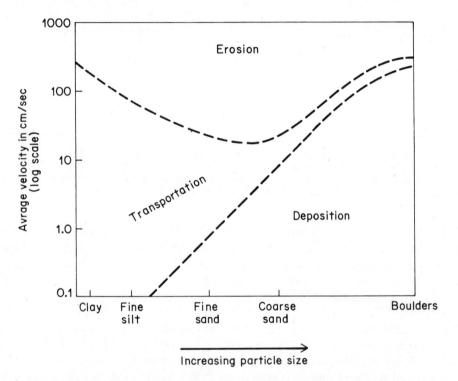

Figure 3.17 Diagram relating erosion, transportation, and deposition of various-sized particles to flow velocity of water. For coarse sand and larger particles, a relatively small increase in velocity will cause the sediment-water interface to change from a surface of net deposition to a surface of net erosion. In sharp contrast, particles of clay-size can be transported by very slow-moving currents; yet once the clay minerals are sedimented, very large currents are required to erode them. (Adapted from Hjulstrom, 1939.)

These furrows, or grooves, are often preserved by the rapid deposition of sand-sized material over the mud surface. In the field, these features are most readily observed as groove casts on the bottom side of sandstone beds; the shales below are usually too friable to produce slabs large enough to observe the actual grooves.

Groove casts can provide valuable information concerning paleocurrent directions in marine sequences.

Flute casts are a somewhat similar phenomena. In this case, the mud at the sediment-water interface is locally disturbed by vortex currents associated with gravity flow down submarine slopes.

Bioturbation and desiccation features. Although moving water and wind impart the most dramatic primary structures to the sedimentary record, other processes can leave important imprints on the fabric of the rock. The burrowing activity of marine benthonic organisms has long been recognized as producing burrow-mottled fabrics that are in themselves diagnostic of specific depositional environments, even though the organisms that produced the burrows are not preserved (see Rhoads, 1967, for example). Similarly, the roots of plants move the sediment around as they grow. This activity will both disrupt any preexisting primary fabrics (see Figure 3.13) and create structures and fabrics diagnostic of the vegetated cover itself. Again, these primary structures may be preserved, although the plants that produce them are totally lost from the geologic record.

Desiccation features, such as mud cracks, are commonly found quite well preserved, especially in intertidal and lower supratidal sediments. In these environments, desiccation features are an extremely valuable environmental datum, for they record those sedimentary environments that were alternately wet during high tide and dry during low tide.

Sequential Relationships in Sedimentary Rocks

We have noted earlier that a single bed, hand specimen, or thin section of rock can yield considerable information concerning the origin of the sediment and the conditions of deposition. Next, we must integrate this information for all of the rock types falling within the scope of the particular area or problem. The basic unit of stratigraphic synthesis is the *measured section* through the lithologic sequence. We would prefer to study each time-rock unit in a map view. We would like to walk across the old depositional surfaces and study the facies changes in much the same fashion as we study facies changes in Recent sediments today. Unfortunately, time-rock units are most typically overlain by other time-rock units, thus precluding random access to the old depositional surfaces. Consequently, we are compelled to direct our attention to specific vertical sequences made available to us by fault escarpments, stream erosion, road cuts, quarry operations, and drill holes.

Thus, the object of most stratigraphic work is to synthesize numerous measured sections into maps and cross sections that portray the paleogeography and earth history recorded within the sediments. It is extremely important that we learn to make maximum use of observations concerning sequential relationships within measured sections of strata.

Unconformities. Where it can be demonstrated that the sedimentary sequences above a surface are not in sedimentological or temporal continuity with the rocks below the surface, the surface is said to be an *unconformity*. The recognition of unconformities is of prime importance in stratigraphic work. These surfaces must bound the major rock units and time-rock units within the area under consideration.

Within a measured section, unconformities may have visible lithologic properties associated with long periods of subaerial exposure (see Figure 3.18).

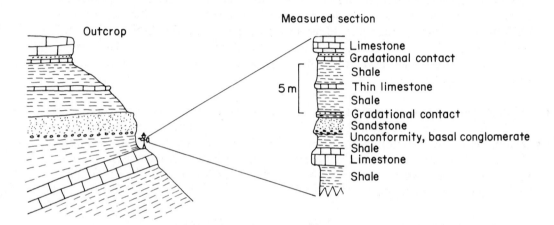

Figure 3.18 Translation of outcrop observations into a description of a measured section. The geologist is measuring the section with a hand level. Holding the level to his eye, he locates the next convenient spot to reach in his traverse up the hill. Then he describes the strike, dip, and lithologic properties of the rocks that lie between his feet and the next spot up the hill. By the time he climbs to the top of the hill, he has the thickness and lithologic description of the sequence. In this schematic representation, the unconformity is easy to recognize. There is an angular relationship between the beds above and below the unconformity, and there is a pronounced basal conglomerate immediately above the unconformity.

Erosion surfaces and fossilized soil zones may be discernible on the material beneath the unconformity. Conglomerates, gravels, or other terrestrial lag deposits may begin the sequence immediately above the unconformity. Where strata above and below the unconformity can be shown to have angular relationships, owing to tectonic activity that occurred during the time represented by the unconformity, the surface is referred to as an angular unconformity. Where angularity cannot be demonstrated, the surface is commonly referred to as a disconformity. Clearly, a disconformity in one set of outcrops may be demonstrated to be an angular unconformity elsewhere. The terms are not genetic, simply descriptive (see Figure 3.19).

Recognition of off-lap and on-lap relationships associated with major unconformities is often a primary key to unraveling stratigraphic complexities.

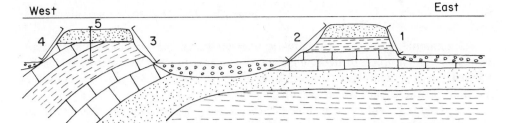

West   East

Figure 3.19 Schematic representation concerning the recognition of a major unconformity that is not readily apparent in individual measured sections. When
the geologist begins to work this area, the unconformity between the upper
sandstone and the shale below is not obvious in measured sections (1), (2),
and (3). As the geologist measures section (4), he notes the occurrence of
limestone where previous sections would have led him to anticipate shale.
Measured section (5) confirms the existence of a major unconformity at the
base of the upper sandstone.

Major regressions and transgressions that produce the intervening unconformity
surface are usually gradual and pulsating events. The last records of marine
deposition below an unconformity, for example, may be much older on the higher
portions of the land area and much younger on its seaward flanks (see Figure
3.20). Conversely, the first appearance of marine sediments over the unconformity surface may be much younger over the former high areas than in the
basin sediments flanking the high areas.

Thus, it is a rational and widely accepted stratigraphic practice to solve difficult
correlation problems by pushing units up or down into major regional unconformity
surfaces.

Figure 3.20 Off-lap and on-lap relationships associated with a major unconformity.
Six biostratigraphic zones are well defined in apparently continuous sedimentation in measured section A. Only zones (1) and (6) occur in measured section
B, separated by an easily recognizable unconformity surface. The apparent
continuity of sedimentation within section A suggests that we are dealing with
off-lap and on-lap relations in the area between sections A and B. As we move
from section A toward section B, more and more of the time represented by
sedimentation in section A will be represented by missing section.

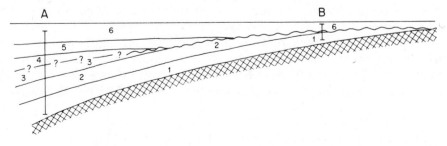

Gradational contacts. Just as the recognition of an unconformity surface demands that we think of everything above as separate from everything below, the recognition of gradational contacts between contrasting lithologies demands that we think of these units as genetically related.

Gradational contacts can take several forms. Sometimes a shale unit, for example, becomes more sandy toward the top as it gradationally passes upward into sandstone. In other situations, the shale unit may begin to show an increasing number of small sandstone beds alternating with otherwise normal shale. Both situations herald the nearby proximity of sandstone deposition. In both cases, the continuing passage of time brings sandstone deposition over a spot that was previously dominated by shale deposition.

Bedding planes. Unconformity surfaces clearly indicate that a portion of geologic time is not represented. On the other hand, gradational contacts demonstrate that sedimentation is more or less continuous throughout the transition from one depositional stage to another. Between these two conditions, there is undoubtedly a vast no-man's-land: situations in which we are not really sure whether we are dealing with continuous deposition or not.

In this connection, bedding planes pose a particularly intriguing problem. Many bedding planes can be attributed to events that make good sedimentological sense. Storm deposits, progradation, channel migration, and ashfalls, for example, produce beds that can be easily understood in terms of observations concerning Recent sedimentation. Contrasting lithology from one bed to the next tells of deposition under differing conditions, similar to environmental fluctuations that we observe today.

On the other hand, some bedding planes separate sediments that appear to be virtually identical. Such bedding planes are particularly troublesome in subtidal normal marine environments. When we observe similar environments today, we find that bioturbation by the marine benthos thoroughly homogenizes the sediment, so that all of the Recent accumulation forms essentially a single bed. Therefore, how can presumably similar sediments in the stratigraphic record exhibit bedding on the scale of centimeters to tens of centimeters?

One possibility is that each bed records a period of deposition and that each bedding plane records a period of nondeposition during which the underlying bed became sufficiently lithified so as to resist the activities of burrowing organisms during the next depositional phase. This possibility leads us to consider a fundamental philosophical point concerning the continuity or discontinuity of the stratigraphic record.

Traditionally, stratigraphers have presumed that measured sections record continuous deposition unless an unconformity can be demonstrated. The fact that some bedding planes were extremely difficult to explain was just one of the messy problems that they learned to ignore. However, a geologist favoring strict interpretation of the evidence available from Recent sedimentation may contend that the very existence of these bedding planes is in itself good evidence of the lack of continuity of sedimentation. Indeed, if we take Recent sediments to represent the formation of a bed, this bed represents 5000 years of geologic time, whereas the bedding plane separating the Recent from Pleistocene deposits commonly represents the passage of more than 100,000 years of geologic time. (See Bloom, 1972, for example.)

For the present, let us simply recognize that these two points of view exist. One stratigrapher may express the opinion that thousands of meters of shallow-marine sediments were deposited more or less continuously and without intermittent sub-aerial exposure. Another stratigrapher may look at the same sequence of rocks and propose subaerial exposure surfaces every few meters. The latter will cite data from the study of Recent sediments, and the former will say that we cannot make that strict an interpretation of Recent sediment data. The accommodation of these two points of view is the subject of Sections III and IV of this book.

Electrical Properties of Sedimentary Rocks

Much effort goes into the study of sedimentary rocks that are below the surface of the earth. This is the region where the petroleum geologist seeks oil and where the engineering geologist tries to determine whether or not a proposed dam will actually hold water. Such studies are accomplished by drilling holes into the rocks. Although it is possible to core the rock and bring a piece back up to the surface for study, it is far more economical to drill the hole with a bit that pulverizes the rock as it drills. This pulverization leaves us, however, with very little to study firsthand.

We therefore rely heavily upon measurement of the electrical properties of the sedimentary rocks after the hole has been drilled. Electrodes are lowered down the borehole on a wire line. As the electrodes are raised slowly up through the hole, a continuous record is made of the electrical properties of the various lithologies encountered.

Resistivity and *spontaneous potential* are the two electrical properties most commonly measured. Taken individually, each set of data is inconclusive; but taken together, these two measurements provide a good indication of some important lithologic distinctions. Shale or porous rock filled with salt water will be a good conductor. Such intervals in the borehole will have low resistivity. On the other hand, porous rocks that are filled with petroleum will have high resistivity, as will impermeable rocks that are tightly cemented with quartz or calcite.

With proper salinity relationships between drilling mud and formation water, spontaneous potential provides a direct indication of the permeability of the rock. If the rock is permeable, a spontaneous potential will be generated; if the rock is impermeable, there will be little or no spontaneous potential. Figure 3.21 il-

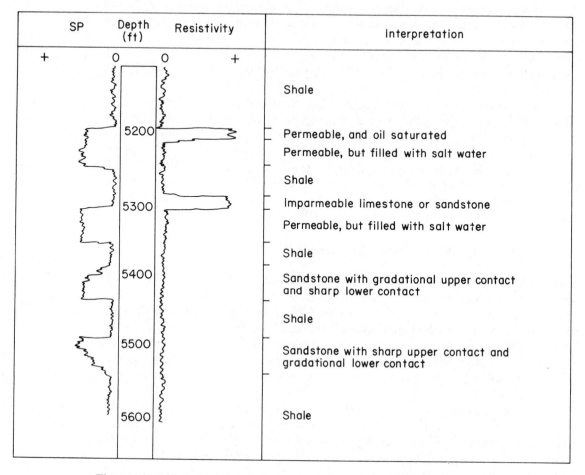

Figure 3.21 Lithologic properties suggested by various combinations of spontaneous potential and resistivity in petroleum exploration boreholes.

lustrates the various possible combinations of resistivity and spontaneous potential data that allow distinction among important lithologic attributes of the sediments encountered in a borehole.

Note further that the contacts between sandstones and shales can be investigated in considerable detail from electric logs. The abruptness of upper and lower contacts of sands may allow us to distinguish marginal marine sands as opposed to alluvial or turbidite sands. Given the availability of an occasional core to check the interpretation, we may be able to map these various sand types over large areas with considerable confidence.

Citations and Selected References

ALLEN, J. R. L. 1963. The classification of cross-stratified units, with notes on their origin. Sedimentology 2: 93–114.

————. 1968. Current ripples: their relation to patterns of water and sediment motion. North-Holland Publishing Co., Amsterdam. 433 p.
Detailed analysis of the how and why of primary structures.

BLATT, H. 1967. Provenance determination and recycling of sediments. J. Sed. Petrology 37: 1031–1044.
An eloquent plea for observational evidence rather than "plausible reasoning" in the discussion of provenance and transportational history of clastic sediments. Extensive bibliography.

————, G. MIDDLETON, and R. C. MURRAY. 1972. Origin of sedimentary rocks. Prentice-Hall, Inc., Englewood Cliffs, N. J. 634 p.

BLOOM, A. L. 1972. Geomorphology of reef complexes, *In* L. F. Laporte (ed.), Reef complexes in time and space. Soc. Econ. Paleont. Mineral. Spec. Pub. 19.

CAROZZI, A. V. 1960. Microscopic sedimentary petrology. John Wiley & Sons, New York. 485 p.

DUNHAM, R. J. 1962. Classification of carbonate rocks according to depositional texture, p. 108–121. *In* W. E. Hamm (ed.), Classification of carbonate rocks. Amer. Assoc. Petrol. Geol. Mem. 1.

FOLK, R. L. 1962. Spectral subdivision of limestone types, p. 62–84. *In* W. E. Hamm (ed.), Classification of carbonate rocks. Amer. Assoc. Petrol. Geol. Mem. 1.
A widely used shorthand system for naming carbonate rocks.

————. 1966. A review of grain size parameters. Sedimentology 6: 73–93.

————. 1968. Petrology of sedimentary rocks. Himpills Bookstore, Austin, Tex. 170 p.
A practical, down-to-earth discussion on how to study sedimentary rocks.

FRIEDMAN, G. M. 1967. Dynamic processes and statistical parameters compared for size frequency distribution of beach and river sands. J. Sed. Petrology 37: 327–354.
Application of statistical parameters to a geologic problem, the distinction between beach and river sands collected in bulk, loose, sample.

GARRELS, R. M., and F. T. MACKENZIE. 1971. Evolution of sedimentary rocks. W. W. Norton & Co., New York. 397 p.

GILBERT, C. M. 1954. Sedimentary rocks, p. 251–384. *In* H. Williams, F. J. Turner, and C. M. Gilbert, Petrography. W. H. Freeman and Co., San Francisco. 406 p.

HAMM, W. E. (ed.). 1962. Classification of carbonate rocks. Amer. Assoc. Petrol. Geol. Mem. 1. 279 p.
Symposium volume presenting various approaches to the classification of carbonate rocks.

HJULSTROM, F. 1939. Transportation of detritus by moving water, p. 5–31. *In* P. D. Trask (ed.), Recent marine sediments. Amer. Assoc. Petrol. Geol.

IMBRIE, J., and E. G. PURDY. 1961. Modern Bahamian carbonate sediments, p. 253–272. *In* W. E. Hamm (ed.), Classification of modern Bahamian carbonate sediments. Amer. Assoc. Petrol. Geol. Mem. 1.

———. 1964. Factor analytic method in paleoecology, p. 407–422. *In* J. Imbrie and N. D. Newell (eds.), Approaches to paleoecology. John Wiley & Sons, New York.

———, and H. BUCHANAN. 1965. Sedimentary structures in modern carbonate sands of the Bahamas, p. 149–172. *In* G. V. Middleton (ed.), Primary sedimentary structures and their hydrodynamic interpretation. Soc. Econ. Paleont. Mineral. Spec. Pub. 12.

JOPLING, A. V. 1966. Some principles and techniques used in reconstructing the hydraulic parameters of a paleo-flow regime. Sed. Petrology 36: 5–49.

KLEIN, G. DEV. 1963. Analysis and review of sandstone classification in the North American geological literature, 1940–1960. Bull. Geol. Soc. Amer. 74: 555–576.

KRUMBEIN, W. C., and L. L. SLOSS. 1963. Stratigraphy and sedimentation. W. H. Freeman and Co., San Francisco. 660 p.

KRYNINE, P. D. 1948. The megascopic study and field classification of sedimentary rocks. J. Geology 56: 130–165.

KUKAL, Z. 1971. Geology of Recent sediments. Academic Press, New York. 490 p.

LAHEE, F. H. 1952. Field geology. McGraw-Hill, New York. 883 p. Standard reference concerning field methods.

MCBRIDE, E. F. 1963. A classification of common sandstones. J. Sed. Petrology 33: 664–669.

MCKEE, E. D., and G. W. WEIR. 1953. Terminology for stratification and cross-stratification in sedimentary rocks. Bull. Geol. Soc. Amer. 64: 381–390.

MIDDLETON, G. V. (ed.). 1965. Primary sedimentary structures and their hydrodynamic interpretation. Soc. Econ. Paleont. Mineral. Spec. Pub. 12. 265 p. Symposium volume.

PETTIJOHN, F. J. 1954. Classification of sandstones. J. Geology 62: 360–365.

———. 1957. Sedimentary rocks. Harper & Row, New York. 718 p.

——— and P. POTTER. 1964. Atlas and glossary of primary sedimentary structures. Springer-Verlag, New York. 370 p.

POTTER, P., and F. J. PETTIJOHN. 1963. Paleocurrents and basin analysis. Springer-Verlag, Berlin. 296 p.

RHOADS, D. C. 1967. Biogenic reworking of intertidal and subtidal sediments in Barnstable Harbor and Buzzards Bay, Massachusetts. J. Geology 75: 461–476.

SHELTON, J. W. 1967. Stratigraphic models and general criteria for recognition of alluvial, barrier-bar, and turbitic-current sand deposits. Bull. Amer. Assoc. Petrol. Geol. 51: 2441–2461.

SIMONS, D. B., E. V. RICHARDSON, and M. L. ALBERTSON. 1961. Flume studies using medium sand (0.45mm). U. S. Geol. Survey Water Supply Paper 1498-A. 76 p.

VISHER, G. S. 1965. Fluvial processes as interpreted from Ancient and Recent fluvial deposits, p. 116–132. *In* G. V. Middleton (ed.), Primary structures and their hydrodynamic interpretation. Soc. Econ. Paleont. Mineral. Spec. Pub. 12.

WALKER, R. G. 1967. Turbidite sedimentary structures and their relationship to proximal and distal depositional environments. J. Geology 37: 25–43.

WELLER, J. M. 1960. Stratigraphic principles and practice. Harper & Row, New York. 725 p.

4
Dynamics
in the Stratigraphic Record

We recognize that depositional environments impart properties to sediments that can easily be preserved in the stratigraphic record. For example, the dynamics of the beach environment commonly produces sediments that are well-rounded, well-sorted sands with abraded marine fossils and well-developed, low-angle cross-stratification. If we examined a hand specimen and thin section of a well-lithified sandstone possessing these properties, we would likely deduce that the sandstone was once beach sediment.

Now, suppose that the hand specimen came from strata that included a 200-meter continuous sequence of that lithology. The modern beach environment generating the sediment type occupies a position within no more than a few meters of mean sea level. How could the geologic record stack up 200 meters of this lithology in a continuous fashion? Clearly, we are dealing with a problem involving dynamics of the surface of the earth. The most logical explanation for the observed stratigraphic thickness would be that this portion of the earth's surface subsided at the time when beach sedimentation was occurring upon it. Thus, relationships between the beach environment and sea level remained constant on a day-to-day basis, while a significant thickness of beach sediments accumulated over the long time period.

Geophysicists, in particular, have gathered a wealth of information concerning the ever-changing surface of the earth. They have studied isostasy, earthquakes, and plate tectonics, to mention a few examples. In a qualitative sense, stratigraphers have long recognized variations in earth history. Without variation on some time scale, all earth history would have been the same from the Precambrian to the Recent. Yet, in the past, stratigraphers have not dealt precisely with change or rates of change in the stratigraphic record, because they did not have adequate control on absolute time in stratigraphy. Now stratigraphy and geophysics have developed to the point that we may be able to relate stratigraphic qualitative observations of change to the dynamics quantitatively described by geophysicists. If we are truly to consider the "present as the key to the past," we must include dynamics in our view of the present. Just as modern beach sedimentation is the key to recognizing beach deposits in the stratigraphic record, modern dynamics of the earth surface are surely the key to understanding change and rate of change in the stratigraphic record. Let us therefore examine some of the generalities of the geophysicist's view of the earth and attempt to relate these to what we might expect to see in the stratigraphic record.

Isostasy and the Stratigraphic Record

The interior of the earth is viscous and the lithosphere of the earth is sufficiently weak that it would collapse under its own weight if it were not supported from below. Thus, the major topography of the earth's surface can be modeled as a

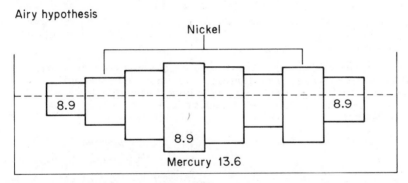

Figure 4.1 Basic concepts of isostasy. Both the Airy and the Pratt hypothesis seek to explain the topography of the earth with disconnected blocks floating in a heavy liquid. The Airy hypothesis suggests that topographic highs are compensated by unusual thicknesses of the same material, whereas the Pratt hypothesis proposes that topographic highs are composed of less dense material than topographic lows. The truth concerning the real earth is a combination of these two concepts. (After Leet and Judson, 1971.)

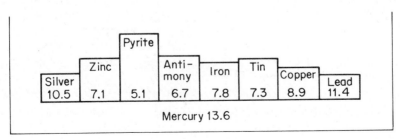

collection of less dense blocks floating upon a more dense liquid. If all the blocks have the same density, thicker blocks will produce higher topography. If the blocks are of differing density, less dense material will float with less of its total volume displacing the heavy liquid below (see Figure 4.1).

In fact, the lithosphere is a complicated combination of these two hypotheses. In either case, if we remove material from a high block and place it on a low block, we have disturbed the previous isostatic equilibrium and the system must adjust to the new distribution of load by a flow of viscous material deep in the earth. Removing material from high blocks and placing it on low blocks is precisely what sedimentation is all about.

Isostasy and basin filling. The stratigraphic record is essentially the material that has accumulated on the low blocks. Let us therefore consider the isostatic adjustments that might accompany two common sedimentological situations.

Consider first the thickness of sediments beneath an alluvial fan adjacent to high topography. Assume that local isostatic equilibrium is maintained throughout the discussion. We shall consider this assumption in a moment. Let us say that the surface of the alluvial fan now stands as much as 300 meters above the floor of the valley. How did things get this way and what lies beneath the alluvial fan?

Figure 4.2 Schematic cross section indicating isostatic subsidence accompanying alluvial fan sedimentation. Steps *A* through *C* indicate how an alluvial fan with 300 meters of topographic relief may actually accommodate 1000 meters of total sediment thickness because of isostatic subsidence accompanying the sedimentation.

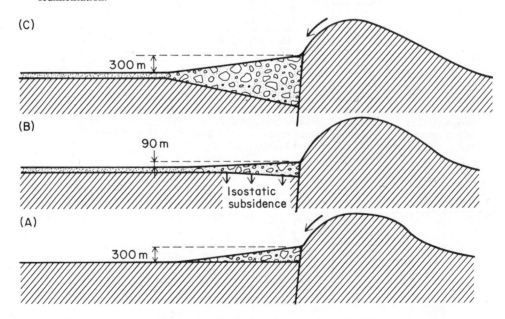

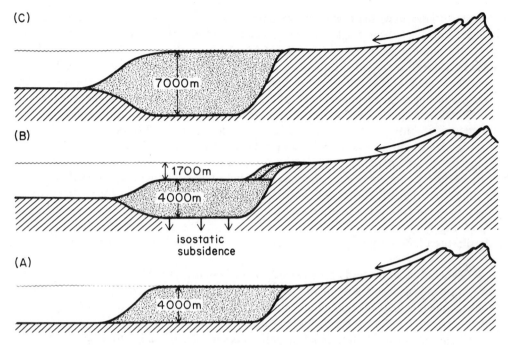

(C)

7000m

(B)

1700m

4000m

isostatic
subsidence

(A)

4000m

Figure 4.3 Schematic cross section indicating isostatic subsidence accompanying marine sedimentation into a basin initially filled with 4000 meters of water. Steps *A* through *C* indicate how isostatic subsidence accompanying sedimentation may allow an ultimate accumulation of sediment thickness that is nearly twice as great as the initial depth of the water column.

Figure 4.2 develops the present-day situation in stepwise fashion. Add the first 300 meters of alluvial fan sediment, and what happens? Sediment of density 2.3 displaces mantle material of density 3.3. This isostatic readjustment depresses the surface of the alluvial fan by 70% of the thickness of sediment added. Thus, after 300 meters of alluvial fan have been deposited, the alluvial fan has only 90 meters of topographic relief. By the time the alluvial fan has 300 meters of topographic relief, the total thickness of sediment is approximately 1000 meters.

The other common situation is the deposition of sediments into a water-filled basin. This process is occurring at the north end of the Gulf of Mexico, for example, where the Mississippi River deposits large volumes of sediment each year. Assumptions are essentially the same as in the preceding model, with the exception of sediment density. For this calculation, we can figure on a sediment density of 2.0.

Let us begin to put sediment into one end of a basin that is 4000 meters deep and initially filled with water (see Figure 4.3). As sediment is deposited in the basin, water of density 1.0 is displaced by sediment of density 2.0. Thus, the net addition of load onto the basin floor is 1.0 grams/centimeter of sediment deposited. This new load will be displacing the mantle material of density 3.3. Therefore, each unit thickness of new sediment deposited will cause basin subsidence amounting to approximately 30% of the thickness of the sediment. Thus, the first 4000 meters of sediment do not fill the basin; instead, isostatic subsidence has generated 1200 meters of new space above the sediment-water interface. The weight of this new water will itself add a component to isostatic subsidence, making the total new space 1700 meters. This process continues until approximately 7000 meters of sediment will be required to fill in the original 4000-meter hole.

In reality, this situation becomes much more complicated by compaction, diagenesis, and perhaps even low-grade metamorphism with increasing burial of the sediment. Whereas each increment added at the sediment-water interface is of rather low density, processes accompanying deep burial transform these loose, watery sediments into dense rock. Ultimately, it is the dense rock that must be isostatically compensated. The total thickness of the sediment required to fill the basin, therefore, may be considerably greater than indicated in Figure 4.3.

Relaxation times and the viscosity of the mantle. In the preceding discussion, we have talked about flow of mantle material induced by loading of new sediment onto a portion of the lithosphere. Inasmuch as mantle material is rather viscous, flow cannot instantaneously achieve a new isostatic equilibrium.

Most geophysical investigations of relaxation time deal with the rebound of continental areas that were previously occupied by Pleistocene glaciers (see McConnell, 1968, for example). Pleistocene history has provided the geophysicist with a natural experiment. Some 12,000 years ago, massive ice sheets existed over Scandinavia and Canada. If we presume that the ice was more or less at isostatic equilibrium, then the weight of the ice must have depressed the lithosphere. As the ice melted, the lithosphere should have risen to achieve new isostatic equilibrium.

Figure 4.4 indicates the present elevation of a 5000-year-old Scandinavian shoreline. Five thousand years ago, this surface was at or near sea level. Today, isostatic rebound has deformed the surface by some 100 meters. Gravity data indicate that an additional 200 meters of uplift is to be anticipated before isostatic equilibrium is achieved. From such data, estimates can be made concerning the time required to achieve a new isostatic equilibrium following a rapid loading or unloading. Times on the order of 10^3 to 10^4 years are to be anticipated. These numbers will become very important to us as we go on to consider isostatic response to rapid eustatic sea-level fluctuations.

Flexural rigidity of the lithosphere. Having recognized that the lithosphere is essentially floating on mantle material, we must consider the structural properties of the lithosphere. Is there separate and independent isostatic compensation for each small piece of lithosphere, or do large portions of the lithosphere have sufficient strength and rigidity to behave as a single unit? If a new load is placed over

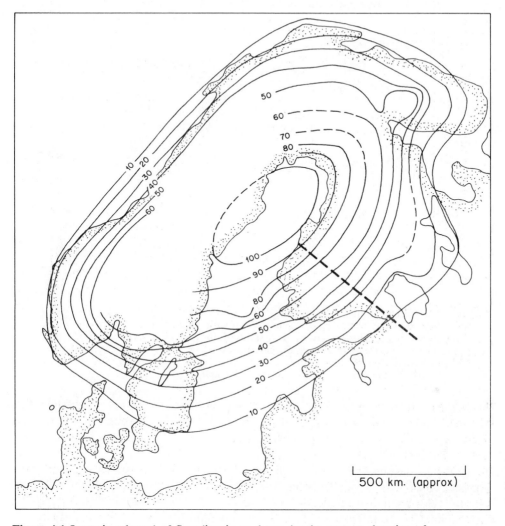

Figure 4.4 Isostatic rebound of Scandinavia as shown by the present elevation of the 5000 B.P. shoreline. [After McConnell, *Jour. Geophysical Research,* **73,** No. 22, 7090 (1968).] Courtesy American Geophysical Union.

a portion of the lithosphere, will isostatic equilibrium obtain directly below that load, or will compensation of the load be distributed over a broad area?

Once again, the geophysicist seeks areas where nature has run an experiment for him (see Walcott, 1970, for example). Proglacial lakes once contained a depth of water that can be estimated and had shorelines that were approximately level at the time the lake existed. Deformation of these shorelines provides information about the area over which the load was distributed. Similarly, volcanic

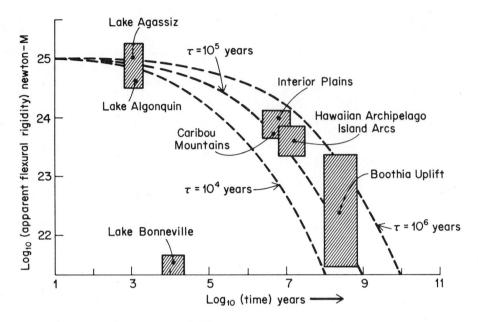

Figure 4.5 Apparent flexural rigidity of the lithosphere as a function of the duration of the naturally occurring experiment from which the flexural rigidity estimates were made. The low flexural rigidity of the late Bonneville area is probably due to extreme fracturing of the crust over broad areas of the western United States. [After R. I. Walcott, *Jour. Geophysical Research,* **75,** No. 20, 3951 (1970).] Courtesy American Geophysical Union.

islands usually have a closely associated bathymetric depression, suggesting that the load represented by the island is distributed over the surrounding lithosphere. Resulting estimates of the apparent flexural rigidity of the lithosphere are summarized in Figure 4.5.

The data suggest that the lithosphere behaves like a Maxwell solid. With rapid application of load ("rapid" being 10^3 to 10^4 years), the material behaves elastically. With continued application of load (a period of 10^6 to 10^8 years), viscous deformation of the lithosphere occurs and apparent flexural rigidity decreases.

The viscosity of the mantle regulates the rate at which isostatic equilibrium is attained, but it is the apparent flexural rigidity of the lithosphere that determines the area over which the new load will become compensated. Figure 4.6 indicates the effect of flexural rigidity upon the area over which a new load may be isostatically compensated.

Our interest in this phenomena centers around various permutations of the following question: Can rapid application of a new load (that is, sedimentation) cause subsidence over broad areas adjacent to the new load? If we look back at Figure 4.2, for example, we would ask whether the deposition of the alluvial fan would cause isostatic subsidence only beneath the fan or throughout the entire

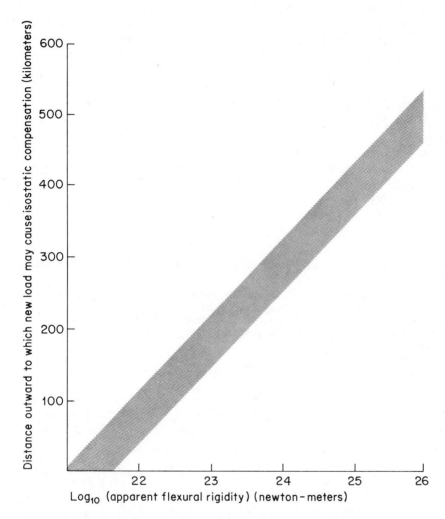

Figure 4.6 High flexural rigidity allows isostatic compensation to be spread over a large area; low flexural rigidity demands local isostatic compensation for new load. The indicated relationship is approximate. [Compiled from data and calculations contained in R. I. Walcott, *Jour. Geophysical Research,* **75**, No. 20, 3951 (1970).] Courtesy American Geophysical Union.

valley. Note that the viscosity of the mantle dictates that isostatic response to a new load will occur within 10^3 to 10^4 years, whereas apparent flexural rigidity of the crust does not undergo significant deterioration until 10^5 to 10^7 years after the application of new load. Thus, if it can be demonstrated that new load was applied within 10^3 to 10^4 years, we would expect apparent flexural rigidity of the lithosphere to be rather high. Moreover, regional downwarping because of the new load should extend approximately 500 kilometers outward from the load. If, on the other hand, it could be demonstrated that the load was applied gradually over a long time span (perhaps 10^7 to 10^8 years), then apparent flexural rigidity of the lithosphere would be much lower and the surrounding region would consequently be less affected by the new load.

Isostatic response to small loads. How small a load will induce isostatic response? The geophysicists have worked primarily with the unloading of continental glaciers. Glacial retreat involves the unloading of approximately 150 bars. Such unloading is clearly a sufficient reduction to initiate significant isostatic response during a reasonably short time. Can similar response be generated by a significantly smaller load?

Bloom (1967) recognized another simple experiment that nature has run for us over the last few thousand years. Most students of the post-Wisconsin sea level rise agree that sea level has been rising rather slowly throughout the last 5000 years. However, various locations give differing estimates concerning the shape of the sea-level rise curve (see Figure 4.7). Bloom proposed that these discrepancies

Figure 4.7 Submergence history of five sites along the eastern coast of the United States. Data points are carbon-14 ages determined on molluscs or peat deposits that accumulated at or near the sea level existing at that time. Each curve records the interaction between rising sea level and local subsidence or emergence. (After Bloom, 1967.)

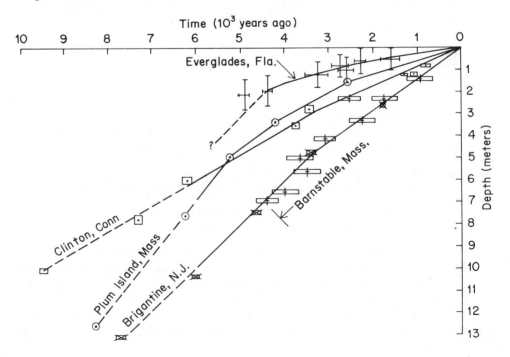

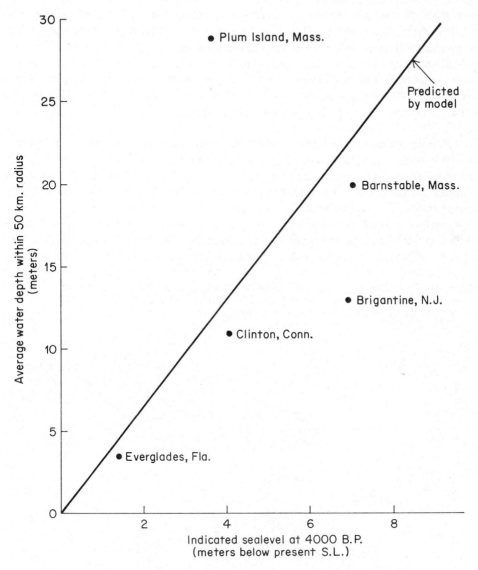

Figure 4.8 Isostatic response to new water load on continental shelves produces a systematic relationship between apparent sea level at 4000 B.P. and average water depth within a 50-kilometer radius of the area where the sea-level estimate was obtained. The solid line is the relationship predicted by the model. (Compiled from data presented in Bloom, 1967.)

are a function of isostatic response of the continental shelf to new water load; the thicker the average depth of water near these shorelines, the greater the isostatic subsidence of that shoreline's sea-level record. Bloom therefore undertook to estimate average water depth within 50 kilometers of the various sea-level data points under consideration. Figure 4.8 summarizes a portion of his results. The data strongly suggest that isostatic response to a load change of less than 10 bars has indeed occurred. Other scientists offer different interpretations of this phenomena (Walcott, 1972, for example), but most now agree that the earth does undergo isostatic response to load changes as small as 10 bars.

Isostatic response to a single eustatic sea-level rise. As we stand on an outcrop and look at a thick sequence of marine strata, we inevitably wish to know the magnitude of the transgression that was responsible for these rocks. Let us therefore consider the relationship between the height of a sea-level rise and the thickness of marine strata that may accumulate in response to that sea-level rise.

Consider a low-lying continental area more or less at sea level. Allow it to be flooded by a 100-meter sea-level rise. Assume that the area is large enough that we can ignore problems related to apparent flexural rigidity of the lithosphere.

Figure 4.9 Estimation of sediment thickness accommodated by a 100-meter eustatic sea-level rise. Steps *A* through *C* summarize the interaction between sea-level rise, isostatic subsidence, and sedimentation.

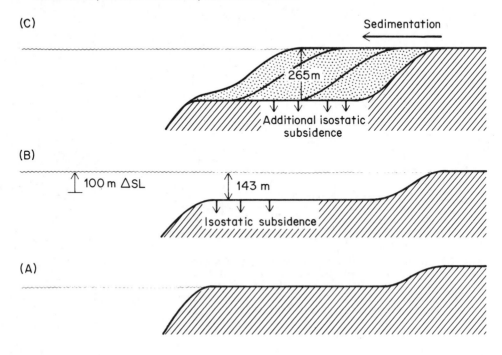

Figure 4.9 summarizes the events that follow logically from a 100-meter sea-level rise. Isostatic subsidence of the continental area due to the weight of new water load would result in a water depth of 143 meters over the former continental surface. One hundred meters of this depth is the result of the sea-level rise; 43 meters is the result of attendant isostatic subsidence of the continental area. If the water is then displaced by sediment of density 2.0 (disregarding complications of com-

Figure 4.10 Magnetic anomaly pattern over the Reykjanes Ridge. The intensity of the present-day magnetic field is anomalously high over the stippled areas and anomalously low over the white areas.

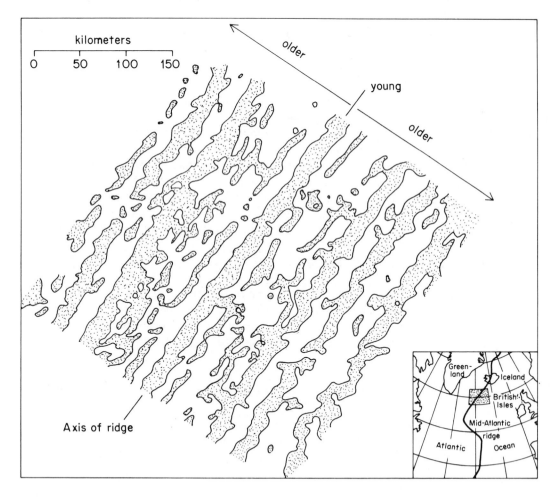

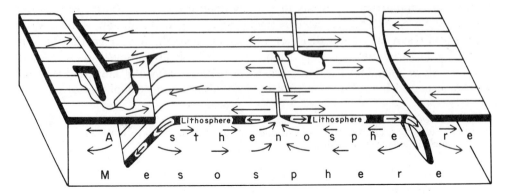

Figure 4.11 Schematic diagram indicating the major elements of plate tectonics. New crust is generated along mid-ocean ridges and old crust is consumed beneath trenches. The major distribution of these features is indicated in Figure 4.12. [After B. Isacks *et al., Jour. Geophysical Research,* **73,** No. 18, 5857 (1968).] Courtesy American Geophysical Union.

paction and lithification), the total accumulation of sediment required to produce a sediment surface at the new high sea-level will be 265 meters. The original depth of water will be displaced by 143 meters of this sediment; the remainder will fill in the resulting isostatic subsidence. Thus, the sea-level rise required to explain any given thickness of shallow-marine sediments over the continental area is approximately 30% to 40% of the total sediment thickness observed. The remaining 60% to 70% of the sediment thickness is accommodated by isostatic subsidence.

Global Dynamics of the Lithosphere

For a number of years, it has been recognized that the earth's magnetic field from time to time was reversed from its present polarity. Scientists have found it particularly convenient to study this phenomenon in layered sequences of volcanic rocks. As lavas cool, they take on magnetic properties indicative of the magnetic

Figure 4.12 (Right) Principal areas of generation of new crust (open bars) and principal areas of consumption of old crust (solid bars). Areas of crustal generation are tectonic highs generally standing 1000 to 2000 meters above surrounding ocean basins. Areas of crustal consumption are commonly topographic lows, as in the case of the circumpacific trenches. Areas of crustal consumption can also form tectonic highs, as in the Alpine-Himalaya mountain belt. The Cordilleran chain of the western Americas is associated with present-day crustal generation in the western United States and with crustal consumption throughout Central and South America. Note the location of Figures 4.10, 4.13, 4.14, 4.15, and the islands of Barbados (1), New Guinea (2), and southern Japan (3), which are mentioned in the text. [After B. Isacks *et al., Jour. Geophysical Research,* **73,** No. 18, 5861 (1968).] Courtesy American Geophysical Union.

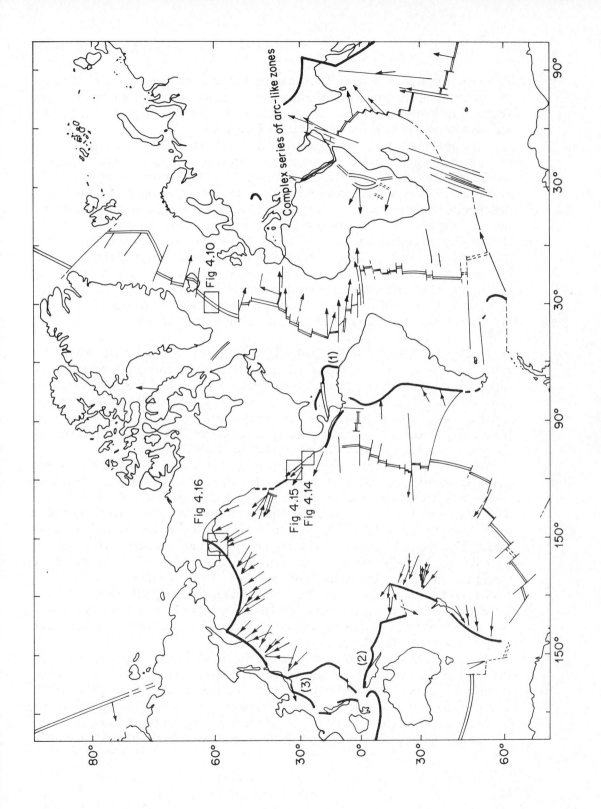

Complex series of arc-like zones

Fig 4.10

(1)

Fig 4.16

Fig 4.15
Fig 4.14

(3)

(2)

field in which they crystallize. Thus, normal and reversed fields are easily discernible in stratigraphic sequences. Furthermore, volcanic rocks are datable by conventional radiometric methods. On this basis, a late Cenozoic magnetic stratigraphy has been developed as is indicated later in Figure 8.1.

Vine and Matthews (1966) proposed that these same patterns of normal and reversed magnetism explain the magnetic properties of oceanic crusts observed along mid-ocean ridges (see Figures 4.10 and 1.2). If new formations of oceanic crust take place along mid-ocean ridges, then the new crust should take on magnetic properties indicating the orientation and intensity of the magnetic field at the time of formation. With continuous generation of new crust beneath these ridges, the ocean floor spreads apart, producing a horizontal stratigraphy of the earth's magnetic history that can be correlated with the vertical stratigraphy developed in layered volcanic rocks on land. This hypothesis has become the cornerstone for one of the most important unifying concepts of modern geology. As indicated in Figure 4.11, new lithosphere is being generated beneath mid-ocean ridges, and old lithosphere is being consumed beneath the deep-sea trenches.

First-order tectonic basins and highs. Deep-sea trenches and mid-ocean ridges, such as indicated in Figure 4.12, constitute the first-order tectonic topography of the earth. The ridges are very broad features that rise a thousand to several thousand meters above the adjacent abyssal plains. These features lie beneath considerable depth of water, so they do not seem like the major tectonic features of the earth. Yet when a ridge system becomes involved with familiar land areas, the magnitude of these tectonic features becomes clear.

Consider, for example, tectonic activity in the southwestern United States, where the East Pacific Rise entangles itself with the North American continent. Figure 4.13 depicts the tectonics of the Gulf of California, where the East Pacific Rise first encounters North America. Baja California is being essentially pulled apart from the rest of Mexico by transform faults active within the Gulf of California. This general rift system extends northward into southern California; we shall discuss it further when we consider second-order tectonic highs and lows within the general concept of sea-floor spreading. To the northeast of the Gulf of California, the whole region from the Sierra Nevadas to the Great Plains has been raised by at least some 1500 meters since the Eocene epoch. It is generally inferred that this broad uplift reflects the interaction between a preexisting continent and the northward extension of the East Pacific Rise.

Just as the areas of crust generation are the major tectonic highs of the earth, the areas of crust consumption are the major tectonic lows. Deep-sea trenches of tectonic origin are today most prominently developed throughout the western Pacific (Figure 4.12). The tectonic depressions associated with underthrusting of the lithosphere range from 10^3 to 10^4 meters, depressed in comparison to associated abyssal plains. Without a doubt, the modern deep-sea trenches closely

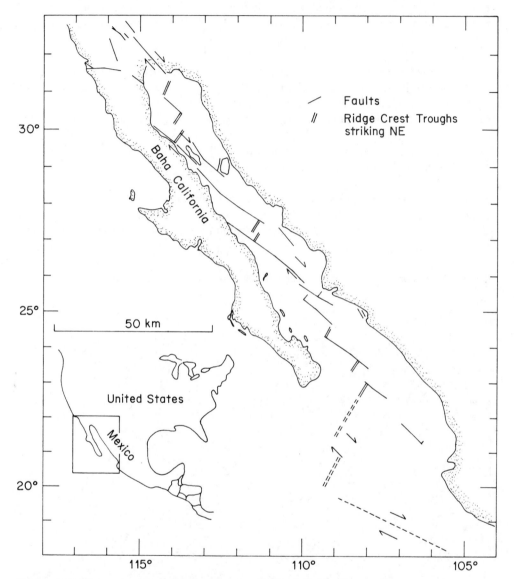

30°

25°

50 km

20°

115°   110°   105°

Faults
Ridge Crest Troughs
striking NE

Baha California

United States

Mexico

Figure 4.13 Major structural features of the Gulf of California. A ridge crest (East Pacific Rise), extensively dissected by transform faults, extends into the Gulf of California. Tectonics associated with the generation of new crust along the ridge crest is pushing Baja California relatively to the northwest. The extension of this tectonic pattern to the north is the subject of Figure 4.14. (After Isacks, 1968, with modification.)

resemble the thick geosynclinal accumulations of sediments long familiar to the geologist from the study of mountain belts. We shall return to this similarity in Chapter 5.

Second-order tectonic basins and highs. It is clear that the strike-slip fault system of southern California originated in transform faults associated with the northward extension of the East Pacific Rise. In Figure 4.14, note that this fault system is in turn a predominant controlling factor on the distribution of local basins and highs throughout southern California. Apparently, the strike-slip fault system has sufficiently sheared the lithosphere under this area, so that each block may behave more or less independently of its neighbor.

Rates of tectonic deformation. When geophysicists discuss motion of the lithosphere, they are usually considering lateral displacement rates of the major plates. For example, if we can make a reliable correlation between the land-derived chronology of magnetic events and the stripes in a map pattern such as in

Figure 4.14 Major strike-slip faults and late Cenozoic basins of southern California.

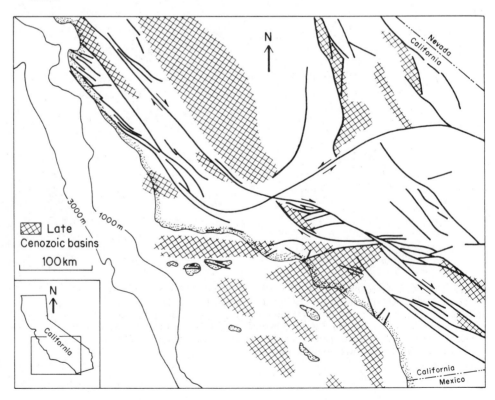

Figure 4.10, then we can estimate the rate at which lithosphere plates are moving away from the crest of the ridge. These lateral velocities are typically 4 centimeters/year; velocities in the range of 15 centimeters/year are not uncommon.

Stratigraphers are much more interested in rates of vertical displacement. For example, tectonic subsidence will generally allow room for new sediment accumulation. Thus, the following discussion deals primarily with rates of vertical displacement.

Earthquakes provide the simplest and most direct opportunity to observe deformation of the earth's surface. Figure 4.15, for example, summarizes vertical displacements that accompanied the magnitude 8.5 Alaska earthquake of 1964 (Plafker, 1965). We can observe (1) specific displacements as great as 8 meters and (2) regional displacements of 1 to 2 meters as much as 75 kilometers to either side of the seismic zone strike. Can we estimate the average rate of vertical tectonic displacement from these data? There are difficulties.

To begin with, these displacements record *discontinuous* adjustment of the earth's surface to *continuous* processes that are occurring in depth. The fact that a piece of land rises 8 meters overnight tells us nothing concerning the *average rate* at which tectonic uplift is occurring. Indeed, Plafker notes that the general history of Prince William Sound immediately prior to the earthquake was one of general subsidence, as indicated by drowned forests and intertidal peat bogs along the shores. Thus, large-scale tectonic deformation occurred overnight, yet similar rapid deformation is not likely to occur again in the near future.

A magnitude 8.5 earthquake is so large and unusual that we have very poor statistical data concerning the likelihood of recurrence within the same rather limited area. Furthermore, as we have previously noted, isostatic adjustment to new load occurs on a time scale of 10^3 to 10^4 years. Thus, a significant portion of the tectonic changes presently observable may be smoothed out with time as isostatic response to the load occurs. But then, on that time scale, there will undoubtedly be a new deformation added upon the deformations presently observed. All in all, it is difficult to estimate an average rate of tectonic deformation over significant time scales from the study of deformation accompanying a single earthquake.

Plafker also reports a radiocarbon date on driftwood associated with the highest of five uplifted beaches on Middleton Island. This uplifted beach dates 4470 ± 250 B.P. and is presently situated 30 meters above sea level. Thus, these data indicate an average rate of tectonic uplift of approximately 8 meters/1000 years over the past 5000 years. Even at this rate, we must note that 5000 years is an extremely short period of time from a sedimentological and stratigraphic point of view. Let us therefore look for some places where nature has run some longer experiments for us.

Approximately 125,000 years ago, the sea stood a few meters above its present level. Where coral reefs grew in association with this sea level, the geologist is provided with both a material (aragonite corals) that can be easily dated (by the

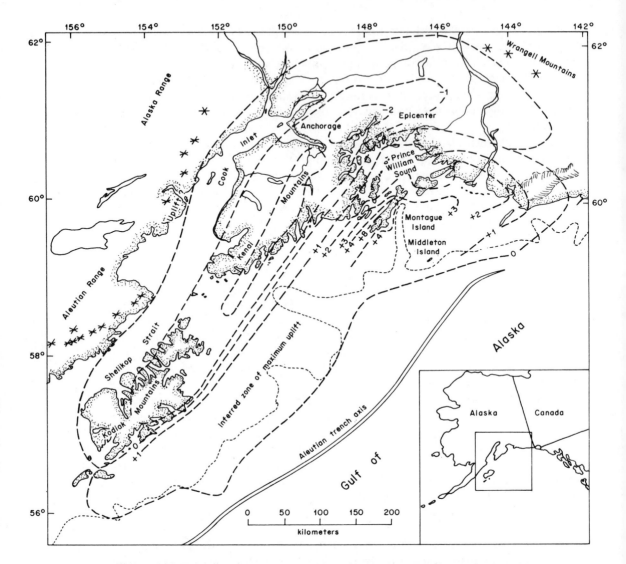

Figure 4.15 Tectonic deformation in southcentral Alaska following the 1964 earthquake. Contours indicate vertical displacement of the land surface; contour interval is 1 meter. [After G. Plafker, "Tectonic Deformation Associated with the 1964 Alaska Earthquake," *Science,* **148,** 1675–1687 (25 June 1965), with modifications.] Copyright 1965 by the American Association for the Advancement of Science.

Th-230 method) and with a datum plane that closely approximates the 125,000 B.P. sea-level surface. Subsequent deformation of this 125,000 B.P. sea-level datum provides a good measure of the average rates of tectonic uplift and tectonic subsidence on the time scale of 10^5 years.

Tectonic uplift would place the 125,000 B.P. coral reefs above sea level, where they would be a conspicuous feature of any landscape containing them. Tectonic subsidence would place the 125,000 B.P. reefs below sea level where they would be inevitably recolonized by younger reefs and thus obscured from easy examination. Therefore, data concerning uplift of this surface are better and more numerous than data concerning the subsidence of this surface.

Uplifted Pleistocene coral reefs have been dated by Th-230 and similar radiometric methods in at least three areas widely separated around the world. The most detailed study is that of Mesolella *et al.* (1969) on the island of Barbados, in the West Indies. The case history of this study is discussed further in Chapter 12 as an example of how we can apply sedimentological data to unravel details of earth history. Similar studies in the Pacific are reported by Veeh and Chappell (1970) and by Konishi *et al.* (1970). The Barbados data suggest average uplift rates of .3 meter/1000 years, with some small portions of the island going up as much as .5 meter/1000 years. The Pacific data, both from the north coast of New Guinea and from the southern Japanese Islands, indicate average rates of tectonic uplift in the range of 1 to 2 meters/1000 years.

In British Honduras, Central America, shallow seismic profiling over the modern coral reefs and associated sediments affords an example of tectonic subsidence of Pleistocene coral reefs (Purdy and Matthews, 1964; Purdy, 1972; see also Figures 12.3 and 12.12). In the northern portion of the study area, Pleistocene coral reefs outcrop on land. The Pleistocene surface gradually slopes southward. In the southern portion of the study area, some 150 kilometers south of the subaerial outcrop, the Pleistocene surface is located 30 meters below present sea level. This deformation apparently reflects tectonic subsidence associated with major left lateral strike-slip motion along the Cayman Trench. An average rate of tectonic subsidence of approximately 30 centimeters/1000 years is indicated.

If we wish to examine the average rates of tectonic deformation over a still longer time scale, we must look beyond sea-level data and radiometric dating techniques. Sea-level events older than 125,000 years B.P. are numerous. However, our usual technique for dating these deposits, the thorium-230 method, approaches equilibrium at around 250,000 years and is in fact of little use on a worldwide scale beyond the 125,000 B.P. sea-level event. Thus, for our longer time scale, we must look to biostratigraphically defined time planes.

Christensen (1965) attempts to estimate vertical displacements in the Coast ranges and San Joaquin valley of central California within the Pliocene-Pleistocene biostratigraphic framework. His results are summarized in Figure 4.16. In the

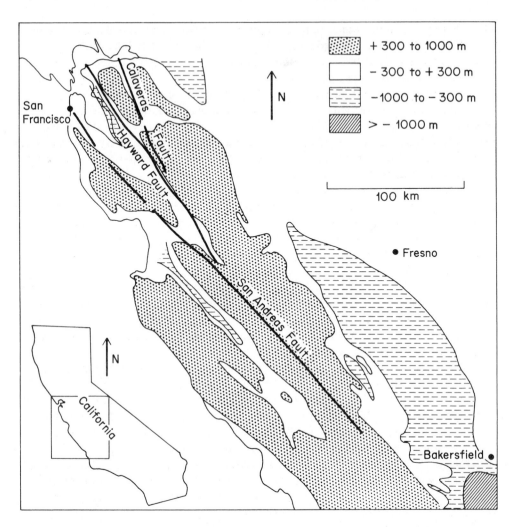

Figure 4.16 Uplift and subsidence in central California during the last 3 million years. (After Christensen, 1965, with simplification.)

southeastern corner of the map area, as much as 3000 meters of subsidence have occurred within the last 3 million years. Over broad areas of the San Joaquin valley, however, 500 meters of subsidence is a more appropriate general figure. Similarly, the Coast ranges have been uplifted in places by as much as 1000 meters over the past 3 million years. Once again, however, a general estimate of 700 meters uplift during 3 million years is a more realistic description of large areas in the Coast ranges. Thus, estimates of the average rate of tectonic subsidence range from the general average of 15 to 20 centimeters/1000 years to as much as 1 meter/1000 years. Similarly, average rates of tectonic uplift range from a general average of 20 to 25 centimeters/1000 years to as much as 35 centimeters/1000 years.

Matsuda *et al.* (1967) present a similar analysis based on the elevation of marine Miocene deposits over a large portion of Japan. They estimate average vertical displacement rates in the range of ±1 meter/1000 years, although local figures may be as high as 5 meters/1000 years.

To summarize, we have discussed documentation of average rates of tectonic uplift that range from 25 centimeters/1000 years to 8 meters/1000 years and average rates of tectonic subsidence that range from 30 centimeters/1000 years to as much as 5 meters/1000 years. Clearly, this summary requires further qualification. Note especially that all the high estimates of tectonic uplift rate come from the major island-arc provinces of the northwest and western Pacific. Geophysical evidence indicates that this region is the most tectonically active area on the face of the earth today. Furthermore, the average rates of tectonic uplift demonstrate that this observation holds true on a time scale of 10^6 years.

It is interesting to observe that the island of Barbados, West Indies, is in a position tectonically similar to the Alaskan uplift area depicted in Figure 4.15. Both areas are broad tectonic uplifts situated 100 to 200 kilometers in front of the volcanic arc. Yet the Aleutian arc-trench system is extremely active seismically, whereas the Caribbean island arc-trench system is the site of only moderate seismic activity. The respective estimates of the average rate of tectonic uplift bear out this contrast: 8 meters/1000 years for Middleton Island, and only 30 to 50 centimeters/1000 years for Barbados. Thus, in areas of extremely active trench development, we may expect vertical tectonic displacement rates toward the high end of the spectrum. In all other areas, we shall probably consider a rate of 1 meter/1000 years, up or down, as the maximum rate to be anticipated. Comparison between this number and the average rate of eustatic sea-level fluctuations will be extremely important in later chapters.

Eustatic Sea Level Fluctuations

At present, the earth has a considerable amount of water stored in glaciers over land areas. When there are fluctuations in the size of continental glaciers, there are corresponding fluctuations in sea level. During Wisconsin time, for example, the sea stood 80 to 100 meters below the present level. If all of the ice presently on the face of the earth were to melt, sea level would rise still another 65 meters.

Fluctuations of sea level resulting from changes in the ice budget or from changes in the volume of the ocean basin are referred to as eustatic sea-level fluctuations. The rate at which sea-level fluctuations occur will be the final, and perhaps the most important, dynamic aspect of our earth today. We shall later consider that rate when we study the dynamics of earth history as illustrated in the stratigraphic record.

The post-Wisconsin eustatic sea-level rise. Relict sediments associated with previous shorelines allow us to reconstruct the history of sea-level rise following Wisconsin glaciation. In particular, samples of bottom sediment taken from water depths as great as 100 meters commonly contain molluscs or peat, which are known to accumulate only under shallow-marine or brackish-water conditions. The presence of these materials *in situ* dictates that the sea was once at that level. Thus, Carbon-14 dates on relict shore deposits of the present-day continental shelf allow the reconstruction of sea-level history.

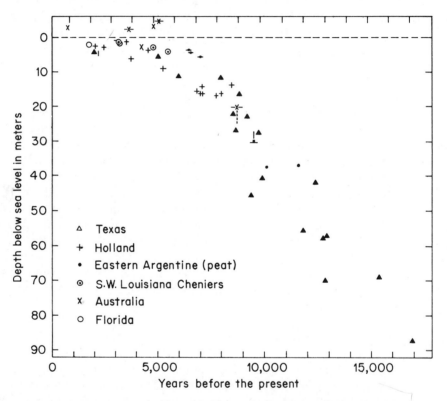

Figure 4.17 Sea-level rise over the last 17,000 years, as determined by carbon-14 dating of peat deposits and brackish-water molluscs. (After Sheppard, 1960.)

Figure 4.17 summarizes Carbon-14 data pertaining to the history of sea-level rise over the past 17,000 years. From 15,000 years B.P. to 6,000 years B.P., it would appear that the average rate of sea-level rise was approximately 7 meters/1000 years. The shape of the curves for Oxygen-18 variation in deep-sea cores (in Figure 15.3, for example) suggest that other sea-level rises during the Pleistocene occurred at similar rates.

Rate of eustatic sea-level lowering. The history of sea-level lowering into the main Wisconsin glaciation lies beyond the range of radiocarbon dating and has been further obscured by the sediment-reworking associated with the post-Wisconsin sea-level rise. We do know that the sea stood a few meters above present levels 125,000 years ago and that it stood at least 80 meters below the present level 20,000 years ago. These data constitute an absolute minimum estimate of the average rate of sea-level lowering: 80 centimeters/1000 years. It must be noted, however, that sea-level lowering after 120,000 years B.P. was undoubtedly discontinuous and interrupted by numerous minor sea-level rises that produced anywhere from 2 to 5 minor terraces in the age range of 40,000 to 105,000 years B.P. (Mesolella *et al.,* 1969; Veeh and Chappell, 1970; Konishi *et al.,* 1970). Over the short term, rates of sea-level lowering in the range of 2 to 8 meters/1000 years are probably to be anticipated from the Pleistocene data.

Transgression and Regression in the Stratigraphic Record

The foregoing discussion outlines the various ways in which the relationship between land and sea level may be altered. Stratigraphers have been dealing with these problems on an *ad hoc* basis for many years. Using the "center of the basin" as a reference point, they have called it a *transgression* when the shoreline moves outward and a *regression* when the shoreline moves inward (Krumbein and Sloss, 1963).

Note that neither term indicates the cause of this dynamic interaction between water and the land. The words "transgression" and "regression," used alone, carry absolutely no connotation concerning the tectonic, eustatic, or isostatic nature of the dynamic event. In this book, we shall use these two words sparingly and only as "inclusive terms of ignorance." If you see continental rocks overlain by marine rocks, there has rather clearly been a "transgression." But go further in your thinking: Was the cause tectonic, eustatic, or isostatic?

The word "regression" is particularly difficult. In addition to possible tectonic, eustatic, or isostatic causal mechanisms, a "regression" may simply record how sediment has filled in an area.

To get around these problems, we must employ new words. It would be senseless to redefine "transgression" and "regression"; they are simply too ingrained into the stratigraphic literature. Let us therefore more accurately define the terms "submergence," "emergence," progradation," and "aggradation."

Submergence and *emergence* refer to changes in the vertical relationship between sea level and a fixed point on or in subtidal sediments. If the thickness of water plus sediments over that point increases, the area is undergoing submergence. Similarly, a decrease will be referred to as emergence. *Progradation* occurs as sediment accumulation builds outward laterally. Continued accumulation of beach sand, for example, will cause the beach to prograde seaward. *Aggradation* refers to the strictly vertical accumulation of sediment. Note that the terms "progradation" and "aggradation" are independent of the terms "submergence" and "emergence." There is no reason why progradation, for instance, cannot occur under conditions of general submergence. Sea level may be rising, but there is sufficient sediment supply to continue building seaward. These are points upon which we shall dwell at some length in later chapters.

Citations and Selected References

BANDY, O. L., and R. E. ARNAL. 1969. Middle Tertiary basin development, San Joaquin Valley, California. Bull. Geol. Soc. Amer. 80: 783–820.
A combination of biostratigraphy and paleoecology of foraminifera allows us to interpret sedimentation rates and basin subsidence.

BLOOM, A. L. 1967. Pleistocene shoreline: a new test of isostasy. Bull. Geol. Soc. Amer. 78: 1477–1494.

CHRISTENSEN, M. N. 1965. Late Cenozoic deformation in the central coast ranges of California. Bull. Geol. Soc. Amer. 76: 1105–1124.

———. 1966. Late Cenozoic crustal movements in the Sierra Nevada of California. Bull. Geol. Soc. Amer. 77: 163–182.

COX, A. V., C. G. DALRYMPLE, and R. R. DOELL. 1967. Reversals of the earth's magnetic field. Sci. American 216: 44–61.

HILL, M. L. 1971. A test of new global tectonics: comparison of northeast Pacific and California structures. Bull. Amer. Assoc. Petrol. Geol. 55: 3–9.
One of the "grand old men" of southern California geology suggests that things are not quite so simple as proponents of "the new global tectonics" might have us believe.

ISACKS, B., J. OLIVER, and L. R. SYKES. 1968. Seismology and the new global tectonics. J. Geophy. Res. 73: 5855–5899.
Excellent summary paper concerning seismology and plate tectonics.

KING, P. B. 1959. The evolution of North America. Princeton Univ. Press, Princeton, N. J. 189 p.
In the context of this chapter, it is particularly interesting to consider how classical descriptive geology holds up in the light of "the new global tectonics."

———. 1965. Tectonics of Quaternary time in Middle North America, p. 831–870. *In* H. E. Wright, Jr., and D. G. Frey (eds.), The Quaternary of the United States Princeton University Press, Princeton, New Jersey, 922 p.

KONISHI, A., S. O. SCHLANGER, and A. OMURA. 1970. Neotectonic rates in the central Ryukyu islands derived from Thorium-230 coral ages. Marine Geol. 9: 225–240.

KRUMBEIN, W. C., and L. L. SLOSS. 1963. Stratigraphy and sedimentation. W. H. Freeman and Co., San Francisco. 660 p.

LEET, L. D., and S. JUDSON. 1971. Physical geology. 4th ed. Prentice-Hall, Inc., Englewood Cliffs, N. J. 687 p.

LUYENDYK, B. P. 1970. Dips on down-going lithospheric plates beneath island arcs. Bull. Geol. Soc. Amer. 81: 3411–3416.

MATSUDA, T., K. NAKAMURA, and A. SUGIMURA. 1967. Late Cenozoic orogeny in Japan. Tectonophysics 4: 349–366.

MAXWELL, A. E. (ed.). 1970. The sea, vol. IV: New concepts of sea floor evolution. Wiley-Interscience, New York. Part I, 628 p. Part II, 672 p.

MESOLELLA, K. J., R. K. MATTHEWS, W. S. BROECKER, and D. L. THURBER. 1969. The astronomical theory of climatic change: Barbados data. J. Geology 77: 250–274.

————, H. A. SEALY, and R. M. MATTHEWS. 1970. Facies geometries within Pleistocene reefs of Barbados, West Indies. Bull. Amer. Assoc. Petrol. Geol. 54: 1899–1917.

McCONNELL, R. K., JR. 1968. Viscosity of the mantle from relaxation time spectra of isostatic adjustment. J. Geophy. Res. 73: 7089–7105.

PLAFKER, G. 1965. Tectonic deformation associated with the 1964 Alaska earthquake. Science 148: 1675–1687.

PURDY, E. G., and R. K. MATTHEWS. 1964. Structural control of Recent calcium carbonate deposition in British Honduras [Abstract]. Geological Society of America annual convention, Program with abstracts: 157.

————. 1972. Reef configurations: some causes and effects, *In* L. F. Laporte (ed.), Reef complexes in time and space: their physical, chemical and biological parameters. Soc. Econ. Paleont. Mineral. Spec. Pub. No. 19.

SHEPARD, F. P. 1960. Rise of sea level along northwest Gulf of Mexico, p. 338–344. *In* F. P. Shepard, F. B. Phleger, and T. H. van Andel (eds.), Recent sediments, northwest Gulf of Mexico. Amer. Assoc. Petrol. Geol.

STUART, J. H. 1971. Basin and range structure: a system of horsts and grabens produced by deep-seated extension. Bull. Geol. Soc. Amer. 82: 1019–1044.

VEEH, H. H. and J. CHAPPELL. 1970. Astronomical theory of climatic change: support from New Guinea. Science 167: 862–865.

VINE, F. G., and D. H. MATTHEWS. 1963. Magnetic anomalies over ocean ridges. Nature 199: 947–949.

VOGT, P. R., E. D. SCHNEIDER, and G. L. JOHNSON. 1969. The crust and upper mantle beneath the sea, p. 556–617. *In* P. J. Hart (ed.), The earth's crust and upper mantle. Geophys. Monogr. 13, Amer. Geophys. Union, Washington, D. C.

WALCOTT, R. I. 1970. Flexural rigidity, thickness and viscosity of the lithosphere. J. Geophys. Res. 75: 3941–3954.

————. 1972. Past sea levels, eustasy and deformation of the earth. Quaternary Research 2: 1–14.

5

Geologic Framework

of Sediment Accumulation

In the two previous chapters, we have looked at the variations among sedimentary rocks and at the dynamics of the present-day earth. Now we must begin to put these two sets of information into the context of earth history. Where have dynamic events happened in the past and what kinds of sediments were deposited? In this chapter, we shall approach these questions on a very general level. Our purpose here is to establish the broad context within which we shall examine, in Sections III and IV, the details of small areas.

The following discussion divides the sedimentary geology of a continent into three general categories: (1) the stable craton, (2) the mobile belts (geosynclines and mountain ranges), and (3) the coastal plains of the continental margin. Our discussion of these geologic provinces will center around the geology of central North America (see Figure 5.1).

The Stable Craton

The fundamental property of this geologic province is that it is underlain at relatively shallow depth by Precambrian basement rocks, which have been worn down to base level.

General characteristics of the interior lowlands of the United States. The interior lowlands of the United States is the sediment-covered area that lies south or west of the Canadian shield, west of the Appalachians, east of the Rockies, and north of the Ouachitas.

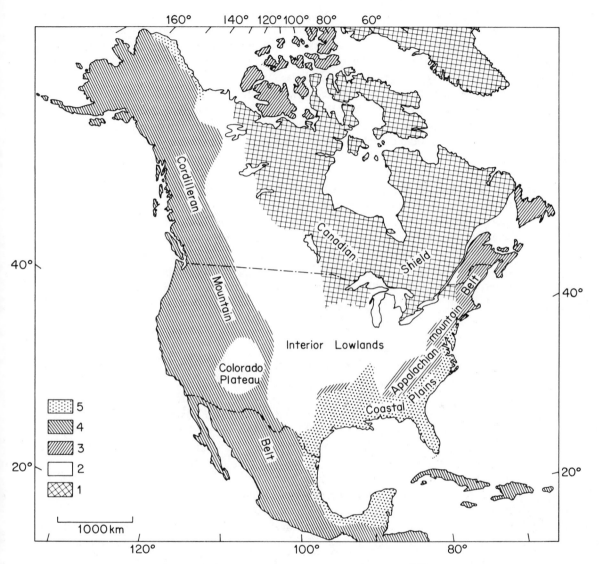

Figure 5.1 Major geologic provinces of North America. Symbols are as follows:
(1) Precambrian basement rocks of the Canadian shield. (2) Flat-lying sediments of the interior lowlands, predominantly Paleozoic but with important
Mesozoic sequences in the west. Sediment thickness in the interior lowlands
generally averages 1 kilometer and seldom exceeds 3 kilometers. (3) Older
Paleozoic sediments of the Ouachita-Appalachian mountain belt. Stratigraphic
thickness of the sediments commonly ranges from 10 to 20 kilometers. (4)
Paleozoic through Cenozoic sediments of the Cordilleran mountain belt. Major
tectonism is late Mesozoic to early Cenozoic. (5) Late Mesozoic to Cenozoic
sediments of the modern continental margin. Sediment accumulations are 5 to
15 kilometers thick and have not undergone tectonic deformation.

The sediments are predominantly Paleozoic and Mesozoic shallow-marine deposits and average 1 kilometer in total thickness. Lower Paleozoic clastic sediment supply was predominantly from the Precambrian rocks forming the core of the continent. Upper Paleozoic and Mesozoic clastic sediment supply came in part from the newly emerging mountain belts.

Low-lying topography and tectonic stability have resulted in the development of classical layer-cake stratigraphy. Geologic events here happened over very large areas and accordingly left sediments that can be physically correlated over literally thousands of kilometers. Limestones, black shales, coal deposits, and orthoquartzite sandstones are common lithologies of large lateral extent.

Intracratonic highs and lows. Within the sediments of the interior lowlands, there are general trends in the thickening and thinning of correlative stratigraphic units. The thickened areas are referred to as *basins* and the areas of thinning are referred to as *domes, arches,* and *uplifts.* Note well that some of these are broad, gentle features. Looking at an outcrop, we would think that these rocks are essentially flat-lying. Yet on the scale of regional mapping, low-angle dips may persist over large areas and thickness relationships show systematic variation. Sediments over the highs may thin to zero; sediments in the lows may thicken to as much as 3 kilometers. The major highs and lows of the interior lowlands are indicated in Figure 5.2.

Transition from stable craton to mobile belts. As indicated in Figure 5.2, the transition from interior lowlands geologic province to adjacent mobile belt province is commonly marked by the development of a foreland basin. The Appalachian basin, the Anadarko basin, and the Alberta basin are such features.

Foreland basins have a unique history. During their early stages, they are intimately related to the stable craton receiving their clastic sediment supply from the craton and accumulating a sedimentary sequence that is correlative with, but thicker than, the cratonic sequence. Later during their history, these basins accumulate thick clastic sequences derived from newly emerging mountain ranges. In addition, structural deformation of the mountain belt commonly spills over into the foreland basin. Thus, the structural complexity within these stratigraphic sequences generally exceeds that of their counterparts in the interior of the continent.

The Mobile Belts

By far the most striking generality of North American geology is the fact that the stable craton is bounded on three sides by mountain chains. It is convenient to recognize on the earth's surface belts that are characterized by tectonic mobility, hence the general term, *mobile belts.* Some of these belts are tectonically active today, whereas others were tectonically active in the past.

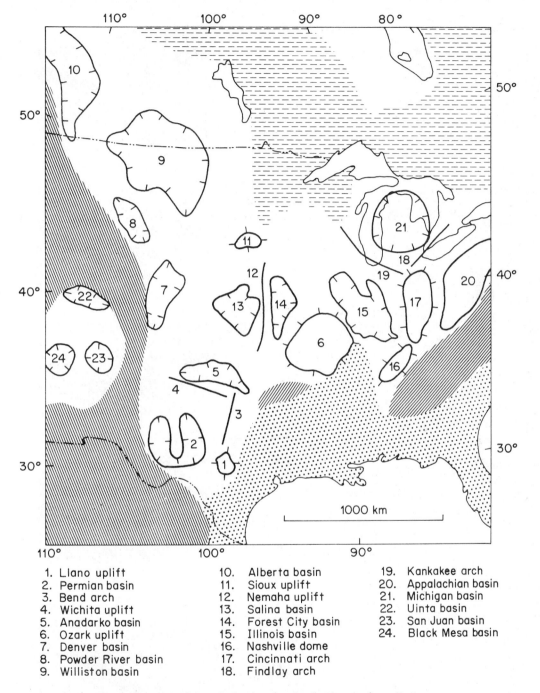

Figure 5.2 Major basins and highs of the interior lowlands province. Patterns indicating surrounding geologic provinces are the same as in Figure 5.1.

1. Llano uplift	10. Alberta basin	19. Kankakee arch
2. Permian basin	11. Sioux uplift	20. Appalachian basin
3. Bend arch	12. Nemaha uplift	21. Michigan basin
4. Wichita uplift	13. Salina basin	22. Uinta basin
5. Anadarko basin	14. Forest City basin	23. San Juan basin
6. Ozark uplift	15. Illinois basin	24. Black Mesa basin
7. Denver basin	16. Nashville dome	
8. Powder River basin	17. Cincinnati arch	
9. Williston basin	18. Findlay arch	

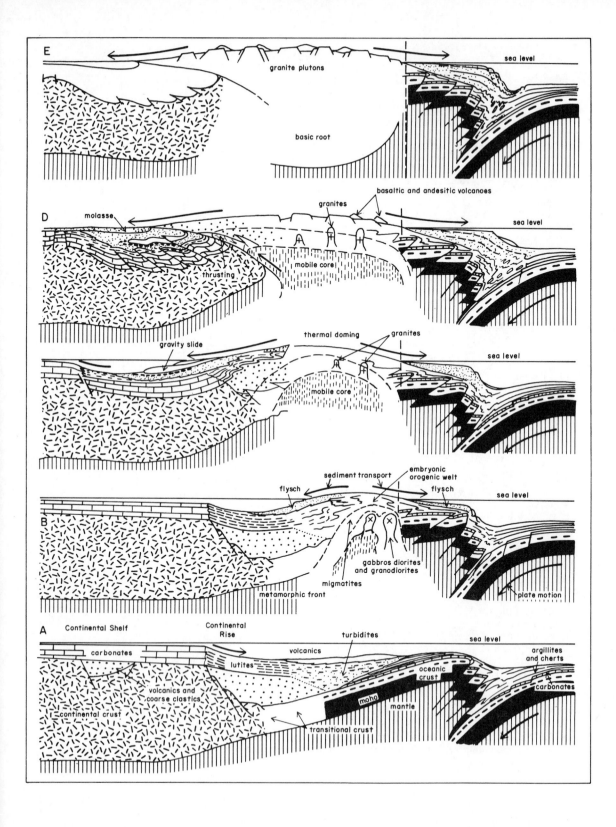

Mobile belts and plate tectonics. In Chapter 4 (Figures 4.10 to 4.14), we noted that most tectonic deformation of the present-day earth surface is ascribable to movement of lithosphere plates approximately 100 kilometers in thickness and of gigantic lateral extent. New oceanic crust is being continuously generated beneath mid-ocean ridges, and old oceanic crust is being continuously consumed in trench areas.

These same generalities appear to apply well back into the Precambrian era. The so-called mobile belts of surficial geologic mapping are but the record of previous tectonic activity at the boundary between two lithosphere plates. As indicated in Figure 5.3, most mobile belts record the consumption of an oceanic lithosphere plate beneath or adjacent to a continental plate. Observe that a number of combinations are possible within the theme of plate tectonics. The example given in Figure 5.3 is intended only to convey a general model, which may be developed in different proportions in different mountain ranges.

Sediments associated with mobile belts. Sedimentological terminology for mobile belts predates plate tectonics. For many years, geologists have recognized that mountain ranges exhibit at least three distinct stages of sediments. Kay (1951) popularized the use of the terms "miogeosyncline," "eugeosyncline," and "exogeosyncline." The words "flysch" and "molasse," from the European literature, are also frequently used. Let us now place these words within the context of plate tectonics (Figure 5.3).

Miogeosynclinal sediments are considered the continental shelf sediments that predate plate consumption and resultant tectonism. These deposits are essentially the thickened seaward extension of stable craton sedimentation. In Figure 5.3, they are characterized as shallow-water carbonate deposits of the continental shelf.

In its broadest context, *eugeosyncline* refers to those sediments that are basinward from and, at least in part, contemporaneous with miogeosynclinal deposits. These deposits would include the fine-grained sediments of the continental rise, turbidites and pelagic sediments of the abyssal plain, and a certain amount of volcanic sediment.

Precise origin of the volcanic input to eugeosynclinal deposits is unclear. This portion of the mobile belt has usually undergone such extreme tectonic deformation and metamorphism that stratigraphers are unable to unravel the precise paleogeography. On the one hand, these volcanics may record the previous existence of island arcs associated with plate consumption. The association of linear volcanic belts with zones of plate consumption is an outstanding generality today, and we may assume that it was a valid generality in the past. On the other hand, it may

Figure 5.3 (Left) Sequence of schematic cross sections illustrating a model for the evolution of a mountain belt developed by consumption of an oceanic plate beneath a continental plate. [After J. F. Dewey and John M. Bird, *Jour. Geophysical Research,* **75,** No. 14, 2638 (1970).] Courtesy American Geophysical Union.

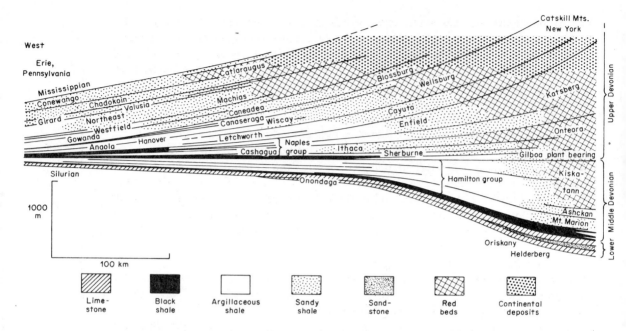

Figure 5.4 Stratigraphic cross section through New York State. The westward advance of the exogeosynclinal clastic wedge during Upper Devonian time is shown. [After Philip B. King, *The Evolution of North America,* (copyright © 1959 by Princeton University Press), Figure 32, p. 58. Reprinted by permission of Princeton University Press.]

be that eugeosynclinal sediments represent tectonic accumulation of abyssal plain sediment at the site of plate consumption. For example, Figure 14.5 maps the occurrence of numerous scattered volcanic seamounts in a portion of the North Atlantic. Consider what would happen to these seamounts if the North Atlantic plate were to consumed beneath the North American plate. We can well imagine the accumulation of a linear belt of extremely thick oceanic sediments and volcanics as these materials became scraped free from a descending slab of oceanic lithosphere.

With continued evolution of the mobile belt (that is, with continued plate consumption), a topographic inversion occurs along the site of the former continental shelf-continental rise. A large mass of sediments, volcanics, and intrusives rises up out of the former basin, and it becomes a source of sediment for redeposition to either side of the newly formed high. Where these sediments accumulate above miogeosyncline deposits, as in Figures 5.3 (C) and (D), they are referred to as *exogeosynclinal* deposits.

The relationship between miogeosynclinal and exogeosynclinal sediments is a hallmark of the Paleozoic stratigraphy throughout the sedimentary Appalachians.

Figure 5.4 presents a schematic stratigraphic cross section through the Devonian period in New York State. From the historical details recorded in these sedimentary sequences, we infer events that must have at one time occurred in the area now occupied by the crystalline Appalachians. In Figure 5.4, note that limestone sedimentation dominates the region in Lower Devonian time and is followed by clastic sedimentation.

Figure 5.5 Schematic map indicating the location and age of major exogeosynclinal wedges of the Appalachians. [After Philip B. King, *The Evolution of North America* (copyright © 1959 by Princeton University Press), Figure 33, p. 59. Reprinted by permission of Princeton University Press.]

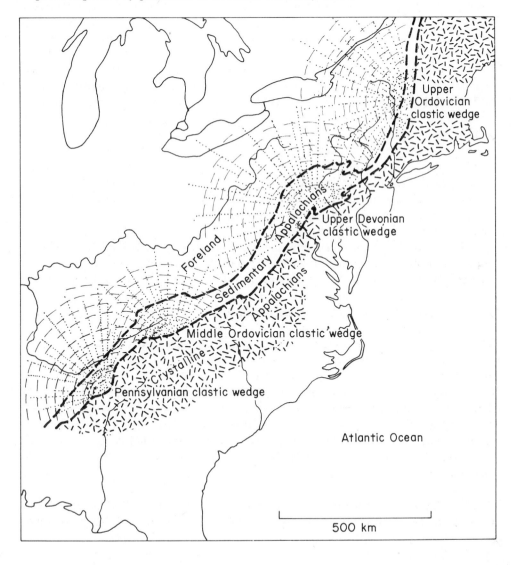

The pre-Devonian stratigraphic section in this area is likewise divided into a lower carbonate miogeosynclinal unit and an upper clastic exogeosynclinal unit. The Taconic orogeny brought clastic sedimentation over Cambro-Ordovician shelf sandstones and carbonates during late Ordovician time. Relative tectonic quiescence throughout the Silurian allowed essentially miogeosynclinal conditions to return to the area by the Lower Devonian period.

In the Middle Devonian Hamilton group, we again see a significant input of clastic detritus from the east. With the continued passage of time, thick sequences of nonmarine clastics built westward several hundred kilometers during Upper Devonian time. In Figures 5.5 and 5.6, we observe that this same general transition from miogeosynclinal Lower Paleozoic to exogeosynclinal Upper Paleozoic is evident along the length of the sedimentary Appalachians from Alabama to New York. Note, however, that the timing of the transition from carbonate sedimentation to clastic sedimentation varies widely along the mountain chain.

The terms "flysch" and "molasse" refer to the same sediments as the words "eugeosyncline" and "exogeosyncline," but they are rooted in a different conceptual framework. Whereas "eugeosyncline" and "exogensyncline" are defined in terms of *position* relative to the mobile belt, "flysch" and "molasse" are defined with relation to the *timing* of major tectonism.

Flysch sedimentation takes place during the major lateral compression of the mobile belt. In theory, flysch sediments not only are the product of newly emerged tectonic lands but also become involved in the continuation of the tectonic deformation. Flysch sediments, therefore, roughly correspond to eugeosynclinal sediments and to those exogeosynclinal sediments that themselves become highly deformed during continuing tectonism.

Molasse refers to those exogeosynclinal sediments that accumulate during the relative tectonic quiescence following major compression of the geosyncline. The Upper Devonian rocks of Figure 5.4 are thus a classical molasse sequence.

The Coastal Plain and Continental Margin

The coastal regions from Mexico northeastward to Cape Cod (Figure 5.1) reveal the accumulation of essentially undeformed sedimentary sequences of Cretaceous and younger age. Viewed in a total context of geologic history, these sediments are analogous to the thickened miogeosynclinal sequences that bordered the North American continent prior to Appalachian and Cordilleran mountain building. Because these sediments are essentially undeformed, however, we can extract from them a much more detailed record of their sedimentation than is usually possible in sedimentary sequences that have become caught up in the mountain-building process.

The Texas-Louisiana Gulf Coast clastic province. In the preceding discussions of foreland basins and exogeosynclines, we noted that newly emergent mountain

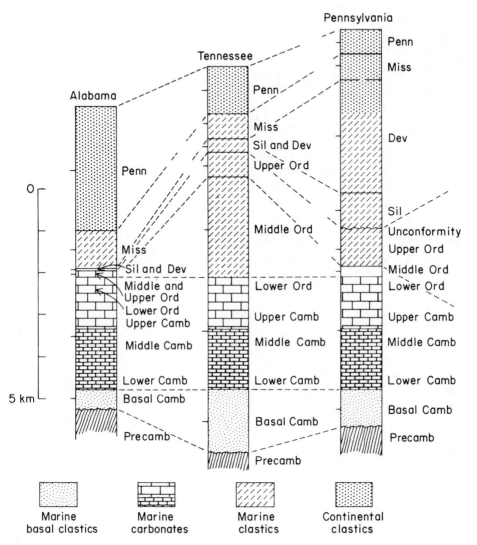

Figure 5.6 Generalized stratigraphic sections of the miogeosynclinal-exogeosyn-
clinal sediment belts of the Appalachians. Note that all sections record essentially the same sequence of events. Basal Cambrian sands give way to Lower
Paleozoic miogeosynclinal carbonate sediments, which in turn give way to thick
exogeosynclinal clastic sequences, culminating in continental deposits. Note
further, however, that the time of transition from miogeosyncline to exogeosyncline sedimentation varies greatly. Similar information is presented in
map form in Figure 5.5. [After Philip B. King, *The Evolution of North
America* (copyright © 1959 by Princeton University Press), Figure 3.4, pp. 61.
Reprinted by permission of Princeton University Press.]

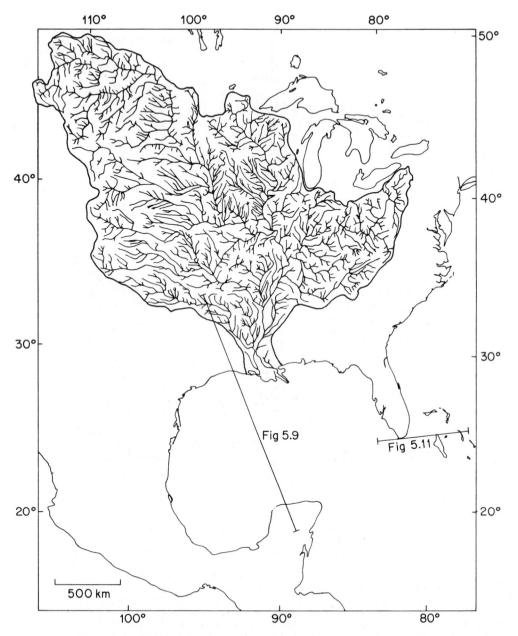

Figure 5.7 The Mississippi River drainage basin. Uplift of the Appalachians to the east and the Rocky Mountains to the west funnels rivers, and their clastic sediments, to the north end of the Gulf of Mexico.

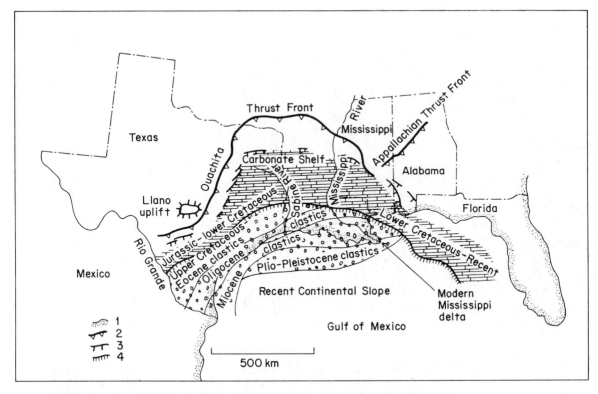

Figure 5.8 Texas / Mexico map labels:

- Texas
- Thrust Front
- Mississippi River
- Appalachian Thrust Front
- Carbonate Shelf
- Alabama
- Ouachita
- Sabine River
- Mississippi River
- Llano uplift
- Florida
- Jurassic-lower Cretaceous
- Upper Cretaceous clastics
- clastics
- Lower Cretaceous-Recent
- Rio Grande
- Mexico
- Eocene clastics
- Oligocene clastics
- Miocene clastics
- Plio-Pleistocene clastics
- Recent Continental Slope
- Modern Mississippi delta
- Gulf of Mexico
- 1 2 3 4
- 500 km

Figure 5.8 Sketch map showing major sediment accumulations in the northern Gulf of Mexico. North and west of the modern Mississippi delta, carbonate sediments accumulated on a broad shallow shelf during Jurassic and Lower Cretaceous time. By Upper Cretaceous time, carbonate sedimentation had given way to predominantly clastic sediments supplied by an ancestral Mississippi River. The position of major accumulation of clastic sediments has prograded seaward 200 kilometers since that time. In sharp contrast, carbonate sedimentation continued to prevail in the area east of the Mississippi River. (After Lehner, 1969, with modification and simplification.)

ranges shed large volumes of clastic detritus toward the stable craton. Some of this sediment is trapped within the exogeosyncline. But eventually an equilibrium is reached, in which large volumes of sediment are carried away from the mountain by rivers and find their way to the continental margin. Figure 5.7 depicts the drainage basin of the modern Mississippi River system. Clastic detritus from the Appalachians and from the Rockies is ultimately funneled down this system to the Gulf of Mexico. It is small wonder that the Gulf Coast clastic province has one of the largest accumulations of Cenozoic clastic sediments in the world.

Figures 5.8 and 5.9 summarize the general features of the Gulf Coast clastic province. During Jurassic and Lower Cretaceous time, a broad shelf was devel-

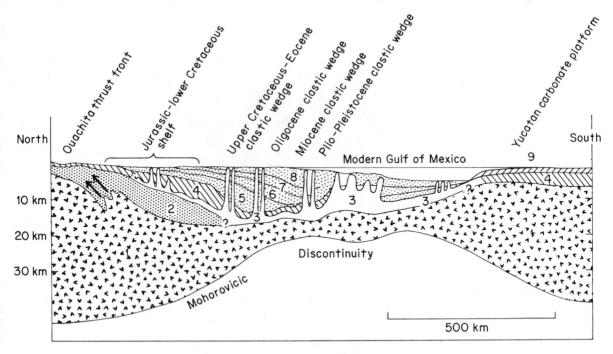

Figure 5.9 Generalized cross section of the Gulf of Mexico geological province. See Figure 5.7 for location. Rock and sediment types are as follows: (1) crystalline basement rocks; (2) Paleozoic sediments of the Ouachita-Appalachian mobile belt province; (3) salt, presumably Jurassic in age; (4) Jurassic and Lower Cretaceous sediments, predominantly shelf carbonates; (5)–(8) clastic sediments of Upper Cretaceous-Pleistocene age; and (9) Upper Cretaceous-Recent shelf carbonates. (After Lehner, 1969, with modification.)

oped; it was characterized by carbonate sedimentation along its seaward margin. With the major uplift of the Rocky Mountains during the Upper Cretaceous-Eocene, clastic sedimentation began to dominate the area. The position of major sediment accumulation during the various epochs of the Cenozoic is indicated in Figures 5.8 and 5.9.

The Gulf Coast clastic sequence records a complex interaction between sediment supply, sea-level fluctuations, and regional isostatic subsidence in response to the weight of new sedimentary load. In Figure 5.9, note that the Cenozoic sediment thickness of 15 kilometers is indicated by seismic refraction data. Oil wells in southeastern Louisiana have been drilled as deep as 8 kilometers and have bottomed in Middle Miocene sediment. The thickness of Cenozoic sediments is presumed to result from synchronous sedimentation and isostatic subsidence. The major tectonic features of the area are normal faults, presumably related to sediment loading and associated subsidence. Salt domes rise upward through the Cenozoic strata, probably from a source layer of Jurassic salt, and provide local

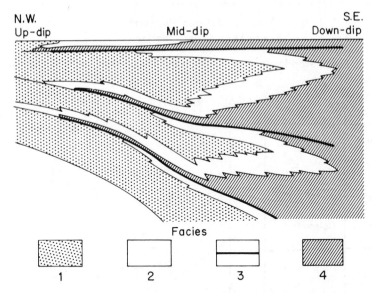

N.W.
Up-dip
Mid-dip
S.E.
Down-dip

Facies

1   2   3   4

Figure 5.10 Schematic stratigraphic cross section illustrating the primary sedimentary cycle of the Gulf Coast clastic province. Sediment types are as
follows: (1) alluvial sediments; (2) marginal-marine sands and shales; (3)
marine shale or carbonate facies marking rapid transgression; and (4) deep-
water marine shale and pelagic marl. The first-order sedimentary cycle of the
Gulf Coast clastic province consists of rapid submergence followed by gradual
progradation. (After Lowman, 1949, with modification.)

structural deformation of the Cenozoic sequence. These domes are extremely
important to the localization of large petroleum reserves.

The fundamental sedimentation cycle of the Gulf Coast clastic province is indicated in Figure 5.10. With the rapid subsidence or with eustatic sea-level rise,
relatively deep water moves landward, and so much of the clastic sediment supply
is temporarily trapped in alluvial environments. With continued passage of time,
marginal marine environments again advance seaward, followed by alluvial facies.
Continued progradation and subsidence accommodates a large clastic wedge that
is finally terminated by yet another episode of rapid transgression of the sea.

The Florida-Bahama carbonate province. In the absence of large clastic input,
continental margin sedimentation may be dominated by shallow water and by
carbonate and evaporite deposition. The Florida-Bahama platform provides a
good example. The eastern portion of Figure 5.8 indicates the paleogeographic
continuity between the Cretaceous carbonate provinces of Texas and Florida.
Whereas the Gulf Coast clastic province received a large clastic sediment supply
from the Mississippi River during late Cretaceous through Cenozoic time, the
Florida-Bahama province continued to accumulate carbonate sediments throughout the gradual subsidence of the region. Figure 5.11 is a cross section through

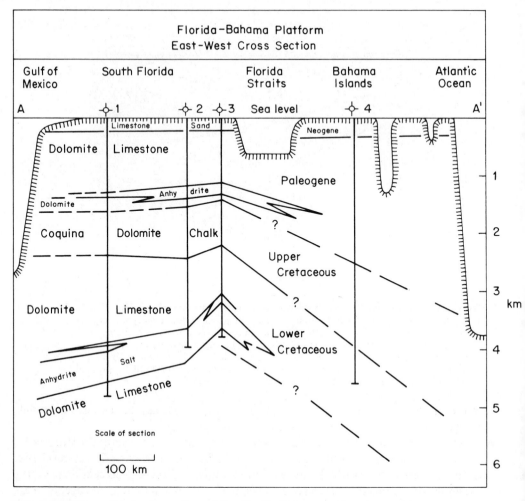

Figure 5.11 Generalized cross section of the Florida-Bahama carbonate province. See Figure 5.7 for location. (After Ginsburg, 1964, with modification.)

the Florida-Bahama carbonate province. Note that Cenozoic carbonate accumulation in this area is a scant 1.5 to 3 kilometers as compared to the 15 kilometers of clastic sedimentation within the Texas-Louisiana Gulf Coast clastic province.

Citations and Selected References

ATWATER, T. 1970. Implications of plate tectonics for the Cenozoic tectonic evolution of western North America. Bull. Geol. Soc. Amer. 81: 3513–3536.

BIRD, J. M., and J. F. DEWEY. 1970. Lithosphere plate-continental margin tectonics and the evolution of Appalachian orogene. Bull. Geol. Soc. Amer. 81: 1031–1060.

DEWEY, J. F., and J. M. BIRD. 1970. Mountain belts and the new global tectonics. J. Geophys. Res. 75: 2625–2647.
A very helpful review of the earth's mountain chains in the light of ocean-floor spreading.

DIETZ, R. S., J. C. HOLDEN, and W. P. SPROLL. 1970. Geotectonic evolution and subsidence of Bahama platform. Bull. Geol. Soc. Amer. 81: 1915–1928.

———. 1972. Geosynclines, mountains, and continent-building. Science 226 (3): 30–38.
Easy reading and well-illustrated.

FISHER, G. W., F. J. PETTIJOHN, J. C. REED, and K. N. WHEELER (eds.). 1970. Studies of Appalachian geology, Central and Southern. Wiley-Interscience, New York. 460 p.

GINSBURG, R. N. 1964. South Florida carbonate sediments. Guidebook for field trip no. 1. Geological Society of America annual convention, Miami Beach, Fla. 72 p.

HALES, A. L., C. E. HELLSLEY, and J. B. NATION. 1970. Crustal structure study on Gulf Coast of Texas. Bull. Amer. Assoc. Petrol. Geol. 54: 2040–2057.

KAY, M. 1951. North American geosynclines. Geol. Soc. Amer. Mem. 48. 143 p.
A landmark in the literature of mountain-belt geology.

KING, P. B. 1959. The evolution of North America. Princeton Univ. Press, Princeton, N. J. 189 p.
A delightful synthesis of the geology of North America. Easy reading.

LEHNER, P. 1969. Salt tectonics and Pleistocene stratigraphy on continental slope of Northern Gulf of Mexico. Bull. Amer. Assoc. Petrol. Geol. 53: 2431–2479.

LOWMAN, S. W. 1949. Sedimentary facies in the Gulf Coast. Bull. Amer. Assoc. Petrol. Geol. 33: 1939–1997.

RODGERS, J. 1970. The tectonics of the Appalachians. Wiley-Interscience, New York. 288 p.

STANLEY, K. O., W. M. JORDON, and R. H. DOTT, JR. 1971. New hypothesis of early Jurassic paleogeography and sediment dispersal for western United States. Bull. Amer. Assoc. Petrol. Geol. 55: 10–19.
Island arc and shallow seas of the western third of the United States as the Atlantic began to open.

WALTHALL, B. H., and J. L. WALPER. 1967. Peripheral gulf rifting in northeast Texas. Bull. Amer. Assoc. Petrol. Geol. 51: 102–110.
Comparison is made between fault zones bounding the northwestern Gulf Coast and the rift valleys of East Africa.

ZEN, E., W. S. WHITE, J. B. HADLEY, and J. B. THOMPSON, JR. (eds.). 1968. Studies of Appalachian geology, northern and maritimes. Wiley-Interscience, New York. 475 p.

6

Physical Stratigraphy

In the preceding chapters, we have looked at the dynamics of the present-day earth and have recognized that the interactions of sedimentation, tectonics, and sea-level fluctuations can be expected to produce and preserve a sedimentary record of earth history. We briefly examined the stratigraphy of North America and saw, at least in general terms, that such a record does exist. In this and subsequent chapters, we shall consider the principles and tools that allow us to construct a more detailed understanding of the earth history as recorded in the stratigraphic record. The first step in this direction is to understand the physical relationships among sedimentary rocks within relatively small areas.

The Basic Rationale of Physical Stratigraphy

Before we discuss the geologic history of an area and the relation of that history to the history of other portions of the earth, we must catalog the contents of that particular area. What rock types are present and what are their relationships to the other rock types in the area?

To answer these questions, we shall be concerned primarily with the physical characteristics of rock units. Time in stratigraphy will be of only secondary concern to us at this level of investigation. Certainly we shall realize that rocks on the bottom of the pile are older than rocks on the top. Similarly, we may recognize times of marine deposition followed by times of continental deposition. The point is that, at this level of investigation, we are not concerned with how these time relationships fit other events on a regional or global scale. For the moment, we want only to make good sense out of the physical events that occurred within a relatively small area. Hopefully, we can quantify the time dimension in subsequent studies.

Rock units: group, formation, and member. The basic mapping unit of physical stratigraphy is the *formation*. The underlying concept for the definition of a formation is convenience. A formation need only be a mappable rock unit. The word "mappable" clearly suggests that a formation should be defined on the basis of characteristics that are discernible under normal working conditions to the geologist making the map. Gross lithology, color, and, possibly, the general fossil content are common parameters upon which to base the definition of a formation. Alternatively, several lithologies may be grouped as a single formation, provided only that the upper and lower contacts of the formation are conveniently recognized under existing mapping conditions.

The legalisms of stratigraphy require the designation of a type section for each formation. Any stratigraphic section that can be shown to have had physical lithologic continuity with the type section during deposition may be properly assigned to the formation name of the type section.

As mapping progresses, it usually becomes apparent that several formations, existing side by side or stacked one upon the other, may be viewed as a single unit for general purposes. Where one formation thickens, they all thicken; where one formation is absent, they are all absent; and so on. Thus, it may be convenient to have a single designation for all of these formations collectively. Such a collection of formations is referred to as a *group*.

On the other hand, it may become important to keep track of some smaller subdivision of a formation. Once again, the definition of the smaller subdivision is commonly based upon some lithologic characteristic easily recognizable under mapping conditions. Smaller subdivisions of a formation are referred to as *members*.

Thus, the basic mapping units of physical stratigraphy—group, formation, and member—are arbitrary units of convenience. The use of these words will vary widely with the area under consideration and with the field conditions under which the area was first mapped.

In a practical sense, this discussion is more a matter of history than it is a matter of present-day working geology. Few modern stratigraphers will have the opportunity to name new formations. The formations have been defined and named already. We may not always like the definition, but we usually learn to live with it. Often it is convenient to get around original poor definitions of formation boundaries by naming new members within the formation. We shall see such an example in Chapter 11 concerning the Oswego sandstone.

Basement, unconformities, and contacts. What constitutes a rational basis for the definition of group, formation, and member? Let us start with the really big items and work our way down. Every area of sediment accumulation has a *basement* of metamorphic or igneous rocks. Clearly, the *contact* between basement and sediment constitutes a starting point for the definition of group and formation. Many stratigraphic sequences contain angular *unconformities*. Sedimentary sequences were deposited over basement; tectonic deformation and erosion occurred; then additional sedimentary sequences were deposited over the erosional surface. Obviously, we should study these naturally occurring discontinuities. We attach group or formation significance to materials above and below major unconformities.

Major unconformities do not always show angular relationships at every outcrop. A major unconformity may at first be passed over as just another bedding plane (see Figures 3.19 and 3.20, for example). If angular relationships do not exist anywhere within the map area, major unconformities may go unnoticed by the local physical stratigrapher. In such cases, the biostratigrapher may later inform him that large biostratigraphic time intervals are missing from his single "formation." On the other hand, if angular relationships do exist somewhere within the map area, the physical stratigrapher may recognize his own error early in the mapping process. At all times, the physical stratigrapher must be on the lookout for subtle evidence of a major discontinuity parallel to bedding planes. Conglomerates, regoliths, and solution breccias are but a few examples of the features to watch for.

The Utility of Physical Stratigraphy

We have already emphasized physical stratigraphy as a basic necessity for understanding the earth history of any particular small area. In the next chapter, we shall go on to see how biostratigraphy places this small area into a regional or global time-stratigraphic framework. However, even at the relatively simple level of physical stratigraphy, examination and cataloging of stratigraphic sequences provide some exciting and useful data.

Recognition of general paleogeography. Consider the basic paleogeographic information contained in Figures 5.4, 5.5, and 5.6. For the moment, disregard the time framework within which these data are cast. Viewed as physical stratigraphy in the absence of any time framework, the major elements of paleogeography remain intact. Figure 5.5 continues to portray thick clastic sections immediately westward from the present-day crystalline Appalachians. Generalized stratigraphic sequences from Alabama to Pennsylvania (Figures 5.4 and 5.6) still indicate that carbonate sedimentation in the lower portions of the section gives way to clastic input from the east. Recognition of the basic paleogeography throughout the entire length of the Appalachian mountain chain is not dependent upon knowledge of detailed time relationships. The fact that the clastic wedge in Alabama is of Pennsylvanian age whereas the clastic wedge of the Catskill Mountains in New York is of Upper Devonian age is an interesting secondary refinement upon our understanding of Appalachian paleogeography. However, the fundamental recognition of the paleogeography need rest only upon physical stratigraphy.

Exploitation and utilization of local areas. Exploration for petroleum best illustrates the utility of approaching sedimentary sequences by means of physical
stratigraphy. When the geologist is looking for oil, his view of geology often becomes very localized. Specifically, he wonders whether the hole he is drilling in
the ground is going to encounter an oil reservoir? Figure 6.1 schematically summarizes how physical stratigraphy might lead to an oil discovery. Petroleum often
accumulates in permeable rocks at the crest of anticlines. Consequently, anticlines

Figure 6.1 Application of physical stratigraphy to a local petroleum exploration
problem. As indicated in *A,* dry holes nos. (1) and (2) were drilled for the
wrong reason, but they provide subsurface information from which the physical
stratigrapher can begin to make more refined interpretations concerning local
geology. As indicated in *B,* the physical stratigraphy of boreholes nos. (1) and
(2) may logically lead to the location of borehole no. (3). See text for further
discussion.

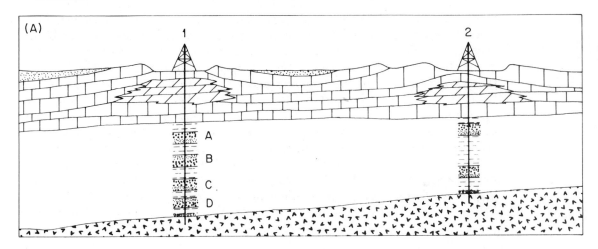

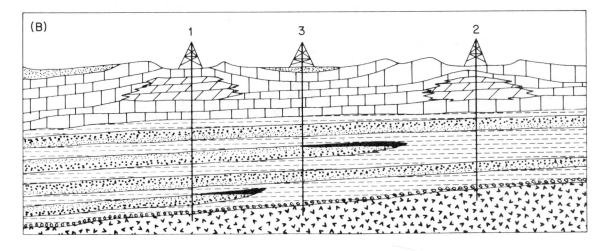

get drilled early in the exploration history of any oil province. Figure 6.1 shows that dry holes (1) and (2) were drilled on surface anticlines. It turned out that these anticlines were not structural features but rather were sedimentary drapeover reefs located very near to the surface. Both wells were dry, but they provided some interesting physical stratigraphic information. Hole (1) even had an oil show in the *D* sandstone, although it was not enough to make an oil well.

According to Figure 6.1, both holes begin in limestone and pass abruptly into shale at about the same depth. An unconformity? Perhaps. At least we should call it a formation boundary. Both wells then continue through a marine sandstone-shale sequence until basement is encountered, somewhat deeper in hole (1) than in hole (2). It is a well-established generality that marine clastic sequences tend to thicken basinward. Thus, we may begin to suspect that the transgressions and regressions responsible for the alternating sands and shales came from a more persistent basin to the west. Note further that borehole (1) encounters four sandstone units whereas borehole (2) encounters only two. Closer examination of the lithologic properties of the sandstone units in hole (1) indicate that sandstones *A* and *C* are finer-grained at the bottom and become coarser upward to the point where they are overlain by sharp contact with shale. In contrast, sands *B* and *D* tend to become fine upward and pass gradationally into the overlying shale. At this point, we probably begin to get excited. Sands *A* and *C* display classical marginal-marine, regressive features (coarsening-upward), whereas *B* and *D* appear transgressive. Could it be that rapid sea-level rise drowned the shorelines represented by sands *B* and *D* and left them with porosity pinchouts somewhere between dry holes (1) and (2)? A quick glance at samples from borehole (2) confirms the correlation of these sands with *A* and *C* of borehole (1), and a new wildcat well goes down to seek petroleum accumulations at the pinchout of sands *B* and *D*. With any luck, hole (3) will be an oil well. Moreover, stratigraphic information gained from it will also suggest more drilling prospects.

Figure 6.2 A "comedy of errors" concerning physical stratigraphy and city planning.

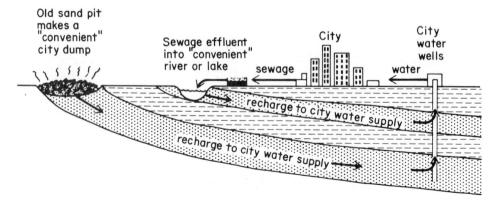

Even though this discussion unravels a small piece of local geology, no one really cares how this geology is related to any other geology, either regionally or in a time-stratigraphic framework. Many of the day-to-day details in petroleum exploration are carried out in precisely this manner.

Similarly, the physical stratigraphy around metropolitan areas may be of only major local importance. City planners need not be concerned with how local geology fits into the global time-stratigraphic framework, but they must be interested in whether or not effluent from the dump and the sewage disposal plant gets mixed up with the city water supply. Figure 6.2 portrays a rather simple comedy of errors in this regard. In this hypothetical example, poor planning has resulted in town dump and sewage treatment plant being situated over the area that recharges the city's aquifer. This example may be exaggerated; yet, the general point is still valid.

Citations and Selected References

AMERICAN COMMISSION ON STRATIGRAPHIC NOMENCLATURE. 1961. Code of stratigraphic nomenclature. Bull. Amer. Assoc. Petrol. Geol. 45: 645–660.
Updated continuously in the AAPG Bulletin.

EICHER, D. L. 1968. Geologic time. Prentice-Hall, Inc., Englewood Cliffs, N. J. 150 p.
Good supplementary reading for Chapters 6 through 9.

KRUMBEIN, W. C., and L. L. SLOSS. 1963. Stratigraphy and sedimentation. W. H. Freeman and Co., San Francisco. 660 p.
Advanced text.

LAHEE, F. H. 1952. Field geology. McGraw-Hill, New York. 883 p.

WELLER, J. M. 1960. Stratigraphic principles and practice. Harper and Bros., New York. 725 p.

7

Biostratigraphy:

Introduction to Temporal Correlation

In the previous chapter, we noted that rock units are the basic starting point for understanding the sedimentary geology of a small area. As we attempt to expand our discussion to larger and larger areas, we run into two major problems. First, the areal extent of demonstrable physical correlation of rock units is often limited by outcrop patterns and availability of subsurface data. Perhaps the formation under consideration is well exposed in a mountain range, but the next mountain range is 300 kilometers away. Can we really trust lithologic correlation across a 300-kilometer gap? Or perhaps this particular formation has been removed from adjacent areas by erosion. It was there once; we could have walked it out for thousands of kilometers. But now it is gone, and so physical correlation is impossible.

Even if we can establish lithologic correlation over large areas, we run into a second problem. Lithologic units are commonly diachronous on this large a scale. For example, if a sandstone originated as beach deposits prograding westward across a shallow-marine environment, the eastern portion of the lithologic unit will be younger than the western portion. In a small mapping area, the fact that beach sand is found to overlie subtidal marine shale portrays a fairly vivid picture of local earth history. Yet, as we view this situation in larger areal extent, we recognize that these lithologic relationships must surely transgress time. Consequently, our view of earth history begins to go out of focus. If we wish to bring large-scale earth history into proper focus, we must pay more serious attention to the dimension of time. It is through biostratigraphy that the geologist traditionally controls this dimension.

Rock Units, Time-Rock Units, and Time Units

Rock units carry with them absolutely no connotation of time. The formation exists as a three-dimensional body of rock. When it was deposited and how it was deposited simply does not enter into the definition of a rock unit. In similar fashion, we accept *a priori* that time existed in the past irrespective of whether or not rocks record the passage of that particular interval of time. For example, we have little doubt that there was "a time"—8 to 10 AM, March 22, 127,442,361 B.C.—Yet, obviously we shall never retrieve the historical record of precisely what happened within that two-hour "time unit." This problem constantly faces us as we attempt to work a time dimension into our understanding of the earth history recorded in sedimentary sequences.

Figure 7.1 Diagram indicating increasing levels of sophistication concerning time in the stratigraphic record. Rock units have no time significance whatsoever. Time-rock units convey relative relationships such as "older than" or "younger than," but they carry with them the implicit recognition that the stratigraphic record of time may have large gaps in it. Time units are the next conceptual step above time-rock units; they designate a continuum of time between two events in the history of the earth. Radiometric dating allows absolute determination of the age of these events within a margin of error. Interpolation between events that have well-known radiometric ages allows still more detailed discussion of absolute time in the stratigraphic record of earth history.

No time significance	*Relative time significance*	*Increasingly sophisticated absolute time significance* $\longrightarrow$
Rock units	Time-rock units	1. Time units
		Era
		2. Radiometric dating of time unit boundaries
Group Formation Member Bed	System - - - - Period	3. Interpolation between radiometric date points in continous sedimentation sequences.
	Series - - - - Epoch	4. Ultimately, "there was a storm the morning of August 23, 467,931 B.C.", etc.; but we cannot hope to really achieve quite that level of sophistication.
	Stage - - - - Age	

We get around this problem by defining an intermediate time-related statement. Instead of attempting to define absolute time units, stratigraphers have defined a large interwoven system of relative time units called *time-rock units*. A sequence of sediments is designated to represent a certain unspecified period of time in earth history. Where other sediments can be demonstrated, usually on the basis of fossil content, to have formed at that same time, they are assigned to the same time-rock unit. Ultimately, we seek to relate time-rock units to the absolute time that they represent. Radiometric dating offers a significant potential for the eventual accomplishment of this task. In the meantime, stratigraphers can still understand earth history on a global scale within the framework of time-rock units. Figure 7.1 illustrates the various levels of sophistication concerning our treatment of time in stratigraphy.

Fossils and Time-Rock Units

Fossil content of the sediments provides the basis for large-scale time-rock correlation. Life on earth has constantly undergone change. By recognizing the changes and putting them in their temporal sequence, we can correlate time and stratigraphy.

If a life-form is to serve as a good biostratigraphic indicator of relative time, it should have four attributes. First, it must produce preservable hard parts; there must be something left for us to find in the stratigraphic record. Secondly, these hard parts should make up a significant portion of the sediment. The more remains there are, the more easily they are found and used by the biostratigrapher. Thirdly, the organism must have had a life style or life cycle that allowed rapid dispersal around the world or at least over very large areas. Pelagic organisms are useful, but many benthonic organisms have a pelagic larval stage and are therefore just as useful. Finally, this life-form will hopefully have had a rather abrupt beginning and a rather abrupt demise that were not too widely separated in time.

The basic bookkeeping unit of biostratigraphy is the range of the species. What are the oldest strata in which the species first appears, and what are the youngest strata in which the species is last found? Beyond this simple level, zones of maximum abundance, zones containing assemblages of various species, and so on, all have time-stratigraphic significance under various circumstances.

Because of the preservability of fossil remains and because of the distribution of major rock types, shallow-marine invertebrate faunas are by far the most widely used biostratigraphic tools. Although the following discussions will deal solely with shallow-marine invertebrate faunas, the principles are equally applicable to pelagic faunas, terrestrial faunas, and so on.

A Biostratigraphic Overview of Evolution

Let us now investigate why fossils should work as indicators of equivalent time. We shall develop the topic of evolution in a very broad fashion and shall point out generalities that will be applicable both at the species level and at the ecosystem level. To us, *evolution* will simply mean change in the character of life on earth as a function of time. Sometimes this change has occurred within a single phylogenetic lineage. For example, an interbreeding population of molluscs may undergo mutation in such a way that individuals become morphologically distinct from their predecessors. In these situations, our use of the word "evolution" is close to the biologist's use of the word.

But we shall be equally interested in the evolution of community structure within specific environments, regardless of the phyla that may be involved. Consider molluscs, as an example. Today they dominate many of the niches that were dominated by brachiopods in the Paleozoic. "Successful" brachiopods did not evolve into "successful" molluscs. The brachiopods instead lost a long-term competitive battle for these niches to molluscs, which independently evolved a more successful life style.

Diversity of life. Two aspects of physiological diversity are important for our studies: (1) diversity among the phyla and (2) diversity within single species. As indicated in Figure 7.2, highly organized invertebrate phyla have existed on the earth for at least the last 550 million years. Because the geologic record of life on earth is heavily biased in favor of those organisms that generate preservable hard parts, we must recognize that complicated communities of highly developed organisms existed even prior to many of the "first appearances" indicated in Figure 7.2.

The existence of a complicated community structure including numerous phyla carries with it a certain guarantee for the continued existence of life within the environment occupied by the community. It is highly unlikely that changing conditions on the face of the earth would prove simultaneously unfavorable to all species within the community. If any one species were to undergo decline in response to unfavorable changes in the environment, it is likely that another species would undergo expansion as it exploited at least a portion of the niche vacated by the declining species.

The second aspect of diversity that interests us is the physiologic and morphologic variability within a single species. A *species* is generally defined as a collection of organisms that interbreed in nature to produce fertile offspring more or less resembling themselves. There commonly exists considerable diversity within a single species. Such diversity can usually be expressed in terms of a normal distribution curve, such as Figure 7.3. No matter what characteristic we choose to measure (shape, heat tolerance, etc.), the majority of the population falls within a certain range of variation. Yet, to either side of this dominant range, there are individuals that may still be an interbreeding part of the population. The gene pool of the total population, therefore, includes characteristics lying beyond the norm of the species. If changing conditions of the environment render these abnormal characteristics valuable, then the ($+$) tail of the normal distribution curve will continue to reproduce, whereas individuals on the ($-$) tail will tend to die without reproducing. Thus, the shape and position of the normal distribution

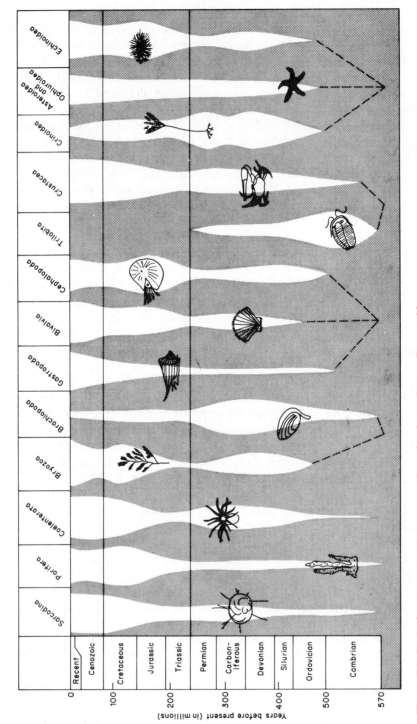

Figure 7.2 Geologic range of major groups of marine invertebrates. Preservable hard parts of most groups of marine invertebrates have existed since the Cambrian. (After McAlester, 1968.)

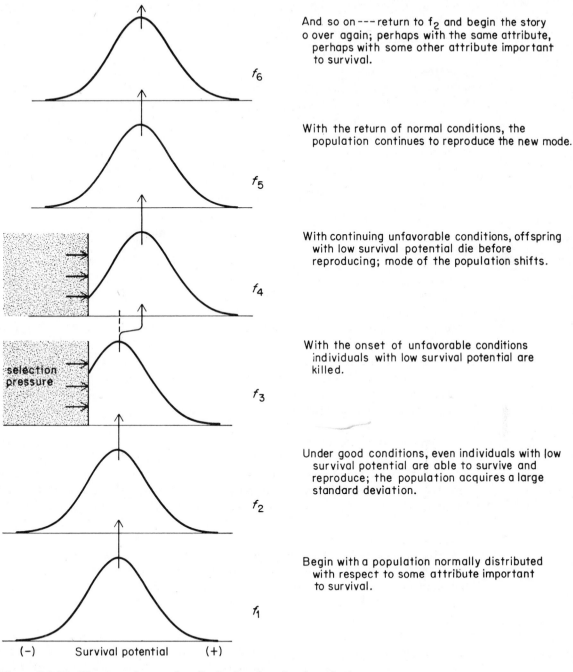

And so on --- return to f_2 and begin the story over again; perhaps with the same attribute, perhaps with some other attribute important to survival.

With the return of normal conditions, the population continues to reproduce the new mode.

With continuing unfavorable conditions, offspring with low survival potential die before reproducing; mode of the population shifts.

With the onset of unfavorable conditions individuals with low survival potential are killed.

Under good conditions, even individuals with low survival potential are able to survive and reproduce; the population acquires a large standard deviation.

Begin with a population normally distributed with respect to some attribute important to survival.

selection pressure

f_6

f_5

f_4

f_3

f_2

f_1

(−)   Survival potential   (+)

Figure 7.3 Modification of a species distribution function by selection pressure. Time passes from f_1 to f_6.

curve with respect to any characteristic may shift with time as various individuals contribute input to the continuity of a gene pool. We shall discuss this point in greater detail later under selection pressure.

Ecological limits of species and communities. Given steady conditions, communities will approach a dynamic equilibrium with their environment. Availability of nutrients will limit the size of the biomass. Predator-prey relationships and physical factors, such as temperature, salinity, and substrate, will determine the relative abundance of the various species within the community. Both total biomass and abundance of individual species will tend to expand to the limit of the average conditions of the environment. As environmental conditions change, the dynamic equilibrium may be disrupted, and species and communities may undergo modifications that will be interesting as biostratigraphic markers. The following discussion particularly concerns shallow-marine benthonic communities and the changes that they may undergo in response to variations in water depth, living space, and temperature.

Numerous environmental parameters vary as a function of water depth. Wave agitation may provide food supply for filter feeders. It may also produce a firm sandy substrate that contrasts sharply with the soupy mud substrate of deeper water environments. Shallow, clear water allows sufficient light penetration for luxuriant development of a benthonic flora. If living space is available in suitable environments, communities will expand to fill the space. Finally, each species has its own range of preferred temperatures. Some range more widely than others, but all species have their limits. If any of these conditions are altered, we must expect the dynamic equilibrium of the community to respond accordingly—be it a positive response or a negative response.

Changing conditions on the earth's surface. In our review of the dynamics of the earth's surface today (Chapter 4) and of the general geology of North America (Chapter 5), we have observed gross variation in all three of the general parameters just cited (water depth, living space, and temperature). Water depth may change rapidly; recall Pleistocene sea-level fluctuations of as much as 80 to 100 meters. The shallow seas that formerly existed over the interior lowlands of the United States surely provided more living space for shallow-marine communities than exists on the North American continent today. In addition, note that Pleistocene glaciers existed adjacent to shallow-marine environments off New England. Undoubtedly, the shallow-marine environments of this area were much colder during the Pleistocene than they are today.

Lacking strong evidence to the contrary, we must assume that variable conditions have been the general rule throughout the geologic record. When a change was favorable to a species or a community, it exploited that change by expanding its population. When a change was unfavorable, the population of the species or of the community underwent alteration in response to the selection pressure.

Selection pressure. If a population is existing at optimum environmental conditions and some aspect of the environment undergoes adverse change, a portion of the population may be adversely affected. In Figure 7.3, we portray the distribution of the generalized variable: *survival potential.* The precise identification of this variable is left open and would presumably vary with the type of stress that the population is about to face. It may be something as simple as temperature tolerance during a three-day cold snap, salinity tolerance when a bay becomes flooded by an unusual amount of fresh water, food-gathering efficiency during times of scarce supply, or perhaps just a willingness to reproduce even during a generally unromantic mating season. Alternatively, survival potential may be an exceedingly complicated collage of all of these and other factors affecting the general maintenance of the population. Indeed, factors contributing to survival potential will change from one crisis situation to the next.

The adverse shift in the environment may be regarded as a selection pressure that will cause the early demise of those individuals having the lowest survival potential. The more serious the environmental crisis, the further into the bell-shaped curve the selection pressure will operate. Thus, selection pressure can substantially modify the gene pool of successive generations. In Figure 7.3, phylogenetic evolution has occurred between f_3 and f_4. If some aspect of this modified distribution shows up in the preservable hard parts of this organism, we may be able to use the shift in population characteristics as a biostratigraphic indicator of time.

Note also how the potential for selection pressure operates among different species. Consider, for example, marine invertebrates *A* and *B,* both with preservable hard parts. They have separate and distinct feeding habits. *A* feeds on *C* and *B* feeds on *D, C* and *D* being species with no preservable hard parts. At this point, *A* and *B* are not competing with each other for food. Then, for reasons that will never be known to us, species *D* simply vanishes from the environment. *A* and *B* now seek to feed on the single species *C.* Species that were formerly not competing with one another are now brought into competition for food and, therefore, for survival. *B,* being the more aggressive food gatherer, continues to thrive. *A,* being the less aggressive food gatherer, undergoes severe selection pressure and is ultimately starved out of the area entirely. Once again, we see the generation of a biostratigraphic shift that must certainly have at least local time-stratigraphic significance; that is, beds containing *A* and *B* give way to beds containing only *B.*

Sea-level fluctuation as a selection pressure on shallow-marine benthonic invertebrates. Fluctuating sea level can produce large variations in the amount of subtidal shallow-marine living space available on or near continents. The Recent epoch provides us with an excellent example of this problem in the Great Bahama Banks (Figure 12.1). Shallow subtidal environments cover the top of the banks today and did so numerous times in the past. During low stands associated with glaciation, however, these banks stood high and dry, forcing subtidal marine or-

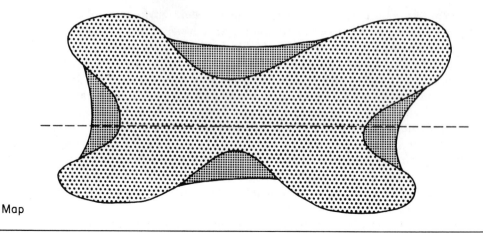

Map

Cross section

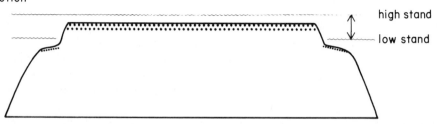

high stand

low stand

Figure 7.4 Sea-level fluctuation as a control of living space and, therefore, of
selection pressure. During low stand, shallow-marine benthonic communities
are forced to occupy the four small areas of low-stand living space. With high
stand of the sea, these four isolated communities can expand onto the much
larger area of the platform.

ganisms to try to eke out survival within the narrow band between the exposed
platform and the deep water surrounding the platform.

Numerous models can be set up to portray the possible interaction between
shallow-marine invertebrate communities and fluctuating sea level. Let us work
through one of these models just to see how things might fit together. Keep
in mind that the possible combinations are more or less infinite.

Figure 7.4 depicts a platform on which high-stand living space grossly exceeds
low-stand living space. You may think of this area as being on the scale of carbonate platform or on the scale of a continent; the principles are similar.

Let us begin with four low-stand populations situated on the various side of
the platform. These populations are isolated from each other by geographic barriers. With rising sea level, the available living space for these communities ex-
pands greatly. At this time, there is indeed a great deal of positive environmental
shift for the species. Space is available for the population increase; and, as an
additional bonus, previously isolated gene pools are now united once again into a

single gene pool. Because selective pressure is low, nearly all combinations of genes produce offspring that will survive. Thus, the population grows not only in numbers but also in diversity.

But then the sea level begins to fall, and so the platform becomes emergent once more. Reduction in living space places severe pressure on the species. It must once again split into four isolated communities around the margins of the platform.

Furthermore, each isolated low-stand population endures its own separate selection pressures. With the high stand, benefits gained by selection pressure acting upon the isolated populations are contributed to a single high-stand gene pool. For example, if selection pressure has caused the west coast population to acquire a beneficial feeding habit, this information can be transmitted to all populations via the next high-stand gene pool. Alternatively, selection pressure on the low-stand west coast community might be so intense as to produce a population that will not interbreed with the other populations during the next high stand. The stage is set for speciation. With the next low stand, some of the west coast variant will undoubtedly end in several of the low-stand platform-margin havens. If they continue to survive and if they continue to breed only among themselves, speciation has occurred.

As another interesting variation on this model, consider the invasion of one of the low-stand havens by a totally different organism (*B*), which possesses superior adaptation to the environment in question. So long as the platform remains emergent, the new superior organism is excluded from the other low-stand havens by geographic barriers. Although this new organism has completely taken over the southern haven, for example, the west, east, and north havens remain untouched. With the next high stand, however, events move rather rapidly. Organism *B* gets his first look at all that platform area with shallow water over it. Those low-stand populations of organism *A* hardly see what hit them. *B* takes over the whole place, bringing about the rapid demise of *A*. In this fashion, biostratigraphic time horizons of large areal extent can be generated. Note that the example involves replacement of a species by a competitor rather than replacement of a species by its phylogenetic progeny.

The Complications of Niche and Biogeography

The preceding discussion outlines the potential for biostratigraphic horizons to be good time horizons. Now we must recognize some of the pitfalls of these generalities.

Variation in depositional environments perpendicular to shoreline. All of the various major depositional environments indicated in Figure 7.5 are capable of accumulating a stratigraphic record of a time interval. Yet the organisms that live in these environments may be quite different from each other and may not occur within the same outcrop.

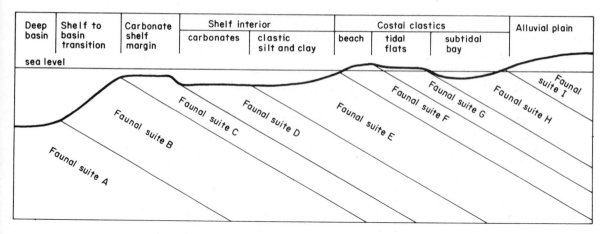

Deep basin	Shelf to basin transition	Carbonate shelf margin	Shelf interior		Costal clastics			Alluvial plain
			carbonates	clastic silt and clay	beach	tidal flats	subtidal bay	

sea level

Faunal suite A
Faunal suite B
Faunal suite C
Faunal suite D
Faunal suite E
Faunal suite F
Faunal suite G
Faunal suite H
Faunal suite I

Figure 7.5 Separate and distinct faunal suites as a function of variation in the depositional environment perpendicular to the shoreline. Variations in water depth, substrate, turbidity, and salinity cause different suites of animals to occupy the various environments within a single time plane. Inasmuch as suite *A* contains no species in common with suites *D* through *I,* demonstration that suites *A* through *I* are indeed time correlative can be a sizable biostratigraphic task.

Two considerations are important to us here. First, each of these environments has its own set of communities that respond independently to ecologic stress. The equation for survival potential within the deep basin need not have any *a priori* relationship to the equation for survival potential in carbonate shelf-interior environments. For example, sea-level fluctuations may have a profound effect on carbonate, shelf-interior, benthonic environments, yet they need have no effect whatsoever on deep-basin pelagic environments.

Secondly, the first appearance of a fauna in a stratigraphic sequence may record local sea-level fluctuation rather than an evolutionary biostratigraphic event. For example, the first appearance of deep-basin fauna over clastic bay fauna means one of two things: Either the deep-basin fauna has suddenly adapted itself to a very successful life in the clastic bay environment, or the clastic bay environment has been replaced by deep water. The former would be an evolutionary event of potential worldwide importance; the latter may record only local submergence. In most cases, we suspect that the latter is true.

Variation in depositional environment parallel to shoreline. If a shallow-marine environment covers a sufficiently large area, the temperature varies significantly over its areal extent. Organisms sensitive to temperature may thus be restricted to certain portions of this time plane.

Figure 7.6(A) depicts the distribution of communities along the eastern and southern margins of a continental land mass. Tropical, temperate, and cool communities are recognized. Except for the temperature factor, these communities

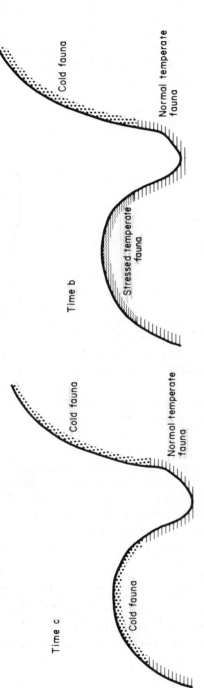

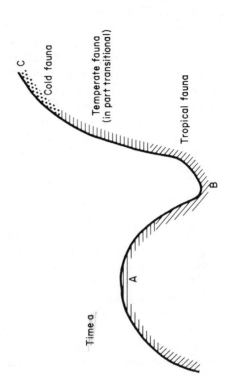

Time c

Cold fauna

Normal temperate fauna

Cold fauna

Time b

Stressed temperate fauna

Cold fauna

Normal temperate fauna

Time a

A

B

C

Cold fauna

Temperate fauna (in part transitional)

Tropical fauna

Figure 7.6 Separate and distinct faunal suites as a function of variation in the depositional environment parallel to the shoreline. As indicated at time (*a*), temperature variation among otherwise similar environments allows faunas that do not resemble each other to occupy similar sediment types within a single time plane. Diagrams (B) and (C) trace some of the biostratigraphic complications that can arise as changing climate interacts with geographic barriers. At time (*b*), cooling has allowed the cold fauna to migrate considerably southward, yet the geographic barrier excludes the cold fauna from the northern portions of area (*A*). At time (*c*), a chance migration or a brief interconnection between the two areas allows the cold fauna to occupy the northwestern area from which it was previously excluded by a geographic barrier. Thus, the first appearance of the cold fauna is time transgressive from north to south along the east coast, but it is a good time plane where the cold water fauna invades the northwestern area (*A*). Figure 7.7 continues this discussion.

are occupying the same depositional environments. Because of the temperature factor, however, the tropical and the cool communities do not have a single species in common. The temperate community likely will be less well defined. A few tropical species may be more tolerant of cold than the majority of the tropical community. Similarly, some members of the cold community may extend into the temperate area. Thus, we face a very involved correlation problem if we are to demonstrate contemporaneity of cool and tropical communities. Even if both communities are undergoing evolutionary shifts that will be of regional biostratigraphic importance, numerous sections will be required in between to relate biostratigraphic horizons in the tropical area to biostratigraphic horizons in the cool area.

Time-transgressive first appearance in response to changing climate. Given the distribution of tropical, temperate, and cold-water species portrayed in Figure 7.6(A), consider next the stratigraphy generated by sea-level fluctuations during a time of general climatic cooling [Figures 7.6(B) and (C)]. With each high stand, sediments are laid down that record the areal distribution of climatic conditions during that high stand.

To make the model even more illustrative, allow a new species X to be introduced to the cold area at time (a) by invasion from another continental shoreline (Figures 7.6 and 7.7). The deep water between the two continents had kept X out of this area for millions of years; but once it was introduced to the area by chance migration, it proved to be far superior to the local inhabitants of its niche and proliferated rapidly throughout the cold community. Thus, the first appearance of X provides an excellent time line throughout an area dominated by the cold community.

As our model continues, the general cooling trend introduces species X into stratigraphic sections farther and farther to the south. Thus, the first appearance of X transgresses time as we move southward.

Crossing geographic barriers. Our model can become further complicated by the introduction of a geographic barrier. In Figure 7.6(B), cold climatic conditions have pushed down into areas that were temperate during time (a). Observe, however, that the continental mass is a geographic barrier to the migration of species X in the western portion of our map area. Ecological conditions are right for X, but it cannot get there; it can neither cross the land nor go around the land because the water to the south is too warm.

Allow a chance migration or a brief interconnection between the two cold water masses, and X will take over the western area very rapidly. Thus, the first appearance of X becomes an excellent regional time line throughout the western portion of our map area. But note that this first appearance may be greatly separated in time from the first appearance in the northeast portion of the map area (see Figure 7.7).

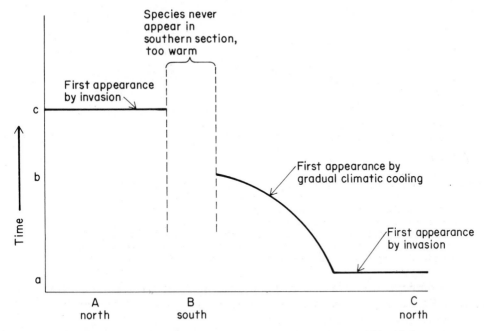

Figure 7.7 Varying significance of the first appearance of species X within Figure 7.6. Locations (A), (B), (C) and times (a), (b), (c) are keyed to Figure 7.6. Because of interaction between earth history (climatic change), niche, and geographic barrier, the first appearance of species X may be strongly diachronous and of changing stratigraphic values.

In looking at this model, we have presumed to know a great many things that the geologist in the field does not know as he begins his work. Identifications of tropical, temperate, and cold faunas might be quite difficult in stratigraphic situations. We really know very little about global climatic patterns during the Paleozoic era, for example. Faunal differences that we have assigned to tropical, temperate, and cold regions might easily be mistaken for evolutionary sequences by the biostratigrapher first undertaking temporal correlation within these stratigraphic sequences. The search for the best time correlation is open-ended. We must remain flexible on this point.

Filtering out the complications. One of the most sensible approaches to this problem is to consciously avoid *a priori* judgments concerning which fossils will be useful for biostratigraphic correlation and which will not. Alternatively, we might demand of ourselves systematic attention to all species present in each stratigraphic section under consideration.

Shaw (1964) has proposed an extremely useful way to establish biostratigraphic correlation between any two stratigraphic sections. His work is based

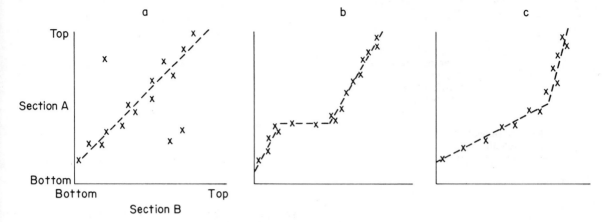

Figure 7.8 Examples of the biostratigraphic correlation technique of Shaw (1964). Data points are the first and last occurrences of species that the measured sections *A* and *B* have in common. The generally good linear fit in diagram (A) suggests precise correlation between the two sections. In diagram (B), the pronounced offset in the otherwise linear trim line suggests that a section is missing out of *A*. In diagram (C), pronounced change in the slope of the correlation line suggests change in relative sedimentation rates.

primarily on data concerning the base and the top of the range for all species that the two sections have in common. We could equally use abundance zones or even such physical time markers as bentonites.

Figure 7.8 provides examples of Shaw's techniques. Section *A* is plotted on the *Y*-axis and section *B* on the *X*-axis. Top of the range and bottom of the range are plotted for each species that is common to both sections. In Figure 7.8(A), the majority of these data cluster around a straight line. The *X* and *Y* intercepts of this straight line constitute precise biostratigraphic correlation between the two sections. Spurious points result from species that are strongly facies-dependent or from species that suddenly cross geographic barriers between the two sections. Note further that the diagram indicates that section *A* has some strata that are older than anything represented in section *B*; conversely, section *B* has some strata that are younger than anything represented in section *A*.

This type of diagram is an extremely powerful stratigraphic tool. For example, in Figure 7.8(B), we see a graph that indicates immediately the absence of strata in the middle of section *A*. A graph such as Figure 7.8(C) suggests abrupt change in relative sedimentation rate, perhaps heralding the onset of deltaic sedimentation in the vicinity of section *A*.

Biostratigraphic Nomenclature

The foregoing discussion gives some idea of the complexity involved in the establishment of reliable biostratigraphic correlation. Such efforts have provided us with a list of recognized names for time-rock units. Table 7.1 is a convenient scorecard containing many of the commonly used names of time-rock units.

Table 7.1 COMMON TIME-ROCK NAMES OF NORTH AMERICA AND EUROPE

			NORTH AMERICA	EUROPE		
C E N O Z O I C	Quaternary	Recent Pleistocene		Calabrian Astian		C E N O Z O I C
	TERTIARY	Pliocene		Zanclian Messinian Tortonian		
		Miocene		Langhian Burdigalian Aquitanian	NEOCENE	
		Oligocene		Chattian Rupelian Tongrian		
		Eocene	Jacksonian	Ludian Bartonian		
			Claibornian	Auversian Lutetian	PALEOGENE or NUMMILITIC	
			Wilcoxian	Cusian Ypresian		
		Paleocene	Midwayan	Thanetian Montain Danian		
M E S O Z O I C	CRETACEOUS		Laramian	Maestrichtian		M E S O Z O I C
			Montanan	Senon. Campanian Santonian Lan Coniacian		
		Upper Cretaceous	Coloradoan Dakotan Washitan	Turonian Cenomanian Albian	CRETACEOUS	
		Lower Cretaceous	Fredericksburgian Trinitian	Aptian Neocomian		
	JURASSIC	Upper Jurassic		Portlandian Kimmeridgian Oxfordian		
		Middle Jurassic		Callovian Bathonian Bajocian	JURASSIC	
		Lower Jurassic		Toarcian Pliensbachian Sinemurian		
	TRIASSIC	Upper Triassic		Hettangian Rhaetian Norian		
		Middle Triassic		Keuper Karnian Ladinian	TRIASSIC	
		Lower Triassic		Muschelkalk Anisian Scythian Buntsandstein		

Table 7.1 (continued)   *NORTH AMERICA   EUROPE*

	Period (N.A.)	Subdivision	North America	Europe	Period (Europe)
	PERMIAN	Upper Permian	Ochoan	Chideruan	PERMIAN
				Kazanian	
		Lower Permian	Guadalupian	Kungurian	
			Leonardian	Artinskian	
			Wolfcampian	Sakmarian	
P A L E O Z O I C	PENNSYL-VANIAN	Upper Pennsylvanian	Virgilian	Uralian	CARBON-IFEROUS
			Missourian	Gshelian	
			Desmoinesian		
		Middle Pennsylvanian		Moscovian	
			Atokan		
			Morrowan		
		Lower Pennsylvanian		Namurian	
			Springeran		
	MISSISSIP-PIAN	Upper Mississippian	Chesteran	Viséan	
			Meramecian		
			Osagian	Tournaisian	
		Lower Mississippian	Kinderhookian	Etroeungtian	
	DEVONIAN	Upper Devonian	Conewangoan	Famennian	DEVONIAN
			Cassadagan		
			Chemungian		
			Fingerlankesian	Frasnian	
		Middle Devonian	Taghanican		
			Tiouchniogan	Givetian	
			Cazenovian	Eifelian	
		Lower Devonian	Onesquethawan		
			Deerparkian	Coblenzian	
			Helderbergian	Gedinnian	
	SILURIAN	Upper Silurian (Cayugan)	Keyseran	Dowtonian	GOTHLANDIAN / SILURIAN
			Tonolowayan		
			Salinan	Ludlovian	
		Middle Silurian (Niagaran)	Lockportian		
			Cliftonian	Wenlockian	
			Clintonian		
		Lower Silurian	Alexandrian	Llandoverian	
P A L E O Z O I C	ORDOVICIAN	Upper Ordovician (Cincinnatian)	Richmondian	Ashgillian	ORDOVICIAN
			Maysvillian		
			Edenian	Caradocian	
			Trentonian		
		Middle Ordovician (Mohawkian)	Blackriveran		
			Chazyan	Llandeilian	
				Skiddavian	
		Lower Ordovician	Canadian	Tremadocian	
	CAMBRIAN	Upper Cambrian (Croixian)	Trempealeauan	Potsdamian	CAMBRIAN
			Franconian	(Lingula Flags)	
			Dresbachian		
		Middle Cambrian	Albertan	Menevian	
				Acadian	
				Solvan	
		Lower Cambrian	Waucoban	Comleyan	
				Georgian	

Nevertheless, we must still face the task of placing these time-rock units into a framework of absolute time. Absolute chronology in geologic history rests heavily upon radiometric dating and is the subject of Chapter 9.

Citations and Selected References

BERGGREN, W. A. 1970. Tertiary boundaries and correlations, p. 693–809. *In* Scientific Committee on Oceanic Research (SCOR) Colloquium (1967), Cambridge, England.

BRETSKY, P. W. 1969. Central Appalachian, Late Ordovician communities. Bull. Geol. Soc. Amer. 80: 193–212.

CUMMEL, B., and C. TEICHERT. 1971. Stratigraphic boundary problems: Permian and Triassic of West Pakistan. University Press of Kansas. 480 p.

GIGNOUX, M. 1955. Stratigraphic geology. W. H. Freeman and Co., San Francisco. 682 p.
Classical reference to European stratigraphy. Strongly biostratigraphic.

HAYS, JAMES D., and W. A. BERGGREN. 1970. Quaternary boundaries and correlations, p. 669–691. *In* SCOR Colloquium (1967), Cambridge, England.

HAZEL, J. E. 1970. Binary co-efficients and clustering in biostratigraphy. Bull. Geol. Soc. Amer. 81: 3237–3252. Biostratigraphic zonation by statistical techniques.

LAPORTE, L. F. 1968. Ancient environments. Prentice-Hall, Inc., Englewood Cliffs, N. J. 115 p.

MCALESTER, A. L. 1968. The history of life. Prentice-Hall, Inc., Englewood Cliffs, N. J. 151 p.

SHAW, A. B. 1964. Time in stratigraphy. McGraw-Hill, New York. 365 p.
An excellent and scholarly discussion concerning the recognition of precise time relationships within stratigraphic sequences.

STANLEY, S. M. 1968. Post-Paleozoic adaptive radiation of infaunal bivalve molluscs: a consequence of mantle fusion and siphon formation. Jour. Paleont. 42: 214–229. Adaptation to living *in* the sediment instead of *on* the sediment afforded competitive advantages to these molluscs.

WALKER, K. R., and L. F. LAPORTE. 1970. Congruent fossil communities of Ordovician and Devonian carbonates, New York. J. Paleont. 44: 928–944.

WILSON, E. O., and W. H. BOSSERT. 1971. A primer of population biology. Sinauer Associates, Inc., Stamford, Conn. 92 p.

8

Time-Stratigraphic Correlation
Based on Physical Events
of "Short" Duration

The search for the best time-stratigraphic correlation is unending. We use all our knowledge but still cannot be absolutely certain that we have events worked out correctly. We must become accustomed to bringing as many lines of evidence as possible to bear on our time-correlation problems. We can develop a real confidence in our correlations only when several lines of evidence begin to agree.

Many physical events in the history of the earth have left us with hints of time-stratigraphic correlation. If we can learn to use these hints, they can help us write the precise history of the earth. Some of the events in the following discussion are global in scope; others are of an extremely local nature. Some of the events may have occurred in a matter of hours or days; others may have occurred over a period of thousands or tens of thousands of years. Whereas worldwide correlation may sound conceptually simple, it is in fact a gigantic undertaking. A geologist may feel that two years of his life were well spent if he has genuinely demonstrated precise time correlation between two sedimentary sections 100 kilometers apart.

Magnetic Stratigraphy

From time to time, the polarity of the earth's magnetic field has become reversed. Precisely why this reversal has happened is a rather difficult problem that we shall leave to the geophysicist. Instead, we shall just use this phenomenon as a time-stratigraphic tool.

Magnetic minerals in rocks preserve a magnetic vector indicative of the orientation and intensity of the global magnetic field within which the rock formed. In the case of igneous rocks, this vector is acquired as the cooling rock passes through the Curie point. In the case of sediments, detrital magnetic minerals may take on

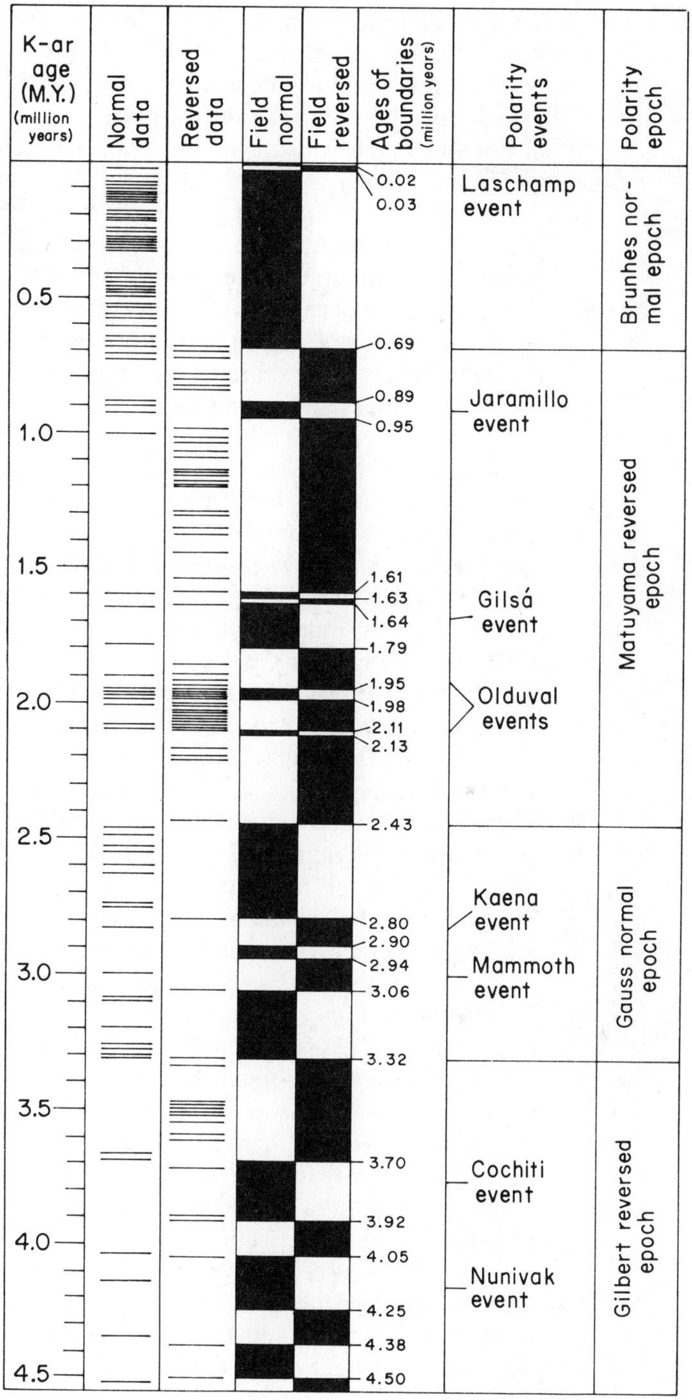

K-ar age (M.Y.) (million years)	Normal data	Reversed data	Field normal	Field reversed	Ages of boundaries (million years)	Polarity events	Polarity epoch
					0.02	Laschamp event	Brunhes normal epoch
					0.03		
0.5							
					0.69		
					0.89	Jaramillo event	
1.0					0.95		
1.5							Matuyama reversed epoch
					1.61	Gilsá event	
					1.63		
					1.64		
					1.79		
					1.95	Olduval events	
2.0					1.98		
					2.11		
					2.13		
2.5					2.43		
					2.80	Kaena event	Gauss normal epoch
					2.90		
3.0					2.94	Mammoth event	
					3.06		
					3.32		
3.5							Gilbert reversed epoch
					3.70	Cochiti event	
					3.92		
4.0					4.05	Nunivak event	
					4.25		
					4.38		
4.5					4.50		

Figure 8.1 Magnetic polarity time scale for the last 4.5 million years. Compiled from magnetic data on volcanic rocks that have been dated by the potassium-argon method. [After A. Cox, "Geomagnetic Reversals," *Science*, **163**, 237–245 (17 January 1969).] Copyright 1969 by the American Association for the Advancement of Science.

a preferred orientation during or shortly after sedimentation, or diagenetic magnetic minerals may record the earth's magnetic field at the time of diagenesis.

Both polarity and intensity of the magnetic field are potentially useful data for time-stratigraphic correlation purposes. To date, most of the emphasis has been placed on correlation of polarity time intervals, and black and white time-stratigraphic diagrams, such as Figure 8.1, have resulted. Clearly, there are certain difficulties when we attempt to relate a new section of rock to one of these diagrams. Numerous times in the history of the earth, the magnetic field has been normal, and numerous times it has been reversed. Correlation can be based only on similarities concerning the spacing of magnetic events. Because this graphic technique does not assign any particular identifying characteristic to any one magnetic interval, magnetic stratigraphy has found its greatest application in relatively young rocks and sediments.

Volcanic rocks. The fundamental documentation of the earth's magnetic history over the last 4.5 million years has been carried out in volcanic sequences of ocean islands and continental areas. Lavas are particularly good for magnetic work because they have formed hard rock before their temperature lowers through the Curie point of the magnetic minerals. Furthermore, these rocks are relatively easy to date by radiometric methods. Thus, we can get good magnetic data on rocks that can be easily placed within an absolute time-stratigraphy.

The one serious disadvantage with developing a magnetic stratigraphy based upon volcanic rocks is the fact that we cannot be certain that volcanism has been continuous through time. Even if we sample every lava flow on a volcanic island— a physical impossibility in itself—we cannot be sure that a complete record of earth history is represented by the suite of samples collected.

Figure 8.1 summarizes the state of our knowledge in 1969 concerning the polarity history of the earth over the last 4.5 million years. It is important to bear in mind that this area of research is rapidly evolving. As indicated in Figure 8.2, the more data we gather, the more magnetic events we recognize.

Oceanic sediments. Fine-grained, deep-ocean sediments also record a magnetic history. Detrital magnetic minerals tend to take on a preferred orientation that is, at least in some statistical sense, related to the magnetic field in which the sediment is ultimately deposited. Reaching this final sedimentation is a more complicated process than it might at first appear. The major problem is the continuing activity of burrowing organisms. As a particle of detrital magnetic mineral first adheres to the sediment-water interface, it is likely very well oriented with respect to the existing magnetic field. Then, along come the burrowing organisms. They ingest the sediment to extract organic matter and excrete the mineral residue. The precise effect of this activity upon the orientation of magnetic minerals is not well understood. Nevertheless, two points seem very clear: (1) Any given sedimentary particle does not achieve its ultimate orientation within the

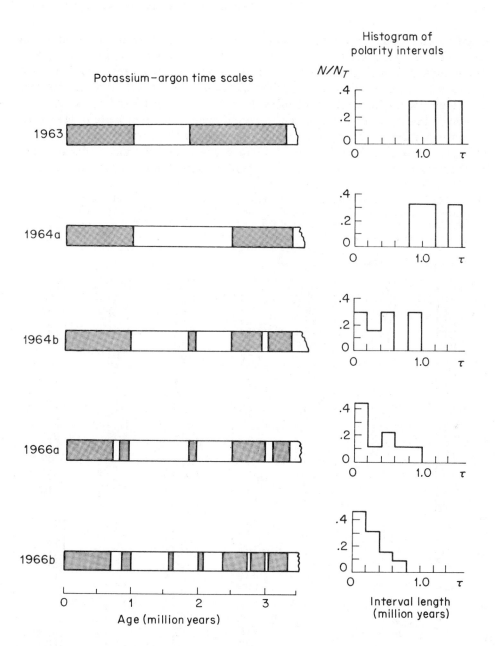

Figure 8.2 Successive refinements of the magnetic polarity time scale. Polarity is either normal or reversed. As more and more data are accumulated, more and more reversed events are found in normal epochs and normal events in reversed epochs. [After A. Cox. "Geomagnetic Advancements," *Science*, **163**, 237–245 (17 January 1969).] Copyright 1969 by the American Association for the Advancement of Science.

sediment until after sediment accumulation has placed that particle out of the reach of burrowing organisms. Clearly, this fact will tend to blur the precise stratigraphic positioning of magnetic events in the deep-sea cores. (2) It must be anticipated that magnetic events of short duration may be lost from the sedimentary record by homogenization of the sediment because of burrowing activity.

Thus, both volcanic rocks and oceanic sediments offer certain assets and certain liabilities concerning the precise recording of the earth's magnetic history. Whereas volcanic rocks afford excellent magnetic data and can be dated readily by radiometric methods, they do not record a continuous history, owing to the

Figure 8.3 Correlation of Antarctic cores by magnetic stratigraphy. Greek letters indicate radiolarian biostratigraphic zones. Note that biostratigraphic boundaries often occur near magnetic events. [After N.D. Opdyke *et al.,* "Paleomagnetic Study of Antarctic Deep-Sea Cores," *Science,* **154,** 349–357 (21 October 1966).] Copyright 1966 by the American Association for the Advancement of Science.

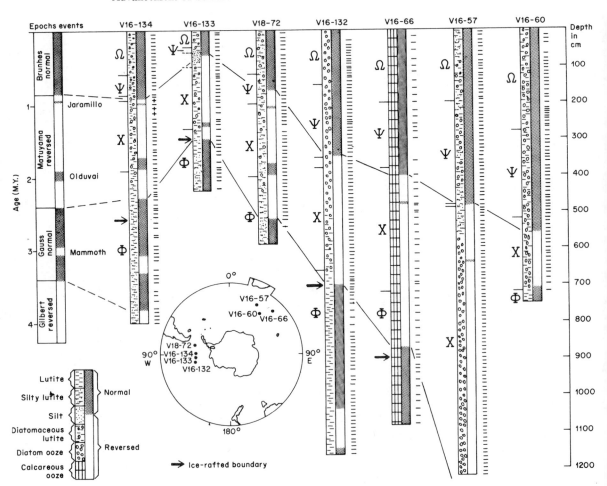

periodic nature of volcanism. In contrast, oceanic sediments offer a continuous record of the earth's magnetic history; but the record is somewhat blurred by burrowing activity, and so precise dating by radiometric methods is often difficult. Where the ultimate goal of scientific research is the establishment of the complete and detailed magnetic history of the earth, scientists move back and forth among magnetic data from volcanic rocks, deep-sea sediment, and mid-ocean ridges. In our present discussion, the main concern is using magnetic reversals as physical events that allow time-stratigraphic correlation.

Figure 8.3 depicts correlation of seven Antarctic deep-sea cores on the basis of their magnetic stratigraphy and relates this correlation to the standard magnetic stratigraphy composite section. Note that the magnetic stratigraphy provides rather convincing correlation among these cores, even though they contain varying types of sediment and are physically separated by as much as 2000 kilometers.

Of further interest to stratigraphers is the fact that biostratigraphic events, such as beginnings and terminations of ranges, appear in some cases to be closely associated with magnetic events (see Hays and Opdyke, 1967; and Phillips *et al.*, 1968, for example). This fact offers the stratigrapher the possibility that a single event in the history of the earth may be expressed in both *physical* (magnetic properties) and *biological* (termination or beginning of a range) data. The event, therefore, becomes a concept that allows correlation among various types of data.

Continental iron-bearing sediments. Alluvial sediments deposited in an oxidizing environment may also record the magnetic history of the earth. The usual problems of detrital versus authogenic origin of the magnetic minerals must be taken into account in these sediments. Furthermore, interaction between stream transport and gravity may produce depositional fabrics unrelated to the magnetic field. Thus, cross-bedded, channel-sand deposits may be a poor choice for magnetic study, whereas associated muddy, flood-plain deposits may yield reliable magnetic data.

Soil zones may likewise contain significant magnetic minerals to be suitable for study. The activity of burrowing organisms in the soil and the movement of sedimentary particles by root growth tend to disrupt primary depositional fabrics resulting from particle shape and gravity. As homogenization removes these strong primary fabrics, the magnetic mineral may take on a weak, but statistically significant, preferred orientation with respect to the magnetic field.

Time-Stratigraphic Correlation Based on Climatic Fluctuations

Physical events, such as magnetic reversals, provide actual *physical* bases for time-stratigraphic correlation. Physical events may also provide *conceptual* bases for precise time-stratigraphic correlation. For example, the distribution pattern of organisms may offer indirect evidence of such events as climatic fluctuations.

Variation in climate usually induces variation in the distribution pattern of organisms. Even if the organisms involved in these shifting patterns remain the same over large intervals of time, changes in their distribution pattern may produce a stratigraphy in which climatic fluctuations provide the conceptual basis for time correlation. A generalized example of this concept is provided in Figure 8.4. Note that, in this conceptual model, precise time correlation is based on ecological shifts within the populations rather than on the first appearances or last appearances of the various elements of the fauna.

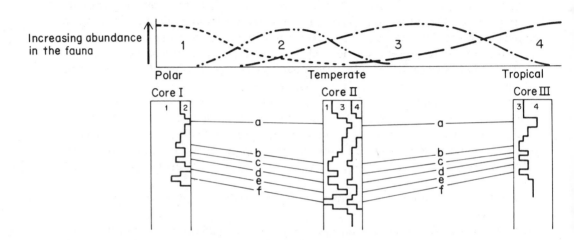

Figure 8.4 Schematic example of time-stratigraphic correlation based on shifts of fauna in response to climatic change. Faunas (1), (2), (3), and (4) occur predominantly in polar, subpolar, subtropical, and tropical regions, respectively. Even though the same organisms may exist throughout a long time interval, variations in their distribution pattern may provide a basis for event correlation. In this fashion, horizon *a* through *f* correlate shifting patterns of "warmer" to "cooler" fluctuations down each of three cores.

Warm-cool foraminifera stratigraphy of deep-sea cores. This correlation technique has been most fully exploited in the study of the Pleistocene history of the ocean basin. Evidence of recurring continental glaciation during the Pleistocene led early workers to hypothesize quite naturally that cyclic climatic fluctuations should provide the basis for precise time correlation of deep-sea cores. Figure 15.1 illustrates correlation between several Pleistocene cores widely spaced throughout the Caribbean and tropical Atlantic. In this example, we assume that relative abundance of the tropical species *Globorotalia menardii* indicates generally warm conditions, whereas relative scarcity of this species indicates somewhat cooler conditions.

Oxygen-18 in Pleistocene planktonic foraminifera. When water evaporates from the oceans to provide moisture to the atmosphere, there is a fractionation of water containing oxygen-16 and oxygen-18. Statistically, water containing O^{18} tends to be left in the ocean because it is heavier. When atmospheric moisture returns to the surface as precipitation, it is likewise O^{18} water with this isotope of oxygen that will tend to precipitate first. Thus, precipitation to the interior of continents usually is significantly depleted in O^{18} when compared to sea water. It follows then that the relative abundance of O^{18} in ocean water as a function of time is dependent upon the amount of ice impounded in continental glaciers. Furthermore, the fractionation of O^{18} between the water and the calcium carbonate of a foraminiferal test is dependent upon temperature. These two factors combine to produce relative enrichment in O^{18} in foraminiferal tests formed during glacial times. These data, therefore, constitute an independent check on the conceptual basis of warm-cool stratigraphies such as just outlined. As will be indicated in Figure 15.5, agreement between *G. menardii* stratigraphy and O^{18} stratigraphy is good in those cores where both techniques are applicable.

Let us note carefully that these are very specific examples of a very general conceptual basis for precise time-stratigraphic correlation. We need not restrict ourselves to either temperature fluctuations or pelagic organisms. It is just as easy to envision similar stratigraphic correlation utilizing terrestrial floras that might be indicative of wet or dry climate.

Time-Stratigraphic Correlation Based on Transgressive-Regressive Events

At this point, our discussion of time-stratigraphic correlation drops down to a much more local level. Our interest in time in stratigraphy does not have to be limited to problems of global scope. A geologist may become quite involved in studying the detailed time relationships within a relatively small stratigraphic interval and not give a single thought to how this area ties into global time-rock or time units. The petroleum geologist, for example, may find a stratigraphic trap simply by unraveling precise time relationships within a limited stratigraphic interval over a small area.

The relation of sediments to sea level forms another important conceptual basis for event correlation. As the sea level changes, the sedimentation patterns change. Sediment types may assume new positions within the paleogeography, or new lithologies may occur over widespread areas.

Symmetrical transgressive-regressive cycles. Lithologic and paleontologic attributes of rock types usually allow generalized reconstruction of a paleogeographic profile from basin margin to basin center. Let us view a measured section in the light of this generalized profile. When more basinward rocks occur over more marginward rocks, we can see that a transgression has occurred. Similarly, when marginward rocks occur over basinward rocks, a regression has occurred. Where we can demonstrate that sediment types within a vertical profile are symmetrically distributed with respect to the most basinward facies present, that surface connecting the most basinward facies among the various vertical profiles approximates time correlation.

This event correlation is similar to our example of polar-tropical faunas in Figure 8.4. In this latter case, it is local environmental conditions related to sea level (Figure 7.5) that control the distribution patterns of faunas (1) through (4).

Asymmetrical transgressive-regressive cycles. Submergence followed by progradation and aggradation commonly results in an asymmetrical cycle. Submergence often occurs with such rapidity that sedimentation does not record the orderly transgression from one facies to another. Rather, we frequently find extremely shallow-water facies or even continental facies abruptly overlain by sediments that indicate deeper water or more open marine conditions. Rapid submergence has produced a discontinuous shift, or *kickback,* in the regional configuration of the generalized paleogeographic profile. In such a case, time correlation can be made just above the discontinuity surface. All lithologies in that stratigraphic position existed almost immediately after the kickback event. Those sediments overlying the kickback surface record the synchronous reestablishment of a paleogeographic profile immediately following the transgressive event. From this point onward, progradation and aggradation may produce a complicated facies mosaic that defies precise time correlation.

Thin limestone beds in clastic marine sequences. Clastic sedimentation in shallow-marine environments is closely related to sediment supply from the land area. Because of the characteristics of meandering-stream sedimentation and deltaic sedimentation (discussed in later chapters), transgressive pulses may cause the clastic sediment supply to be momentarily trapped in alluvial and deltaic environments. "Momentarily" is used here in the geologic context; probably thousands of years would be involved. Such disruption of clastic sediment supply to shallow-marine environments may allow carbonate sedimentation to occupy large areas for a short time; thus, there is opportunity for a special case of kickback correlation. The subsequent return of normal conditions of clastic sediment supply will cause the limestones to be replaced in time by renewed clastic sedimentation. The net result of this overall sequence of events will be the deposition of thin beds of limestone throughout areas that are normally dominated by clastic sedimentation. Such limestones have been recognized historically as valuable marker beds for *physical* stratigraphic correlation. By tracing their relationship to transgressive events, we have a conceptual basis for considering their significance in *time-*stratigraphic correlation.

Rapid Depositional Events

Once again, we return to our opening remarks: The search for the best time-stratigraphic correlation is unending. Some sedimentary events have long been acknowledged as useful in time-stratigraphic correlations; other events appear quite reasonable but have not been widely used.

Volcanic ashfalls (called *bentonites* by stratigraphers) are valuable aids in time-stratigraphic correlation studies. Major eruptions of volcanoes are discrete events of relatively short duration and widely separated in time. The ash from such an event may blanket large areas and therefore provide a time marker within the sedimentary sequence.

Turbidity current deposits may also be catastrophic events that blanket large areas. Other examples are sandstorms, which may lavishly deposit petrographically or mineralogically identifiable material. The possibilities are numerous. The ultimate completeness of a list of such events depends only on keen observation and a mind sufficiently alert to make sense out of what has been observed.

Citations and Selected References

BUREK, P. J. 1970. Magnetic reversals: their application to stratigraphic problems. Bull. Amer. Assoc. Petrol. Geol. 54: 1120–1139.
Application of magnetic stratigraphy to the problem of defining the Paleozoic-Mesozoic boundary.

COX, A., G. B. DALRYMPLE, and R. R. DOELL. 1967. Reversals of the earth's magnetic field. Sci. American 216: 44–61.
A popular summary including a discussion about methods of measurement.

———. 1969. Geomagnetic reversals. Science 163: 237–245.
Excellent summary paper.

EMILIANI, C. 1966. Paleotemperature analysis of the Caribbean cores, P6304-8 and P6304-9, and a generalized temperature curve for the past 425,000 years. J. Geology 74: 109–126.

ERICKSON, D. B., and G. WOLLIN. 1968. Pleistocene climates and chronology in deep-sea sediments. Science 162: 1227–1234.

HAYS, J. D., and N. D. OPDYKE. 1967. Antarctic radiolaria, magnetic reversals, and climatic change. Science 158: 1001–1011.
Relates the disappearance of some radiolaria to magnetic events.

———. 1971. Faunal extinctions and reversals of the earth's magnetic field. Bull. Geol. Soc. Amer. 82: 2433–2448.

IMBRIE, J., and N. G. KIPP. 1971. A new micropaleontological method for quantitative paleoclimatology: application to a late Pleistocene Caribbean core, p. 71–181. *In* K. Turekian (ed.), The Late Cenozoic glacial ages. Yale Univ. Press, New Haven, Conn.

IRVING, E. 1964. Paleomagnetism and its application to geological and geophysical problems. John Wiley & Sons, New York. 399 p.

IRWIN, M. L. 1965. General theory of epeiric clear water sedimentation. Bull. Amer. Assoc. Petrol. Geol. 49: 445–459.

ISRAELSKY, M. C. 1949. Oscillation chart. Bull. Amer. Assoc. Petrol. Geol. 33: 92–98.
Proposes a symmetrical cycle for transgression and regression based on benthonic foraminifera assemblages.

MALKUS, W. V. R. 1968. Precession of the earth as the cause of geomagnetism. Science 160: 259–264.

McELHINNY, M. W. 1971. Geomagnetic reversals during the Phanerozoic. Science 172: 157–159.
Generalized magnetic history of the earth from the Cambrian to the Recent.

NAGATA, T. 1961. Rock magnetism. Plenum Press, New York. 350 p.

OPDYKE, N. D., B. GLASS, J. D. HAYS, and J. FOSTER. 1966. Paleomagnetic study of Antarctic deep-sea cores. Science 154: 349–357.

PHILLIPS, J. D., W. A. BERGGREN, A. BERTELS, and D. WALL. 1968. Paleomagnetic stratigraphy and micropaleontology of three deep-sea cores from the central North Atlantic ocean, p. 118–130. *In* Earth and planetary science letters, vol. 4.
Relates the evolutionary transition from one foraminifera species to another during the Olduvai event.

WATKINS, N. D., and A. ABDEL-MONEM. 1971. Detection of the Gilsa geomagnetic polarity event on the islands of Madeira. Bull. Geol. Soc. Amer. 82: 191–198.

9

Absolute Time in the Stratigraphic Record

In previous chapters, we have explored the methods that the stratigrapher uses to place rocks in their proper sequential relationships within the history of the earth. The ultimate conclusions drawn from these exercises are time-rock statements: These rocks are the same age as those rocks; these rocks are younger than those rocks, and so on. Now we shall look at the ways in which we can obtain estimates of absolute age within the time-rock sequence.

Radioactive Isotopes: The Clocks of Geologic Time

Radiometric dating techniques have been a highly significant fallout of man's interest in the exploitation of nuclear energy. Radioactive isotopes undergo systematic change with time by gaining or losing subatomic particles. All these reactions proceed as an exponential function of time, which can be characterized by the half-life of the reaction. In the case of decay reactions, for example, after one half-life, the abundance of the parent nuclei is one-half what it was at the beginning; after two half-lives, one-quarter the original amount; after three half-lives, one-eighth; four half-lives, one-sixteenth; and so on. Table 9.1 summarizes some pertinent information concerning the utility of various reactions for radiometric dating. Details of applying these methods are discussed elsewhere (Hurley, 1959; Faul, 1966; and Eicher, 1968, for example).

In the following pages, we shall address ourselves to the stratigraphic aspects of radiometric dating. We assume that the geochemist is smart enough to give us a reliable date if we are smart enough to give him a sample that will have genuine stratigraphic significance once it has been dated.

Table 9.1 The Chief Methods of Radiometric Age Determination

Parent Nuclide	Half-life (in years)	Useful Age Ranges (years B.P.)	Daughter Nuclide	Minerals and Rocks Commonly Dated
Carbon-14	5730	< 25,000	Carbon-12	Wood, peat, $CaCO_3$
Uranium-235	—	< 150,000	Protactinium-231 [1]	Aragonite corals Deep-sea sediment
Uranium-234	—	< 250,000	Thorium-230 [1]	Aragonite corals Deep-sea sediment
Uranium series	—	200,000 to tens of millions	Helium 4	Aragonite corals
Uranium-238	4,510 million	> 5 million [2]	Lead-206	Zircon Uraninite Pitchblende
Uranium-235	713 million	> 60 million [2]	Lead-207	Zircon Uraninite Pitchblende
Potassium-40	1,300 million	> 50,000 [2]	Argon-40	Muscovite Biotite Hornblende Glauconite Sanidine Whole volcanic rock
Rubidium-87	47,000 million	> 5 million [2]	Strontium-87	Muscovite Biotite Lepidolite Microcline Glauconite Whole metamorphic rock

(1) Intermediate daughter products.

(2) These methods may give useful information below the minimum indicated, but the methods become increasingly subject to serious error with decreasing age.

Radiometric Dating of the Stratigraphic Record

A glance at Table 9.1 confirms that we may have some difficulties when we attempt to apply radiometric dating methods to the stratigraphic record. The rocks and minerals on our list are, in general, igneous and metamorphic. Indeed, there are relatively few situations in which we can directly tell the age of a sediment. More often, we must determine the age of associated igneous rocks and estimate the age of the sediments by stratigraphic inference based on the relationship of the sediments to the igneous rocks.

Direct dating of the sediment. Carbon-14 dating on materials in the sediment has proven invaluable in the study of the last 25,000 years of earth history. The dating is usually performed on wood, mummified organic matter, peat, or the skeletal remains of carbonate-secreting organisms. Although the analytical equipment used in C-14 determinations is capable of measuring apparent ages as old as 40,000 years, the possibility of contamination by Recent carbon makes ages beyond 25,000 years highly suspect.

Various methods involving thorium-230 and protactinium-231 are directly applicable to sediments in the time range of 1000 to 200,000 years B.P. Once again, we note that these sediments are relatively young. Yet, a precise understanding of this interval of earth history is extremely important to our understanding of earth history as a whole. Thorium-230 and protactinium-231 are not detectable in seawater, whereas their parent nuclides, uranium-238 and U-235, are relatively abundant. Thus, thorium-230 or protactinium-231 growth in the skeletons of such marine organisms as corals provides an excellent clock, which can be easily read. We need only determine the abundance of either pair of nuclides to estimate the age of an aragonite coral sample. Because the daughter products were not present in the seawater in which the coral grew, their presence in the coral skeleton records only the decay of the parent after the coral skeleton formed. Thorium-230 and protactinium-231 undergo decay themselves. Thus, the ratios of these intermediate daughter products to their parents approach an equilibrium value with the continuing passage of time. Thorium-230 growth reaches steady state after approximately 250,000 years; protactinium-231 reaches steady state after 150,000 years.

Because thorium-230 and protactinium-231 are produced by their parents at different rates, the ratio of thorium-230 to protactinium-231 changes systematically with time. The ratio method can be applied to deep-sea sediments that initially contained measurable amounts of detrital thorium and protactinium. If we can assume constant isotopic composition of the detrital input to the sediment, the age of sediment from deep-sea cores can be calculated from the analytical data. In some cases, this assumption seems valid; in other cases, the assumption is almost certainly invalid. Thus, the ratio method on deep-sea cores yields age estimates that are inferior to estimates based on thorium-230 and protactinium-231 growth methods on corals.

The growth of helium concentration in aragonite coral skeletons provides a dating method potentially applicable to sediments as much as tens of millions of years old. Uranium-series nuclides produce helium as a decay product. The helium is trapped within the aragonite skeleton, and its concentration provides an estimate concerning the age of the coral.

All these methods for dating calcium carbonate require unaltered skeletal material. If the skeleton has undergone recrystallization, the system has been opened at some unspecified time and under some unspecified environmental conditions. The apparent date on such material will be invalid. Inasmuch as carbonate skeletons commonly undergo recrystallization, the specification of unaltered skeletal material places serious limitations upon the application of these techniques to the stratigraphic record.

In certain special cases, potassium-argon dates on clay minerals provide direct evidence for the time at which the sedimentary rock was deposited. However, we must approach the direct dating of clay minerals with extreme caution. For example, many "illites" in the stratigraphic record are nothing more than finely divided detrital grains of igneous or metamorphic muscovite. Other common clay minerals may have a complex origin involving chemical alteration of igneous and metamorphic materials followed by interaction with the water chemistry of the depositional environment. Thus, the dates for these minerals may not record the time at which the sediment was deposited but may indicate instead the time of origin for the igneous or metamorphic rock from which the detrital sediment was derived. Such a date provides information concerning the maximum possible age of the sediment. The sediment, however, may be much younger.

We can easily infer that some clay minerals are indeed the authogenic product of the marine environment in which the sediment was deposited. Potassium-argon dates on the mineral glauconite, for example, are commonly in fair agreement with other lines of evidence concerning the precise age of the sediments containing them. In fact, the glauconite data tend to give a somewhat younger date in comparison to other presumably reliable methods.

Contemporaneous volcanic sediments. In general, volcanic sediments are relatively easy to date, particularly by the potassium-argon method. Occasionally, we may be so fortunate as to find volcanic sediments interbedded with otherwise continuous sedimentary sequences. Ashfalls, or bentonites, are particularly valuable because the observant stratigrapher can often demonstrate conclusively that the ashfall and the surrounding sediments are truly contemporaneous.

If the ashfall occurred within the sedimentary environment, then the date on the ash is a date on the biostratigraphic interval containing it. If, on the other hand, the ash simply fell upon an exposed rock surface of some ancient paleogeography, then the date on the ash simply tells us that everything below the ash is older and everything above the ash is younger than the date on the ash.

There are many things we can look for to infer the contemporaneity of the ash and the surrounding sediment. First and foremost, biostratigraphic zones are the main working tool of the stratigrapher. Are the rocks above and below the ash assignable to the same biostratigraphic zone? Furthermore, can it be demonstrated that the ash fell on soft sediment? Burrowing organisms frequently carry portions of the ash bed down into the underlying sediment. They also may carry overlying sediment down into the ash layer. Distribution of the ash layer with respect to the paleogeography may provide another important clue. Within shallow-marine environments, for example, the ash should accumulate within quiet-water environments, but it would presumably be disrupted and removed from agitated environments. Thus, lagoons and offshore areas should have the ash, whereas beach deposits should not. If the ash does cut across lagoons, beaches and offshore areas, we should suspect that the underlying paleogeography was not the site of active, shallow-marine sedimentation at the time the ash was laid down. Perhaps, for example, the entire paleogeography was lithified rock sitting well above sea level at the time the ash fell.

Lavas interbedded with sediments provide another opportunity to date volcanic material that may be synchronous with the interbedded sediment. The deposition of lavas into a stratigraphic sequence is a much more violent process than the deposition of ashfalls. Contact relationships between the sediments and the lavas will not be so conclusive as the evidence concerning ashfalls. In addition, we must be careful to distinguish lava flows from injected sills, a problem that we shall take up shortly.

Age brackets on sedimentary sequences from radiometric dates on related igneous rocks. Radiometric dating techniques have much more general application to igneous and metamorphic rocks. If we can demonstrate a relationship between sediments and nearby datable igneous rocks, then we have information concerning the absolute time of deposition of the sediment. Usually this information takes the form of an "older than" or "younger than" statement. If a sediment unconformably overlies metamorphic rocks and the sediment itself shows no signs of metamorphism, a date on the metamorphic rocks clearly tells us that the sediments are younger than that absolute age. At this point, we have no idea how much younger; the age of the metamorphic rocks simply puts a lower limit on our estimation of the age of the sediment.

If we continue to be concerned about the precise age of these sediments, we must find stratigraphic relationships with datable materials to provide an "older than" statement. For example, if we can demonstrate igneous intrusion into these same sedimentary beds, then the date of igneous intrusion tells us that the sediments are older than that age. Thus, in this example, two dates on unrelated igneous and metamorphic materials yield absolute ages that must bracket the age of the sediment.

Note that we are obliged to take very special care in documenting contact relationships between the sediments and associated igneous or metamorphic rocks

Table 9.2 GEOLOGIC TIME SCALE

Relative Durations of Major Geologic Intervals	Era	Period	Epoch	Duration in Millions of Years (Approx.)	Millions of Years Ago (Approx.)
Cenozoic		Quaternary	Recent	Approx. last 5,000 years	—0
			Pleistocene	2.5	2.5
Mesozoic			Pliocene	4.5	7
			Miocene	19	26
			Oligocene	12	38
			Eocene	16	54
	Cenozoic	Tertiary	Paleocene	11	65
Paleozoic		Cretaceous		71	136
		Jurassic		54	190
	Mesozoic	Triassic		35	225

(Millions of Years Ago scale markers: —0, —50, —100, —150, —200, —250)

Era	System	Period	Age (10⁶ yr)	Duration (10⁶ yr)	Millions of Years (scale)
Paleozoic		Permian	280	55	—300
	Carboniferous	Pennsylvanian	325	45	
	Carboniferous	Mississippian	345	20	—350
		Devonian	395	50	—400
		Silurian	430	35	—450
		Ordovician	500	70	—500
		Cambrian	570	70	—550
Precambrian		Precambrian		4,030	—4,600

Radiometric ages after Harland et al., 1964. (After Eicher, 1968.)

that are to be dated. Incorrect interpretation of stratigraphic relationships results in patently wrong inference concerning the age of the sediment. If the sediment is either "older than" or "younger than" a given date, there is clearly no room to be halfway right.

The Geologic Time Scale

The various techniques discussed in the last three chapters have been applied for approximately 100 years. The result has been the continuing development and improvement of the geologic time scale. Tables 7.1 and 9.2 are a convenient summary of the words and absolute ages that are used today in discussing the history of the earth. Nevertheless, we should anticipate further modification and addition.

Citations and Selected References

BENDER, M. L. 1970. Helium-uranium dating of corals. Ph.D. dissertation, Columbia University. 149 p.

BROECKER, W. S., D. L. THURBER, J. GODDARD, T. L. KU, R. K. MATTHEWS, and K. J. MESOLELLA. 1968. Milankovitch hypothesis supported by precise dating of coral reefs and deep-sea sediments. Science 159: 297–300.

DALRYMPLE, G. B., and M. A. LAMPHERE. 1969. Potassium-argon dating. W. H. Freeman and Co., San Francisco. 258 p.

EICHER, D. L. 1968. Geologic time. Prentice-Hall, Inc., Englewood Cliffs, N. J. 150 p. Good supplementary reading for Chapter 7 through 9.

EVERNDEN, J. F., D. E. SAVAGE, G. H. CURTIS, and G. T. JAMES. 1964. Potassium-argon dates and the Cenozoic mammalian chronology of North America. Amer. J. Sci. 262: 145–198.

———, and G. T. JAMES. 1964. Potassium-argon dates and the Tertiary floras of North America. Amer. J. Sci. 262: 945–974.

FANALE, F. P., and O. A. SCHAEFFER. 1965. Helium-uranium ratios for Pleistocene and Tertiary fossil aragonites. Science 149: 312–317.

FAUL, H. 1966. Ages of rocks, planets, and stars. McGraw-Hill, New York. 109 p. Convenient discussion of radiometric dating methods.

HAMILTON, E. I., and R. M. FARQUHAR. 1968. Radiometric dating for geologists. Wiley-Interscience, New York. 506 p. Collection of papers at the research level.

HART, S. R. 1964. The petrology and isotopic-mineral age relations of a contact zone in the Front Range, Colorado. J. Geology 72: 493–525. Metamorphism affects radiometric clocks to varying degrees.

HURLEY, P. M. 1959. How old is the earth? Doubleday and Company, Inc. (Anchor books), New York. 160 p.
Colorful easy reading concerning radiometric dating and the history of the earth.

KU, T. L. 1968. Pa^{231} method of dating corals from Barbados Island. J. Geophys. Res. 73: 2271–2276.

KULP, J. L. 1960. Geologic time scale. Science 133: 1105–1114.
Absolute ages of the periods.

LIBBY, W. F. 1955. Radiocarbon dating. 2nd ed. Univ. Chicago Press, Chicago. 124 p.

MESOLELLA, K. J., R. K. MATTHEWS, W. S. BROECKER, and D. L. THURBER. 1969. The astronomical theory of climatic change: Barbados data. J. Geology 77: 250–274.
Application of the thorium-230 method.

𝑥

Sedimentary Environments from the Mountains to the Deep Sea

In the previous section, we looked at the stratigraphic record in a very general fashion. Now we shall learn to recognize specific sedimentary environments within the stratigraphic record in the hope that we may develop a more accurate understanding of how stratigraphic sequences originated.

Each of the following chapters deals with a major suite of sedimentary environments. Each chapter begins with a review of facts based on the observation of Recent sediments accumulating in these environments. Then these facts are formed into a general model of (1) what the sediments of these environments would look like if we saw them in the stratigraphic record and (2) how the environment would react to dynamic events such as submergence and emergence. Finally, with a model based on Recent sedimentation firmly in mind, we shall examine selected Ancient sequences in order to test the model and in order to get some feeling for the adequacy of this approach to the stratigraphic record.

Our discussion of Recent environments is oriented toward the development of principles that will have general application. Specifically, our discussion of Recent environments is *not* oriented toward the detailed analysis of every little facet of sedimentation in this or that area. However, we shall pick up specific points that will be useful to us in the formulation of a general model. The basis for our formulation is usually not at random; rather, it depends on where a study was done that developed a certain point in which we are interested.

The following treatment of sedimentary environments and the stratigraphic record, therefore, focuses on the development of general models. A model with constant sea level is first constructed from Recent sedimentation data. Then it is expanded to a theoretical dynamic model showing how the constant sea-level model should react to conditions of emergence or submergence. The models are of value as a *tool* for solving problems, not as an *answer* to our problems. We must organize our thoughts, recognize our misconceptions and inconsistencies, improve our models, and so work forward in an iterative process that leads to better understanding of the stratigraphic record through the use of ever-improving models.

10

Clastic Sedimentation

in Stream Environments

Our look at the details of sedimentary environments begins with the action of running water on terrigenous clastic debris derived from mountainous terrains. Physical and chemical weathering processes yield boulders, sand, silt, and clay from the mountainsides. These materials immediately begin to accumulate in sedimentary environments that we hope to recognize in the geologic record.

Recent Clastic Sedimentation in Stream Environments

In the following pages, we shall primarily study four examples of stream sedimentation. The Donjek River, a tributary of the Yukon, provides an example of the braided-river environment in mountainous terrain (Williams and Rust, 1969). The Platte River of Colorado and Nebraska is a braided river that extends some 800 kilometers outward from the mountain region (Smith, 1970). The Brazos River of southeastern Texas is an exceedingly well-studied meandering stream (Bernard *et al.*, 1970). Finally, the Red River of northwestern Louisiana is an interesting example of a complicated meandering-stream pattern (Harms *et al.*, 1963).

   The Donjek River. Figure 10.1 indicates the location of the Donjek River. Note that the study area is only a few tens of kilometers downstream from the glaciers that form the headwaters of the river. The modern braided river is 2 to 4 kilometers in width and occupies a valley cut into glacial tills, outwash sands and gravels, and loess.

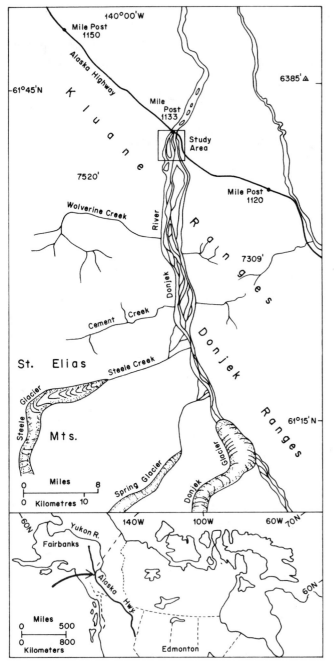

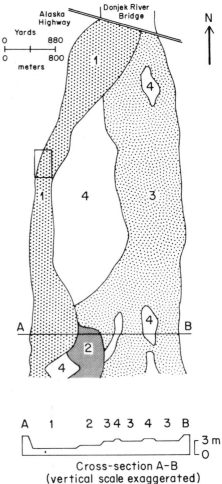

Figure 10.1 (Left) Locality map of the Donjek River and study area. The Donjek River is an example of a braided stream in mountainous terrain very near the source area. (After Williams and Rust, 1969.)

Figure 10.2 (Below) Areal distribution and diagrammatic profile section of stream channels and levels within a portion of the Donjek River. Note location of Figure 10.3. (After Williams and Rust, 1969.)

Cross-section A–B
(vertical scale exaggerated)

Figure 10.2 presents the generalized physiography of the study area, and Figure 10.3 illustrates the complex system of channels and bars that exists within the various levels. Most of the water discharged today is carried by channel level (1). Figure 10.3 shows the major elements of bar and channel geometry in a portion of level (1). The first-order topographic highs are longitudinal bars, separated from one another by the large channels that carry the majority of the water at intermediate- and low-river stage. Deposition and migration of first-order longitudinal bars occur primarily during flood stage. Because they are deposited at flood stage, longitudinal bars are the most poorly sorted sediment within the braided-stream environment. They commonly consist of pebbles and cobbles mixed with sand. As flood waters recede, first-order bars are usually dissected by smaller channels.

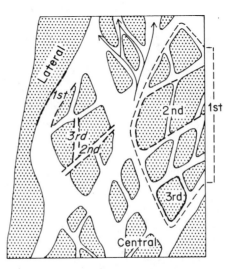

Figure 10.3 Map demonstrating the hierarchal organization of channels and bars within a portion of level (1) of the Donjek River. First-order longitudinal bars, deposited during flood stage, are later dissected by second-order and third-order channels as the river stage subsides. (After Williams and Rust, 1969.)

1st, 2nd, 3rd, = Channel (Orders)
1st, 2nd, 3rd, = Bar (Orders)

During low stage of the river, sedimentation on the longitudinal bars more or less ceases. Continued transport and redeposition of sand-sized material continue to occur within the channels. Transverse sandbars commonly have steep avalanche faces in the downstream direction and small-scale asymmetrical ripples on the top of the sandbar. Both features tend to suggest an almost continuous transportation of sand-sized material across the top of the transverse bar, with sand deposition occurring along the avalanche slope at the front of the bar. Migration of such bars produces large-scale planar cross strata.

Figure 10.4 shows size distribution within a number of samples from the Donjek River. Note the incredible variation in both size and size distribution

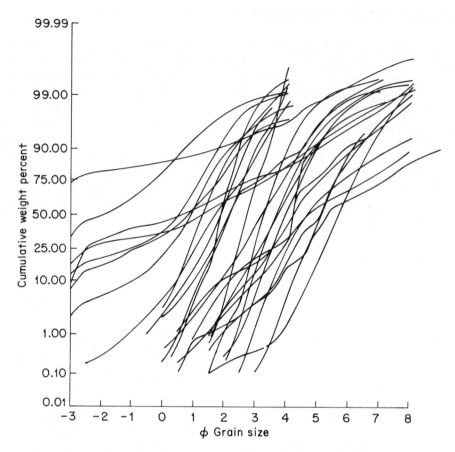

Figure 10.4 Cumulative size-frequency curves of sediment samples from the Don-jek River. Note the extreme variation in grain size and sorting that occurs in samples from a relatively small area within a braided stream. (After Williams and Rust, 1969.)

within the confines of this braided-stream environment. This variation results from the wide range of sedimentation conditions that predominate in the area at varying times throughout the year. During flood stage, sediment supply is large and the accumulation beneath the main flow of water is coarse-grained and extremely poorly sorted. This is also the time of occasional flooding of the higher terrace levels of the river valley. Flood waters reaching these higher levels are relatively slow-moving and deposit only silt and clay. At low stage of the river, sediment supply is also low. Stream power is insufficient to transport the gravels of the longitudinal bar deposits. Whereas the flood waters were extremely muddy, low-stage water is likely to be crystal clear. At this time, only sand-sized material is being transported and resedimented. As indicated in Figure 10.4, some of the medium to fine sands are the best-sorted sediments to be found in the environment.

Figure 10.5 attempts to portray lateral facies relationships within any single bar-channel complex. Observe that transitions from poorly sorted bar gravels to well-sorted channel sand can occur within a matter of a few meters. Where channels lie dormant, they will eventually be filled with silt and clay. Then the entire area will become vegetated unless fluvial processes reoccupy the area within a year or so.

A good indication of bar mobility within the various levels is the colonization of the bar crests by trees. In level (1), bar mobility associated with annual flooding has not allowed trees to establish themselves.

Level (2) is the site of widespread fluvial activity just during flood stage. At other times, only the major first-order channels contain a slight flow of water. Bar mobility during flood stage is not so pronounced in level (2) as it is in level (1).

Figure 10.5 Three-dimensional model illustrating rapid variation in lithology within an area of dissected bars and small channels. (After Williams and Rust, 1969.)

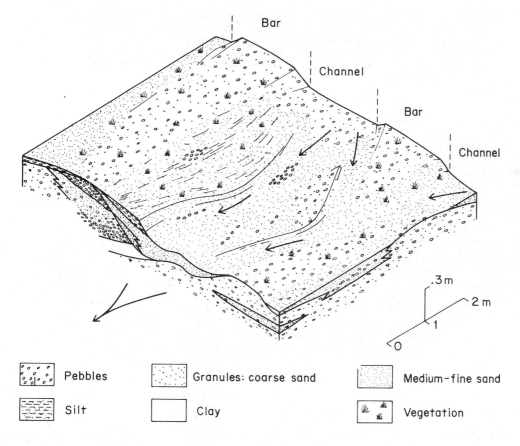

Frequently, willow trees have established themselves on the crests of bars; some are as much as 12 years old (as estimated by tree-ring analysis).

Although levels (3) and (4) retain geomorphic features that are clearly channels and bars, these levels have not been sites of active bar migration for many years. Today, level (3) is covered with water only during maximum stage; at that time, sedimentation is predominated by slack-water settling of suspended muds. Willow trees on level (3) are as old as 27 years, and spruce development on level (4) indicates that the topographic surface there is at least 250 years old.

The ages of vegetation on the various topographic levels of the braided river valley provide a good demonstration of a fundamental characteristic of the braided-stream environment: From time to time the site of major fluvial activity shifts from one area of the river valley to the other in an unsystematic fashion. Level (3) provides the most graphic example of this shift. At some time in the not too distant past, fluvial activity was sufficiently intense in this area of the river valley to develop the system of bars and channels quite similar to those now de-

Figure 10.6 Three-dimensional model indicating the complex relationships among the various stream-channel systems and levels within the Donjek study area. Note especially the extreme lack of lateral continuity, both on the scale of the relationship of one level to the other and even within a single level. Cut and fill and cut again is the ever-recurring story of braided-stream sedimentation. (After Williams and Rust, 1969.)

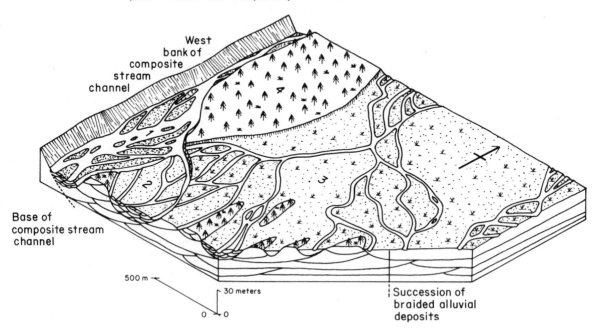

veloping in level (1). Today, these bars and channels are largely inactive, and they are being covered over with a drape of mud deposited only during maximum flood stage. If we may judge from other braided-rivers studies (such as Doeglas, 1962), the eastern side of the river valley will likely become the site of energetic braided-stream activity again at some time in the near future. The present active channel system need only become glutted with sediment or blocked by a rock slide or a log jam; then the site of active transportation and recent sedimentation would rapidly shift back to the east.

Figure 10.6 gives a composite stratigraphic view of the entire study area. Consider whether or not you could untangle this stratigraphic complexity if it faced you in an outcrop of Ancient rocks. We have already noted that lithologic variations occur on a very local scale. The various levels of the river valley that are now so clear in maps and cross sections would appear only as zones of cut and fill in an outcrop of Ancient rocks. The relationship between level (3) and level (2) is well illustrated in Figure 10.6.

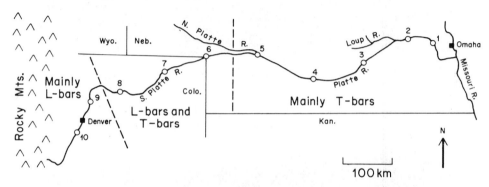

Figure 10.7 Map of the South Platte and Platte rivers showing sample localities in the generalized distribution of longitudinal bars and transverse bars. The Platte River provides an example of braided-stream sedimentation out to considerable distances beyond the mountainous source area of the sediments. (After Smith, 1970.)

The Platte River. Whereas the Donjek River exemplifies braided-stream sedimentation within mountainous regions, the Platte River shows braided-stream sedimentation extending as far as 900 kilometers away from the source area. As indicated in Figure 10.7, Smith (1970) has studied portions of the Platte River system from the front ranges of the Rocky Mountains to the junction of the Platte River and the Missouri River in eastern Nebraska.

As with the Donjek River, the story of Platte River sedimentation is written in terms of *longitudinal bars* and *transverse bars*. Yet this study incorporates a sufficiently long stretch of river so that important differences are noted from west to east. Figure 10.7 indicates that longitudinal bars predominate near the source area, whereas transverse bars predominate in the downstream area. Figures 10.8 and 10.9 depict typical Platte River longitudinal bars and transverse bars.

Figure 10.8 (Above) Longitudinal bar near Fort Morgan, Colorado, South Platte River. Bar is about 8 meters wide. (Photo by N. D. Smith.)

Figure 10.9 (Below) Transverse bars near Valley, Nebraska, lower Platte River. Note especially dissection of the point bar in the left foreground. (Photo by N. D. Smith.)

As in the previous example, deposition of longitudinal bars is primarily a flood stage phenomenon. The internal characteristics of these bars are: (1) a relatively large mean grain size, (2) relatively poor sorting, (3) crude development of horizontal stratification, and (4) trough cross-stratification near the top of the bar.

In contrast, transverse-bar migration may be active at a lower river stage. The internal characteristics of transverse bars include: (1) relatively smaller mean grain size, (2) relatively better sorting, and (3) a preponderance of high-angle, planar cross-stratification (Figure 10.10) topped by small-scale, trough cross-stratification. During low stage of the river, the predominant process is the dissection of transverse bars (see Figure 10.9). Such dissection results in extreme lateral discontinuity of individual sand bodies, a characteristic that will aid our recognition of braided streams in the stratigraphic record.

Figure 10.10 Planar cross strata in a dissected transverse bar, Platte River. Accretion on the avalanche slope causes transverse-bar migration and leaves this sedimentary record. (Photo by N. D. Smith.)

Figure 10.11 provides a graphic summary of mean grain size and standard deviation of sands from longitudinal bars and transverse bars. Because the relative proportion of longitudinal bars and transverse bars varies from west to east, these individual parameters likewise show systematic trends downstream. Figure 10.12 plots mean grain size and standard deviation as a function of distance downstream.

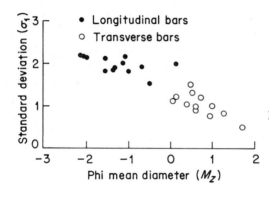

Figure 10.11 Plot of grain size versus sorting for 25 samples of Platte River bar sand. Observe the sharp distinction between coarse-grained, poorly sorted, longitudinal bars and finer-grained, better-sorted, transverse bars. (After Smith, 1970.)

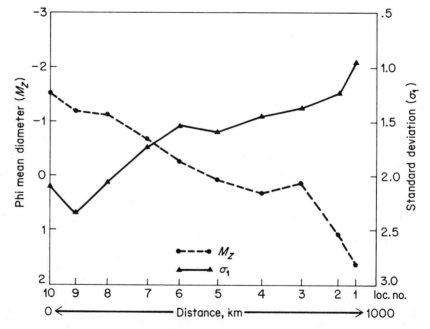

Figure 10.12 Downstream changes in mean grain size and sorting in the South Platte–Platte River. Grain size decreases and the sediment becomes better sorted with increasing distance from the source area. (After Smith, 1970.)

Size decreases and sorting becomes better downstream. Likewise, stratification types can be recast as a function of distance downstream (Figure 10.13). Trough cross-stratification occurs typically as the upper few tens of centimeters on top of both longitudinal bars and transverse bars. The relative importance of trough cross-stratification shows little variation downstream. On the other hand, high-angle, planar cross-stratification occurs predominantly in transverse bars and thus increases systematically downstream. Conversely, horizontal stratification is more typical of longitudinal bars and shows a systematic decrease in importance downstream.

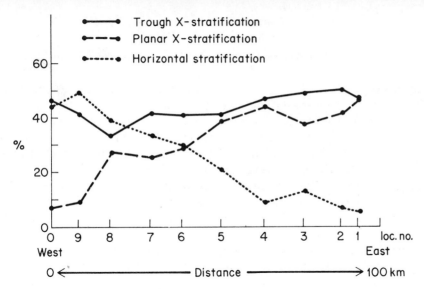

Figure 10.13 (Above) Downstream variation in relative amounts of the principal stratification types, South Platte–Platte River. Whereas trough cross-stratification is uniformly present throughout the study area, planar cross-stratification (associated with transverse bars) is observed to become more abundant downstream, and horizontal cross-stratification (associated with transitional flow regime and longitudinal bars) is most prevalent near the source area. (After Smith, 1970.)

Figure 10.14 (Below) Map showing location and geologic setting of the Brazos River, southeastern Texas. The Brazos is an extremely well-studied example of a meandering stream. Note the numerous abandoned meander loops throughout the coastal plain of southeast Texas. (After Bernard *et al.,* 1970.)

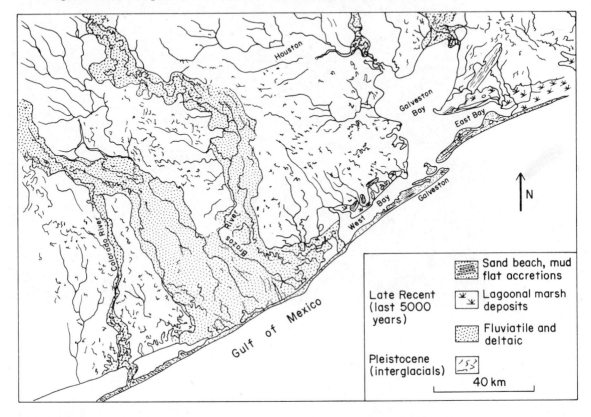

The Brazos River. The Brazos River, southwest of Houston, is an exceedingly well-studied example of a recent meandering stream (Figure 10.14). Whereas the braided stream is characterized by numerous small, shallow channels dissecting a river bed composed of an irregular arrangement of longitudinal and transverse bars, the meandering stream is typified by a single, relatively deep channel in which erosion and sedimentation occur very systematically.

The major elements of the meandering-stream system are indicated in Figure 10.15. The basic subsystem is the individual *meander loop*. The deepest portion of the river channel lies very close to the outer bank of the loop. The outer bank of the loop is the site of active erosion and is commonly referred to as the *cutbank*. The inside bank of the meander loop is the site of active deposition. Sandbars deposited in this position are referred to as *point bars*.

Figure 10.15 Oblique aerial photograph exhibiting the major elements of meandering-stream erosion and sedimentation. Erosion occurs on the outside of the bend (cutbank) and deposition occurs on the inside of the bend (point bar). (After Bernard *et al.,* 1970.)

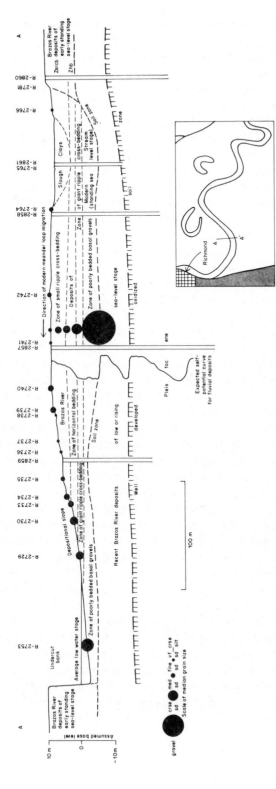

Figure 10.16 Cross section of a Brazos River point-bar deposit. Traversing from the deepest portion of the river channel (left) up onto the point bar (center), note the systematic decrease in grain size as well as a sequence of distinctly different primary structures. This sequence of variation in grain size and primary structures is repeated in vertical sequence in the borehole near the center of the diagram. The point-bar deposit has accumulated by a lateral migration from right to left as the river eroded the cutbank to the outside of the meander loop and deposited sediment on the point bar to the inside of the meander loop. (After Bernard *et al.*, 1970).

Figure 10.16 is a cross section through the river channel and point-bar deposit. A well-developed oxidized soil zone provides a convenient lower boundary for studying Recent meandering-stream deposits. Above the soil are approximately twelve meters of river deposits not related to the deposition of the modern point bar under investigation. Consider first the surficial sediments of the river channel and point bar. Note that the deepest portion of the channel lies at the base of the cutbank. Sediment that accumulates in the deep channel consists of poorly bedded gravel and coarse sand. Grain size decreases systematically, with the coarsest material in the deep channel and the finest sand on top of the point bar. Furthermore, shallow pits dug in the surface of the point bar reveal systematic variations in primary structures. Near the low-water level, there is large-scale, high-angle cross-stratification. Somewhat higher on the bar, stratification is nearly horizontal. The top of the bar is characterized by small-scale, trough cross-stratification. Observe that cores taken through the point bar confirm a vertical sequence of variation in grain size and primary structures. This observation is consistent with observations in the modern river channel and surface of the point bar. Channel migration and concomitant point-bar sedimentation develop a laterally extensive sand body that shows a strong tendency to systematic variation in grain size and primary structures from bottom to top.

Clearly, most of the point-bar deposition process occurs during high stages of the river. Variations in water depth and velocity produce systematic sorting of sediment that is eroded from the cutbank as well as brought down the river from upstream. Coarse sands and gravels remain as a channel-lag deposit in the floor of the deepest portion of the river. In sharp contrast, the upper portions of the point bar record deposition from shallow, relatively slow-moving water. Fine sand accumulates here. The middle portions of the point bar are intermediate between these two extremes of grain size. Similarly, variations in velocity, water depth, and grain size lead to systematic variations in primary structures. In deep, fast-moving water, coarse-grained sediments typically form large-scale, high-angle cross strata. In somewhat shallower water, sand-sized sediments form near horizontal stratification indicative of transitional low regime. Still higher on the bar, fine-grained sand accumulating in relatively slow-moving shallow water forms small-scale, ripple cross-stratifications. The point-bar deposit will be a major portion of the fining-upward unit that we shall come to associate with meandering-stream deposition.

As indicated in Figure 10.17, the present meander belt of the lowstage Brazos River occupies but a small portion of the area dominated by the river during flood stage. During flood stage, large amounts of fine-grained sediment may be deposited throughout the flood plain. Such deposits are commonly referred to as *backswamp* or *overbank* deposits. On the Brazos flood plain, these deposits are generally brown silty clay. Inasmuch as the flood plain is dry most of the year, evidences of soil development and mottling by burrowing animals and plant roots are common.

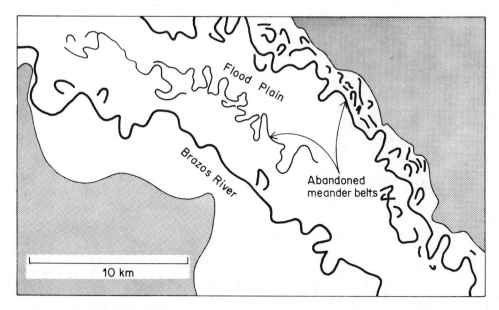

Figure 10.17 Map of the Brazos alluvial valley, illustrating the modern meander belt, abandoned meander belts, and flood plains. During flood stage, water occupies the entire alluvial valley. At such time, overbank deposits of silty clay are deposited throughout the flood plains. (After Bernard *et al.,* 1970.)

The Red River. Whereas the Brazos River provides a simple and pleasing picture of the systematics of meandering-stream sedimentation, the study of point-bar sedimentation in the Red River of northwestern Louisiana (Harms *et al.,* 1963) provides some insight into the complexities that can arise from sedimentological processes operating during various stages of the river. Figure 10.18 indicates the location of the study area. Note once again that the main channel of the present meandering stream occupies just a small portion of the total flood plain. Two topographic surfaces are apparent in the construction of the Beene point bar (Figure 10.19). Both levels are constructed by point-bar deposition of silt, sand, and gravel in a fashion quite similar to that described for the Brazos River. As indicated in Figure 10.20, the upper-level point bar appears to represent sediment accumulation while the river was at or near flood stage, whereas the lower level represents point-bar sediment accumulation while the river was at an intermediate level, still well confined within the banks of the main channel. During the entire year represented by river gauge data in Figure 10.20, the river did not reach the top of the upper level. Thus, upper-level point-bar sedimentation must be regarded as an extremely sporadic process.

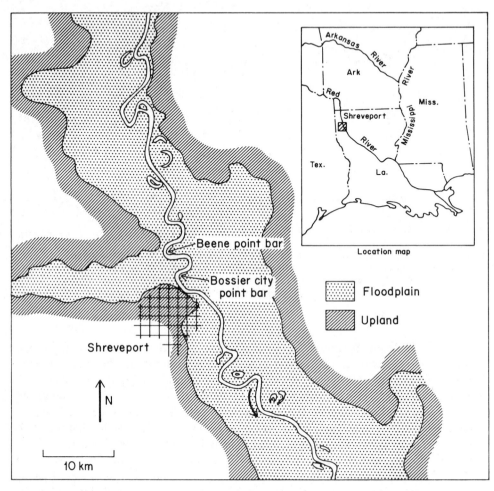

Figure 10.18 Map showing location of the study area, Red River, northwestern Louisiana. (After Harms *et al.*, 1963.)

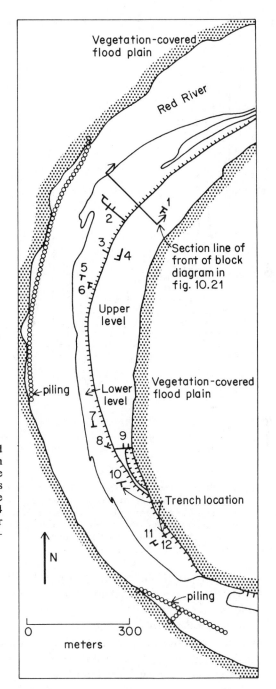

Figure 10.19 Map of Beene point bar, Red River. This point bar consists of an upper level and a lower level. The upper level stands some 7 to 9 meters above the main channel of the river; the lower level stands approximately 4 meters above the main channel. [After J. C. Harms *et al.*, *Journ. Geol.*, **71**, 566–580 (1963).]

Within the figure:

Vegetation-covered flood plain

Red River

Section line of front of block diagram in fig. 10.21

Upper level

Lower level

Vegetation-covered flood plain

piling

Trench location

piling

N

0   300

meters

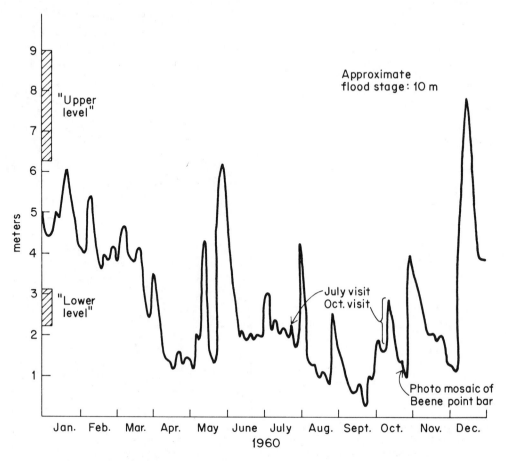

Figure 10.20 Stages of the Red River at the Beene point bar during 1960. Note that the position of the lower level point bar was covered throughout the January–March spring floods and on numerous other brief occasions throughout the remainder of the year. On the other hand, the upper level was only partially flooded for a brief time during December. [After J. C. Harms *et al.*, *Journ. Geol.*, **71**, 566–580 (1963).]

Figure 10.21 summarizes the present distribution of sediment types in a profile of the Beene point bar. According to grain size and type of cross-stratification, the internal construction of the upper level is fairly familiar to us from the Brazos River. Trough cross-stratification in gravelly sands is overlain by horizontal laminations in sand-sized material, which in turn are overlain by large-scale and small-scale trough cross strata in sand or silty sand. In the lower level, cross-stratification in sand and silty sand predominates. Note that we know little about stratification relationships between the lower level and the upper level.

Figure 10.21 illustrates some rather complicated sequences of primary structures that developed during the continued growth and westward migration of the Beene point bar. It would appear that point-bar migration of the lower level is a common process occurring during a net three to four months out of each year (Figure 10.20). Point-bar migration of the upper level seems to occur only during an unusual flood stage. At this time, large-scale, cross-stratified gravel deposits may be expected to migrate across the area formerly occupied by the lower-level point bar, thus forming a complicated vertical sequence of grain sizes and primary structures.

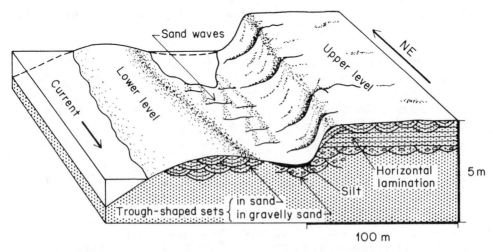

Figure 10.21 Block diagram showing the complex relationships between grain size and primary structures of upper and lower levels of the Beene point bar. [After J. C. Harms *et al., Jour. Geol.,* **71,** 566–580 (1963).]

Alternatively, the upper level may be essentially inactive today, a relict surface recording a time in the near recent past when the Red River flood stage usually reached higher levels than are commonly attained now. At any rate, the complexities of this point bar serve to warn us that a certain amount of flexibility must be built into our generalities concerning meandering-stream sedimentation.

A Basic Model for Clastic Sedimentation in Stream Environments

In the preceding pages, we have noted braided-stream sedimentation in mountainous regions (the Donjek River), braided-stream sedimentation on the plains far out in front of mountain ranges (the Platte River), and meandering-stream sedimentation of the coastal plains (the Brazos and Red rivers). We must now attempt to draw conclusions from these various local situations.

Variation in grain size and sorting. In the Donjek River, we saw that longitudinal bars have large grain size and are rather poorly sorted. In contrast, transverse bars generally consist of sand-sized material and are rather well sorted. In the Platte River, we noted the same generalities as well as the fact that longitudinal bars tend to dominate the upper regions of the river, whereas transverse bars tend to dominate the lower reaches. Thus, we can expect grain size to decrease and sorting to improve away from this source area.

The transition from braided stream to meandering stream is accompanied by further regimentation of grain size. The coarsest material tends to accumulate as channel-lag deposits; finer-grained, but relatively well-sorted, sands tend to accumulate on the top of the point bar; and poorly sorted, silty clays tend to accumulate in the overbank position.

Variation in composition of the sediment. Transportation in the braided-stream environment during flood stage is an extremely violent process. This fact is demonstrated by the compositional variation found away from the sediment source area. For example, we can see large boulders of a poorly cemented sandstone in braided-stream deposits near the outcrop of sandstone. At some distance downstream, however, the rigors of transportation will have smashed other sandstone

Figure 10.22 Effect of transport on rock constituents in the gravel of Rapid Creek, South Dakota. As the ratio of chert to (chert + x) approaches 1, the component x has been removed from the gravel-size fraction by mechanical destruction during the transportation. Component (A) is a rather poorly cemented sandstone; (B), a limestone; (C), Precambrian metamorphic rock fragments; and (D), quartz and quartzite. [After W. J. Plumley, "Black Hills Terrace Gravels: A Study in Sediment Transport," *Jour. Geol.,* **56,** 526–577 (1948).]

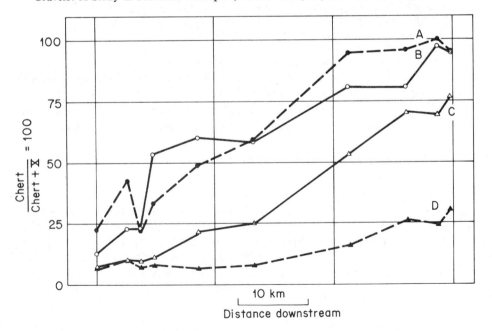

boulders into their constituent sand particles. Similarly, less durable rock types, such as shale or schist, will be ground to a finer powder by continued transport. In general, the shorter the transportation history, the more unstable the rock fragments contained by the sediment. With longer transport, unstable rock fragments are destroyed. These relationships are exemplified particularly well in the data of Plumley (1948). (See also Figure 10.22.)

Primary structures. Structures characteristic of unidirectional transport distinguish stream environments. Large-scale and small-scale, high-angle cross strata predominate. Horizontal laminations of transitional flow regime are noted in high-gradient regions of braided streams and at intermediate levels in point-bar deposits. In these locations, a combination of stream velocity, water depth, and grain size combine to produce transitional conditions. Upper-flow-regime primary structures (antidunes) are not common. If they are found anywhere, they will be in shallow-water environments of high-gradient braided streams.

Figure 10.23 Schematic map and cross section indicating idealized facies relationships within meandering-stream deposits. Sediment types are as follows: (1) channel-lag deposit, commonly gravel; (2) point-bar deposits; (3) natural levee (may be poorly developed). (4) overbank or backswamp deposits; (5) undifferentiated, previously deposited, meandering-stream sediments, now subjected to erosion and resedimentation downstream; (6) older sediments or rock.

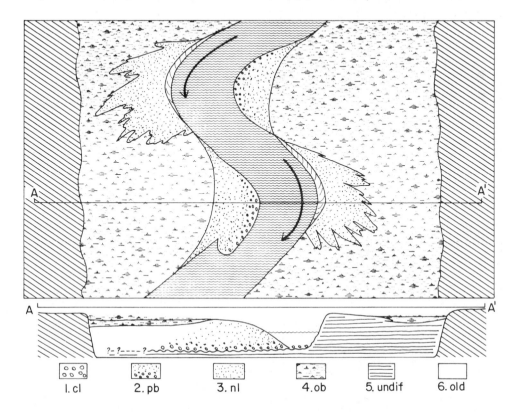

In the meandering-stream environment, we would commonly expect to find a systematic vertical sequence of grain size and primary structures, the so-called *fining-upward unit* (Figures 10.23 and 10.24). As illustrated in Figure 3.15, the lower portion of the fining-upward unit results from the systematic lateral migration of a point bar. Both erosion of the cutbank and scouring in the deeper portions of the river channel produce the sharp lower boundary of the unit. The erosional surface is overlain by the coarsest material present, typically channel-lag gravel and coarse sand. These sediments usually display medium-scale, high-angle cross strata (Figures 3.12 and 3.13). The sequence becomes finer-grained toward the top of the point bar, with primary structures changing from medium-scale, high-angle cross strata to low-angle or horizontal stratification and finally to small-scale, high-angle cross strata. This vertical sequence through the point bar constitutes the lower portion of the fining-upward unit. The horizontal stratification in the middle of the point-bar sequence is not universally developed, but some form of gradation from coarse-grained, medium-scale, high-angle cross strata to fine-grained, small-scale, high-angle cross strata is the invariable hallmark of the point-bar deposit. The upper portion of the fining-upward unit consists of overbank deposits.

Figure 10.24 The basic fining-upward unit of meandering-stream deposition.

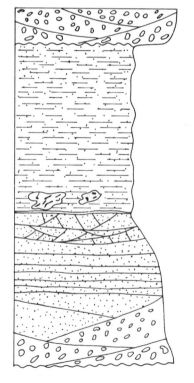

Sharp contact, begin next unit

Poorly sorted sandy silt and clay, highly organic, burrowed, and mottled if deposited in a generally humid climate; mudcracks and evaporite minerals if deposited in a generally arid climate.

Contact highly variable; discontinuous cross stratified sand lenses common.

Fine sand to silt

Small scale high-angle cross-stratification

Gradation

Low-angle or horizontal stratification

Medium scale high-angle cross-stratification

Gravel and coarse sand

Sharp basal contact

In contrast to the systematic variation of grain size and primary structures in the vertical sequence deposited by meandering streams, braided streams would be expected to deposit highly variable sequences. Whereas the deposition of a point bar occurs by systematic lateral migration, sediment accumulation in braided-stream environments occurs largely by vertical aggradation during times of flood. Thus, each new deposit in the braided-stream vertical sequence is not necessarily related to depositional conditions of the sediment that it overlies.

Braided versus meandering habit of streams. For physical reasons beyond the scope of this book, streams naturally should meander. Experiments involving constant gradient of the stream bed and constant discharge rate tend to confirm this theory. Thus, we can consider that a stream showing a braided pattern has not reached the meandering stage.

On the basis of field observations, we can offer several reasons why braided streams do not achieve a meandering pattern. For example, if the banks of a stream are composed of easily eroded material, intermittent collapse of the cutbank may block the main channels of the river and force the stream to follow numerous smaller channels, which may likewise become blocked by slumping sediment. Similarly, it is frequently observed that braided streams tend to have more rapid and more extreme fluctuations in discharge rate than do meandering streams. Abundant sediment load tends to produce a braided-stream pattern.

This situation may be particularly important in those areas where stream gradient changes rapidly from high to low. A high-gradient stream delivers a large volume of sediment, whereas the lower-gradient stream can carry away only a portion of that sediment. The net result is an oversupply of sediment, which tends to glut the channels and causes a braided pattern. Finally, steep-stream gradients generally characterize braided streams; low-stream gradients tend to characterize meandering streams.

The concept of a graded stream profile. So far we have limited our discussion of stream sedimentation to the characteristics of a single bed of sand as it is laid down by the stream. We must now begin to think of stream sedimentation on a more regional scale.

The ability of a stream to transport sediment is dependent upon the gradient of the stream. Other factors being equal, streams flowing down a steep surface will flow faster and therefore transport a larger suspended load and bed load. Thus, if we imagine an initial condition in which a stream of high gradient suddenly becomes a stream of low gradient, we can expect load to be deposited at the change in gradient. Such a situation might occur where a mountain stream comes out onto the prairie. As the net result of this sediment deposition, the transition from high-stream gradient to low-stream gradient will be smoothed out. Eventually, a stream profile will be obtained in which the capacity of the stream to transport sediment will be uniform throughout the length of the profile. Such a stream is referred to as a *graded* stream.

It is convenient to visualize a graded stream in terms of any two stations, *A* and *B,* where sediment flux may be measured. If the stream is graded, the average sediment flux passing *A* and *B* over some appropriate time interval will be the same. Although it is true that both deposition of sediment and erosion of previously deposited sediment have occurred within the region between *A* and *B,* the amount of new sedimentation is precisely balanced by the amount of concurrent erosion. Although we could go to a graded stream and observe newly deposited sandbars, we would realize that ultimately all deposits in the graded stream are eroded and carried a little further down stream. To a sedimentologist, this situation may be very interesting; but a stratigrapher discovers that no net stratigraphic record is accumulating in a graded stream.

Response of graded streams to dynamics of the earth's surface. The statement that graded streams do not accumulate a stratigraphic record carries with it a corollary: Streams accumulate a stratigraphic record only when the equilibrium

Figure 10.25 Schematic example depicting the response of a graded stream to dynamic events. Given an initial topographic perturbation (time *a*), a stream will seek to fill in relatively low areas and establish a smooth graded profile from source area to base level. Given rising base level (time *b*), the entire graded profile is raised and sedimentation will occur to maintain equilibrium grade. Similarly, sediment will accumulate in response to subsiding substrate at the distal end of the profile in response to rising source area. Given lowered base level (time *c*) or lowered source area, the graded stream will erode deeply into its substrate and there will be no net accumulation of sediment throughout the length of the graded profile. Numerous combinations are possible on a regional scale.

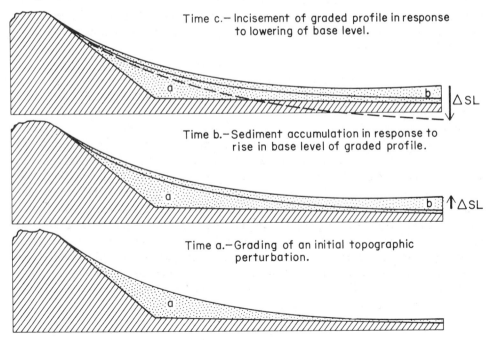

Time c.– Incisement of graded profile in response
to lowering of base level.

Time b.– Sediment accumulation in response to
rise in base level of graded profile.

Time a.– Grading of an initial topographic
perturbation.

profile is undergoing change (Figure 10.25). For example, if the source area is undergoing tectonic uplift, the equilibrium profile will be raised from its source area end. Meandering-stream or braided-stream sedimentary sequences should accumulate to raise the floor of the river valley up to the new equilibrium profile. Alternatively, if sea level rises, the equilibrium profile will be raised from its seaward end. Meandering-stream deposits should accumulate to fill the river valley up to the new equilibrium profile. Similarly, local tectonic subsidence may drop the floor of the river valley below the equilibrium profile. A vertical sequence of alluvial sediments should accumulate to maintain the equilibrium profile throughout such tectonic subsidence.

A second corollary to the concept of a graded-stream profile is the fact that a stratigraphic record will not be accumulated at a time when the equilibrium profile is being lowered. If the equilibrium profile is lowered, the stream will become incised into the surrounding rock types. Net removal of material from that stretch of the river will occur, not net deposition. Similarly, a graded profile will tend to incise into an area that is undergoing gradual tectonic uplift.

All of these relationships between a graded stream and its geographic context are fascinating to a stratigrapher. It is his job to unravel the detailed history of tectonism and eustatic sea-level fluctuations that is recorded by the interaction of these events with the graded profile.

Progradation by meandering-stream deposits. In Chapter 4, we noted that regression is a general term meaning simply that the shoreline appears to have moved basinward for some reason. The reason may be eustatic or tectonic emergence of the coastal area relative to the level of water in the basin. Or the basinward movement may result from progradation of coastal sediments outward into the basin and be independent of submergence-emergence considerations. Note especially that progradation may occur even at a time of submergence, provided only that sediment supply is rapid enough to offset the submergence effect (see Figure 11.17).

In the geologic record, we often observe meandering-stream deposits overlying marine deposits in a more or less continuous fashion. Such a situation is portrayed in Figure 5.4 and is commonly referred to as a *regressive relationship*. Can we say more about the precise earth history recorded by these relationships? Must we use the general term "regressive" or can we use more specific terms such as "emergence," "submergence," or "progradation"? In view of our discussion of the graded profile, we must emphatically state that this relationship could not have been generated by eustatic or tectonic emergence of the coastal area. In either of these situations, the stream valley would tend to become a site of net *erosion* as the stream incised a new bed in response to lowered sea level (eustatic emergence) or to rising land surface (tectonic emergence). Widespread meandering steam deposits would not be preserved under these conditions.

Consideration of the dynamic modification of the graded-stream profile (Figure 10.25) suggests three alternatives for the generation of stratigraphic relationships indicated in Figure 5.4. Perhaps we are looking at the effect of the combination of progradation accompanied by tectonic or eustatic submergence of the coastal region. Alternatively, the equilibrium profile may be raised from the other end by uplift of the source area.

We can debate whether there might exist stratigraphic relationships that would allow distinction among the three possibilities just outlined. Let us simply note now that adherence to models based on the modern stream allows us to totally discredit one of the possible interpretations that had been previously hidden in the background behind the general term "regression."

Isostatic considerations. As stream sediments accumulate into a new equilibrium profile, they are adding new load to that portion of the face of the earth. As we have seen in Chapter 4, isostatic subsidence in response to this new load is to be expected. Thus, the thickness of sediment accumulated will likely be two or three times as great as the initial perturbation of the relationship between the graded-stream profile and the area that it traverses.

An Ancient Example: The Shawangunk Conglomerate and Tuscarora Quartzite, Silurian of Pennsylvania, New Jersey, and New York

In our discussion of geosynclinal sequences in Chapter 5, we noted the common occurrence of thick, clastic wedges marginal to the present position of the crystalline core of the Appalachians. Even on the broad scale of that discussion, it was clear that the thick accumulation of sediment marginal to the crystalline core recorded the uplift and erosion of a large volume of rock in the area now occupied by that core. What we said earlier about clastic wedges sounds very similar to what we are saying now about modern braided streams: Large volumes of debris are derived from the rapid erosion of mountains and deposited relatively nearby. What better place to look for braided-stream deposits than in the clastic wedges of the Appalachians! A modern attempt at paleogeographic reconstruction has been carried out by Smith (1967, 1970). Let us review his work and see if we can agree with his conclusions.

Earlier stratigraphic work. The section under consideration is comprised primarily of the Shawangunk conglomerate, the Tuscarora quartzite, and the Clinton formation of Silurian age in Pennsylvania, New Jersey, and New York (Figures 10.26 and 10.27). As with most of the rocks in the Appalachians, early stratigraphers named the formations of the area on the basis of gross lithology, regional unconformities, and a few fossils. When Smith began his studies, the geologists' understanding of the section had been summarized as appears in Figure 10.27.

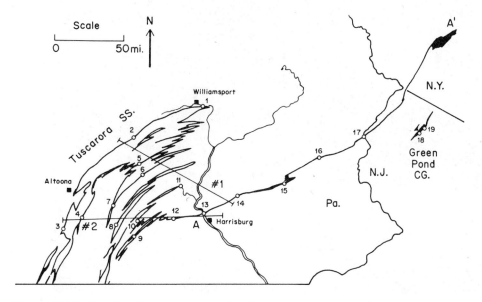

Figure 10.26 Map showing outcrop and distribution of Tuscarora sandstone and Shawangunk conglomerate, northeastern United States. *AA,* denotes the location of stratigraphic cross section Figure 10.27. (After Smith, 1970.)

Figure 10.27 Stratigraphic cross section of Middle Silurian formations in the northcentral Appalachians. Major lithologies are as follows: (1) Ordovician shales and sandstones of the Martinsburg formation, unconformably overlain by (2) basal conglomerates; (3) quartzites and conglomeratic sandstones; (4) interbedded shales and sandstones with some *in situ* marine fossils; and (5) red beds. (After Smith, 1967.)

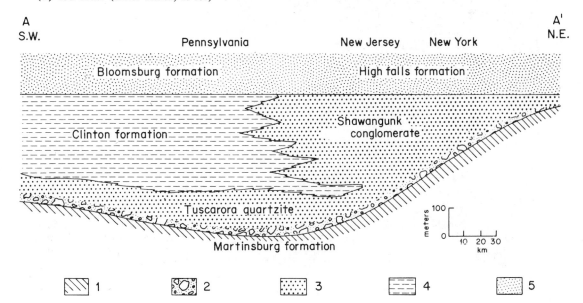

Figure 10.28 (Above) Planar cross-stratification within the Tuscarora sandstone. Compare especially with Figure 10.10, which showed planar cross-stratification in a transverse bar of the modern Platte River. (Photo by N. D. Smith.)

Figure 10.29 (Left) Irregular bedding surfaces in the Tuscarora sandstone. Strata are now nearly vertical; depositional up is to the left. Observe the cut-and-fill structures and extreme irregularity of bedding thickness. (Photo by N. D. Smith.)

The Tuscarora quartzite and the Shawangunk conglomerate are very hard rocks; therefore, they form the ridges of the area and are a starting point for stratigraphic correlation throughout the outcrop belt. Below these quartzites and conglomerates, there is usually a shale or shale and fine-grained sandstone (the Hudson River shales and the Martinsburg formation). Fossils of the Martinsburg formation are decidedly Ordovician. However, these fine-grained sandstones and shales contrast sufficiently with conglomeratic quartzites so that a formation distinction can be made on the basis of lithology alone.

In Pennsylvania, the Tuscarora quartzite is overlain by the Clinton formation. The distinction between the two formations is based on gross lithology and fossil content. The Tuscarora is predominantly sandstone. As we proceed up the section, shale beds begin to come in. Along with the alternation of shale and sandstone, we begin to find more and more marine fossils. Thus, the early geologists found it convenient to set apart the shale and sandstone sequence as a new formation, the Clinton formation.

In this discussion, we need not concern ourselves with the section above the Clinton, other than to say that the Bloomsburg and High Falls formations are similarly defined on the basis of lithologic contrast.

Sedimentological observations within the Shawangunk and Tuscarora. The Shawangunk conglomerate catches the eye of anyone looking for braided-stream environments in the stratigraphic record. Here are 500 meters of conglomeratic sandstone apparently contemporaneous with the marine sands and shales of the Clinton formation. Obviously, the Shawangunk is coarse debris derived from the erosion of a paleogeographic high area to the southeast.

Figures 10.28 through 10.31 offer various sedimentological observations within the Tuscarora and Shawangunk. Figure 10.28 illustrates a rather straightforward example of medium-scale, high-angle, planar cross strata in Tuscarora sandstone. Figure 10.29 shows extremely local irregularities in bedding within Tuscarora sandstone. Some individual beds have irregular thickness, whereas other beds are truncated by scour surfaces. Figure 10.30 presents schematic diagrams of six different types of sedimentation units that recur within these rocks. Variations in grain size and in primary structures suggest extremely local variations in both flow conditions and sediment sorting. Clearly, these variations are reminiscent of the extremely local variations that we discovered in the Donjek River (Figure 10.4). In Figure 10.30, observe that some sedimentation units record vertical decreases in flow regime, whereas others record vertical increases in flow regime. Furthermore, intraclast of shale and siltstone are common in these rocks. These particles are locally derived by erosion of overbank deposits and other desiccated muds. Their abundance indicates the alternating processes of sedimentation, erosion, and resedimentation that we have seen to be so typical of braided-stream environments. Within each stratigraphic section, sedimentation units are seldom more than 1 or 2 meters thick. There is no discernible repeated

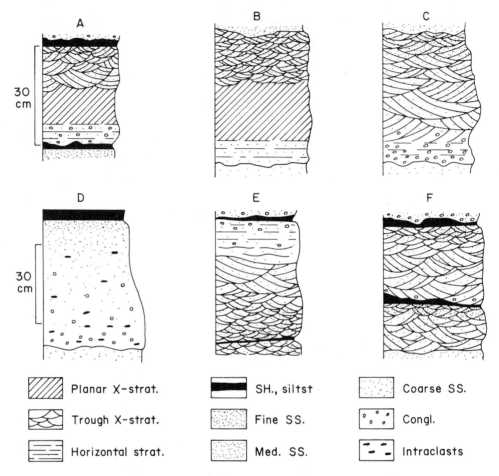

Figure 10.30 Schematic diagrams of six sedimentation units that recur within the Shawangunk conglomerate. Note that numerous combinations of variation in grain size and in primary structures are observed. (After Smith, 1970.)

sequence in which the various sedimentation units may be expected to occur. The deposition of each successive bed appears practically unrelated to the deposition of beds above or below it.

There is systematic variation in the type of cross-stratification that is perpendicular to the general strike of the paleogeography. Figure 10.31 plots the relative abundance of various types of cross strata into traverses from east (near the source area) to west (basinward). Comparison of these data with similar data for the modern Platte River (Figure 10.31) led Smith (1972) to conclude that there was a Silurian geographic distribution of longitudinal bars and transverse bars similar to that observed in the modern Platte River.

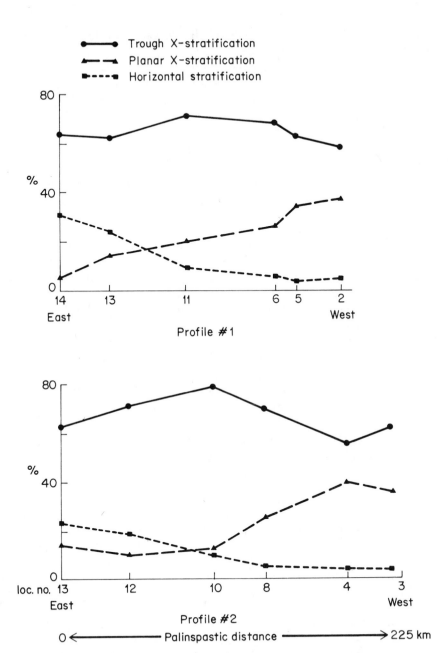

Figure 10.31 Downslope changes in principal stratification types along two east-to-west profiles in the western portion of the outcrop area depicted in Figure 10.26. As noted in Figure 10.13, the relative importance of planar cross strata increases away from the source area and the importance of horizontal stratification decreases away from the source area. (After Smith, 1970.)

Interpretation of earth history recorded by the Shawangunk, Tuscarora, and Clinton formation. Smith (1967) provides a paleogeographic and paleoenvironmental interpretation for each of the formations displayed in Figure 10.27. The Tuscarora and the Shawangunk are braided-stream deposits; the High Falls formation, meandering-stream deposits; and the Clinton and Bloomsburg formations, marginal-marine to subtidal, open-shelf, marine clastics. In this chapter, we have emphasized Smith's evidence for the recognition of braided-stream deposits. Let us assume for the sake of discussion that his interpretations concerning the other formations are correct. Likewise, the Martinsburg formation appears to consist largely of turbidites that presumably accumulated in marine waters of significant depth.

The Tuscarora and Shawangunk formations record activation of major uplift to the southeast. The intertonguing of marine and marginal-marine Clinton formation rocks with Shawangunk conglomerates offers us an additional puzzle concerning the earth's history as recorded by these sequences. Specifically, does the Clinton formation record a major submergence of the North American continent, or does it record local submergence in response to local tectonic conditions, or does it record isostatic submergence in response to deposition of new gravitational load onto this portion of the earth's crust? Recall that calculation of isostatic adjustment for similar situations was treated in Chapter 4. A quick calculation of the isostatic adjustment attendant to Shawangunk and Clinton deposition (use data in Figure 10.27, for example) demonstrates that the local isostatic component of the Clinton transgression is probably quite large relative to any eustatic or tectonic component that may be present.

An Ancient Example: Interaction between Meandering Stream Cycles and Regional Submergence, the Catskill "Delta," Devonian of New York State

Utilizing the concept of the equilibrium profile of meandering streams, McCave (1969) has put together an interesting proposal for event correlation between marine and nonmarine units. He has taken the unifying concept of a model depicting what *should* happen sedimentologically and made seemingly unrelated lithologies the basis for time correlation between marine and nonmarine rocks.

Regional setting and earlier stratigraphic work. McCave's work is primarily in the Middle Devonian relics of the Catskill Mountains. He seeks to correlate these nonmarine rocks with marine rocks to the west.

As indicated in Figure 5.4, the basic history of this area has been understood in general terms for a long time. You can consult numerous papers in Shepps (1963) and Klein (1968) for additional regional detail.

During the Lower Devonian period, shallow carbonate seas covered the area. By the Middle Devonian period, thick accumulations of clastic sediments began to pour into the area from a source to the east. By Upper Devonian time, continental clastic sedimentation extended well into western New York. Marker beds of black shale and limestone allow physical stratigraphic correlation within the marine units. But how can these correlations be extended into a nonmarine unit of the Catskill front?

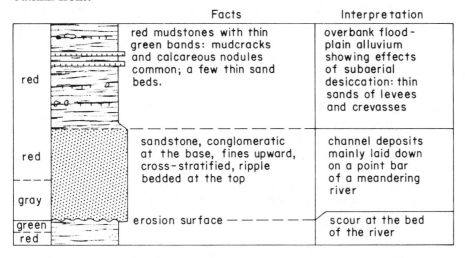

	Facts	Interpretation
red	red mudstones with thin green bands: mudcracks and calcareous nodules common; a few thin sand beds.	overbank flood-plain alluvium showing effects of subaerial desiccation: thin sands of levees and crevasses
red / gray	sandstone, conglomeratic at the base, fines upward, cross-stratified, ripple bedded at the top	channel deposits mainly laid down on a point bar of a meandering river
green / red	erosion surface — — — — —	scour at the bed of the river

Figure 10.32 (Above) Principal sedimentological features of the Catskill fining-upward unit. (After McCave, 1969.)

Figure 10.33 (Below) Histograms showing distribution by thickness of overbank shale and channel sandstone in the Catskill fining-upward units. Whereas the thickness of channel sands is more or less normally distributed, overbank deposits of 20 to 40 meters appear to be anomalously thick. (After McCave, 1969.)

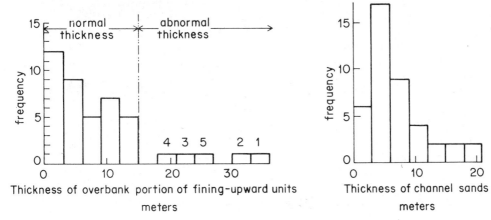

Field observations. While measuring sections in the Catskill Mountains, Mc-Cave came to recognize the basic fining-upward unit that is familiar to us from earlier discussions. Figure 10.32 presents his version of the basic fining-upward unit of the Middle Devonian of the Catskills. Measured sections through these sequences can easily be divided into a lower channel sand and an upper overbank shale. Figure 10.33 is a histogram of the thickness of channel sands and overbank deposits in the sequences measured by McCave. The thickness of channel sands is more or less normally distributed. The thickness of overbank deposits, however, show marked deviations from normal distribution. The high abundance of very thin overbank deposits is easily explained by erosional removal of this material as the next meander belt crossed the area. The occurrence of five units having overbank deposits 18 to 36 meters thick cannot be so easily explained.

Sedimentological interpretation. McCave proposes that the accumulation and preservation of abnormally thick overbank deposits reflect a rise in equilibrium profile resulting from a rise in sea level. With this concept in mind, he proposes the correlations outlined in Figure 10.34.

Figure 10.34 Correlation chart for the upper part of Middle Devonian and the lower part of Upper Devonian in New York State, indicating the event correlation between limestone units within the marine section and abnormally thick overbank deposits in the alluvial section. (After McCave, 1969.)

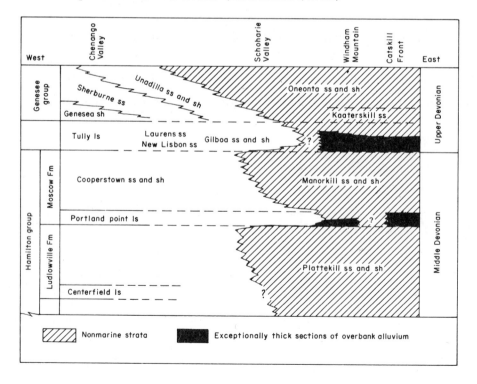

Specifically, an abrupt rise in sea level is a single event that produces different consequences in the marine environment and in the meandering-stream environment. In the marine environment, rising sea level momentarily traps clastic sedimentation in the heads of drowned estuaries. This situation causes brief episodes of clear-water sedimentation (Portland Point limestone and Tully limestone) in areas previously dominated by deposition of marine sandstones and shales.

In the meandering-stream environment, rising sea level has lifted the equilibrium profile of the meandering stream, allowing for the accumulation and preservation of abnormally thick overbank deposits. Thus, the conceptual model dictates correlation of Portland Point limestone and Tully limestone with zones of thickened overbank deposits in the meandering-stream facies. Seemingly unrelated rock types, limestones in the marine environment and thickened overbank deposits in the meandering-stream environment, are both the logical consequences of a single *event:* an abrupt rise in sea level.

Citations and Selected References

ALLEN, J. R. L. 1964. Six cyclothems from the lower Old Red sandstone, Anglo-Welch Basin. Sedimentology 3: 163–198.
Cyclic repetition of sand and shale units is shown to record meandering-stream deposition.

————. 1965. A review of the origin and characteristics of Recent alluvial sediments. Sedimentology 5: 89–191.
Extensive compilation of the literature into the early 1960's.

BERNARD, H. A., C. F. MAJOR, JR., B. S. PARROTT, and J. R. LeBLANK, SR. 1970. Recent sediments of southeast Texas. Univ. of Texas at Austin, Bureau of Economic Geology, Guidebook 11. 16 p. (plus figures and appendices).

BULL, W. B. 1964. Geomorphology of segmented alluvial fans in western Fresno County, California, p. 79–129. U. S. Geol. Survey Prof. Paper 352-E.

DOEGLAS, D. J. 1962. The structure of sedimentary deposits of braided rivers. Sedimentology 1: 167–190.

ECKIS, R. 1928. Alluvial fans of the Cucamonga District, Southern California. J. Geology 36: 224–247.

FAHNESTOCK, R. K. 1963. Morphology and hydrology of a glacial stream, p. 1–70. U. S. Geol. Survey Prof. Paper 422-A.

GIGNOUX, M. 1955. Stratigraphic geology. W. H. Freeman and Co., San Francisco. 682 p.

HARMS, J. C., D. B. MacKENZIE, and D. G. McCUBBIN. 1963. Stratification in modern sands of the Red River, Louisiana. J. Geology 71: 566–580.

HOOKE, R. L. 1967. Processes on arid region alluvial fans. J. Geology 75: 438–460.

JOHNSON, K. G., and G. M. FRIEDMAN. 1969. The Tully clastic correlatives (Upper Devonian) of New York State: a model for recognition of alluvial, dune (?), tidal nearshore (bar and lagoon), and offshore sedimentary environments in a tectonic delta complex. J. Sed. Petrology 39 (2): 451–485.

KLEIN, G. DEV. (ed.). 1968. Late Paleozoic and Mesozoic continental sedimentation, northeastern North America. Geol. Soc. Amer. Spec. Paper 106. 309 p.
Numerous papers concerning Paleozoic stream sedimentation in the Appalachians.

LEOPOLD, L. B., and M. G. WOLMAN. 1957. River channel patterns: braided, meandering, and straight. U. S. Geol. Survey Prof. Paper 282-B. 85 p.

————, M. G. WOLMAN, and J. P. MILLER. 1964. Fluvial processes in geomorphology. W. H. Freeman and Co., San Francisco. 522 p.

McCAVE, I. N. 1969. Correlation of marine and non-marine strata with example from Devonian of New York State. Bull. Amer. Assoc. Petrol. Geol. 53: 155–162.
Event-correlation between marine limestone and thickened overbank deposits in meandering-stream facies.

PLUMLEY, W. J. 1948. Black Hills terrace gravels: a study in sediment transport. J. Geology 56: 526–577.

RIGBY, J. K., and W. K. HAMBLIN. 1972. Recognition of Ancient sedimentary environments. Soc. Econ. Paleont. Mineral. Spec. Pub. 16. 340 p.
Papers from a 1969 symposium dealing with a wide variety of sedimentary environments. Each paper discusses the recognition of a specific environment or cluster of environments. A good starting point for many topics. Generally well-referenced

SHEPPS, V. C. (ed.). 1963. Symposium on Middle and Upper Devonian stratigraphy of Pennsylvania and adjacent states. Pa. Topog. Geol. Surv. Gen. Geol. Rep. G-39. 301 p.

SMITH, N. D. 1967. A stratigraphic and sedimentologic analysis of some Lower and Middle Silurian clastic rocks of the north-central Appalachians. Ph.D. dissertation, Brown University. 195 p.
Braided-stream, meandering-stream and coastal clastic sediments in classic Appalachian geology.

————. 1970. The braided stream depositional environment: comparison of the Platte River with some Silurian clastic rocks, North-central Appalachians. Bull. Geol. Soc. Amer. 81: 2993–3014.
Excellent discussion of cross-stratification and grain-size variation. Includes recognition of the proximal and distal portions of the braided-stream environment.

WILLIAMS, P. F., and B. R. RUST. 1969. The sedimentology of a braided river. J. Sed. Petrology 39: 649–679.

YEAKEL, L. S. 1962. Tuscarora, Juniata, and Bald Eagle paleocurrents and paleogeography in the central Appalachians. Bull. Geol. Soc. Amer. 73: 1515–1540.

11

Clastic Sedimentation
in Coastal Environment

In stream sedimentation, the driving force is gravity acting upon the water. As gravity causes the water to run downhill, the movement of the water causes sediment transportation and deposition. At sea level, a whole new set of processes begins to act on the sediment supplied by streams. Gravity no longer provides the major energy input to sedimentation. Instead, tidal motions and wind-driven currents are the primary transporting agents in coastal environments.

In the fluvial environments, the key word for water motion is "flow." As we begin considering clastic sedimentation in coastal environments, the key word for water movement becomes "swash." The flow of a fluvial stream is more or less constant and unidirectional. In contrast, marine water is continually moving onto and off of clastic coastal sediments. There is the rise and fall of each tidal cycle, and there are the endless waves that roll up onto a beach, only to roll off again. The oscillatory motion of waves usually generates persistent longshore currents, but the oscillatory components of sediment transport also make major imprints on the sediments accumulated in these environments.

In addition, as we study the coastal clastic environments, we see that fossils begin to play an important role in our recognition of depositional environments.

All in all, there are more and more factors to consider in coastal areas. Changes in the relationship of the sea to the land are among the most profound changes recorded in the strata. Coastal clastic environments offer an excellent opportunity to come to grips with the dynamics of earth history.

Recent Clastic Sedimentation in
Coastal Environments

A *delta* forms as the basic sediment accumulation where the river meets the coastal area. It is here that stream processes end and marine processes (tides and currents) begin to operate.

It is convenient to think of the delta as a loading dock. Just as the loading dock separates the manufacturing activity of a factory from the distribution and sale of the product, the delta separates the sediment factory (the continental areas) from the distribution processes of the shallow-marine environment. Continuing the analogy, we can write a budget for sediment inventory on the delta in terms of the capability of the streams to deliver sediment to the delta and the capability of the shallow-marine processes to transport sediment away from the delta. If the stream can deliver a large volume of sediment and the marine processes can remove only a small volume of sediment, a large amount of sediment will accumulate in the delta area. On the other hand, if the stream can deliver only a small amount of sediment and the marine processes are capable of moving a large amount of sediment, the stream may build no delta at all. It may be that all of the sediment delivered by the stream is transported away from the mouth of the stream as soon as it is delivered.

In the following pages, we shall look at coastal environments that represent a wide spectrum in the balance between the sediment supply capability of the streams and the sediment reworking and transport capability of the marine environment. Several of our examples will be from the northwest Gulf of Mexico. In the Mississippi delta, the sediment supply far exceeds the reworking capabilities of the marine environment. A very large and unique type of delta is accumulating there today. On the other hand, in the Niger delta, sediment supply and reworking capabilities of the marine environment are in much closer balance. Although a large delta is accumulating, the sediment is being significantly reworked and locally transported. The result is a delta that is quite different from the Mississippi. As an example of longshore sedimentation, we shall return to southwest Louisiana and the Texas Gulf Coast. Here the sediment has escaped the delta area and is being transported and deposited by processes acting in the shallow-marine environment. Finally, to round out our model of clastic sedimentation in coastal environments, we shall look at tidal-flat sedimentation.

It is often convenient to discuss clastic sedimentation in coastal environments in terms of topset deposits, foreset deposits, and bottomset deposits. The delineation of these terms in this chapter involves the *strand line,* the general area of interaction between open water and the shallow-water sediments associated with the coastal area. The beach, of course, most clearly exhibits the strand line. This line is less well defined in the rough outline of a delta like the Mississippi. *Topset sediments* are those sediments that accumulate landward from the strand line. Sediment accumulation here is essentially vertical.

Bottomset sediments are the offshore, deeper-water marine sediments that are not deposited by the wave and tidal processes dominating the coastal and shallow-marine areas. These sediments are typically clay muds with a normal marine benthonic fauna and perhaps a sizable representation of pelagic organisms. Accumulation of bottomset sediments is essentially vertical.

Foreset sediments occupy that position between the strand line and bottomset sediments. It is a position of almost continual activity. The top of the foreset deposits are the river-mouth bars and the barrier beaches. Seaward from these features, bathymetric gradients are pronounced. Because of this slope to the foreset bathymetry, sediment accumulation is essentially lateral or progradational.

The major task facing the stratigrapher is recognizing and understanding relationships within foreset deposits in clastic sedimentation in coastal environments. Topset and bottomset beds are useful primarily as stratigraphic markers, valuable for setting apart the discrete sedimentary packages that are the stratigrapher's chief aim.

The Mississippi delta. A brief glance at Figure 11.1 confirms that the modern Mississippi delta is a major physiographic feature of the northwest Gulf of Mexico. From the Rio Grande to western Louisiana, modern coastal sedimentation more or less parallels the strike of older sedimentary units and extends 15 to 30 kilometers seaward from the Pleistocene outcrop surface. In southeastern Louisiana, modern Mississippi delta sediments have built outward onto the continental shelf as much as 100 kilometers seaward from the Pleistocene outcrop surface. This area is indeed a site of major deltaic accumulation of clastic sediments.

Sedimentation in the Mississippi delta particularly interests us because of the economically important oil and gas accumulations in similar deposits throughout southern Louisiana. The Mississippi River has been dumping large volumes of sediment into this region from Oligocene time to the Recent epoch. Sedimentation and internal stratigraphy of the modern Mississippi delta is undoubtedly a valuable clue in prospecting for oil and gas in older deposits within this sedimentary basin.

Although the entire Mississippi delta has accumulated since the post-Wisconsin sea-level rise, active delta sedimentation is occurring today only in the modern Birdfoot delta (Figure 11.2). This area has been studied in detail (Gould, 1970, for example) and constitutes in large part the basic model for our understanding of the Mississippi delta as a whole.

Sediments of the modern Birdfoot delta can be divided into three general categories. The mouth of the distributary is the main site of sedimentation. Here the coarsest material brought down the river becomes deposited in a river-mouth bar. Outward from this bar in all directions, the sediment becomes more fine-grained (Figure 11.3). The fine-grained sediment accumulates both in front of the river-mouth bar and also between river-mouth bars. Finally, wherever fine-grained material fills in near sea level, marsh deposits develop. These marsh deposits prograde over shallow, subtidal, fine-grained sediments and form the laterally extensive, and most visible, upper unit of the delta sequence. These marsh deposits are, however, generally thin.

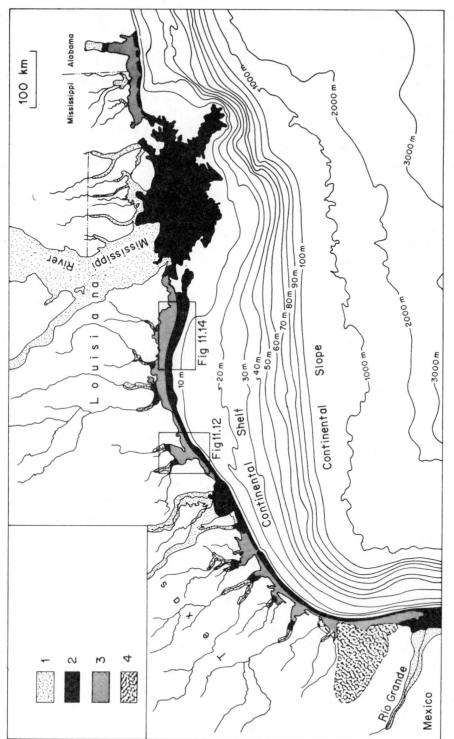

Figure 11.1 Generalized features of Recent coastal clastic sedimentation, northwest Gulf of Mexico. In this chapter, we shall look at Recent sedimentation in the Mississippi delta (Figure 13.2), the Galveston barrier bar-lagoon complex (Figure 13.12), and the Chenier Plain of southwestern Louisiana (Figure 13.14). Generalized sediment types are as follows: (1) alluvial plain deposits, predominantly from meandering streams; (2) subaerial coastal clastic sediments, deltaic complexes, and progradational barrier bars and cheniers; (3) lagoons and marshes behind progradational features; and (4) eolian sand transported inland from the barrier beach environment.

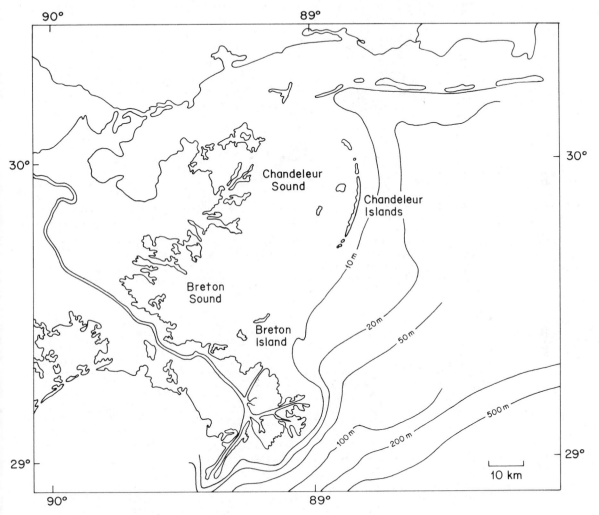

Figure 11.2 Physiography of the modern Mississippi River Birdfoot delta and environs.

(A)

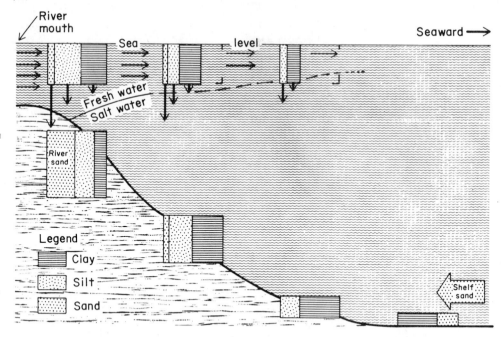

Figure 11.3 Sedimentation seaward from the mouth of the delta distributary. Figure (A) traces the composition of suspended sediment and deposited sediment from the river mouth out to sea. Figure (B) schematically summarizes progradational relationships resulting from this deposition. (After Scruton, 1960).

(B)

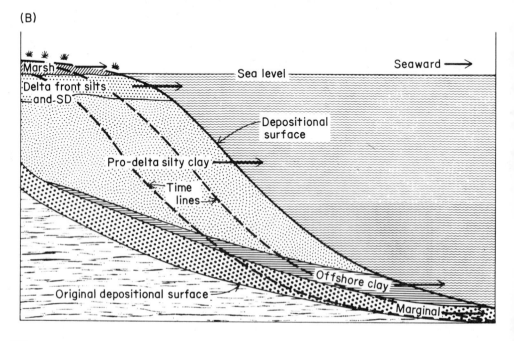

Note that the coarsest sediment accumulates in the shallow water of the river-mouth bar. Grain size becomes progressively finer offshore and between river-mouth bars. Primary structures also reflect the decreasing carrying capacity from river mouth outward into the Gulf. High-angle cross-stratification is common in the shallow-water sands of the river-mouth bar but is uncommon seaward.

The transition from river mouth to open marine is also apparent in the fauna contained in the sediments. Sediments accumulating in front of the river-mouth bar will commonly contain a burrowing marine fauna. The shells of these organisms not only become an important attribute of the sediment, but their burrowing activity also reworks and mottles the sediment. If primary structures were formed in the subtidal environments in front of the delta, they would likely be destroyed by the burrowing activity of these organisms.

Historical data, combined with coring of the modern Birdfoot delta, indicate that the mouths of the river distributaries have been building seaward at a very rapid rate. The net result of this seaward progradation of the distributary channels is the construction of *bar-finger sands* (Figure 11.4). The characteristics of present-day sediments, previously discussed in relation to areal distribution from the distributary mouth, are repeated in the vertical profile of any single bar-finger sand [Figure 11.3(B)]. The coarsest material in shallow water and the finest material in deeper water in front is translated into the coarsest material at top of the bar-finger sand and the finest material at its base. This process continues for the various characteristics of these sediments.

As indicated in Figure 11.4, coarse-grained material in the Mississippi delta accumulates only in the bar-finger sands, resulting in progradation of major distributaries. Areas between major bar-finger sands are filled in with fine-grained sediment, not unlike the muds accumulating in front of the mouths of passes. The basic geometry of modern Birdfoot delta sedimentation is, then, the radial arrangemen of bar-finger sands surrounded by finer-grained sediments. As the area between bar-finger sands becomes filled in with muddy marine sediment, marsh deposits prograde seaward to fill in between the "spokes" of the radially-arranged bar-finger sands. Although these marsh deposits are areally extensive, they are generally thin. Thus, they are not volumetrically significant as a lithology characteristic of delta sedimentation. They may prove useful for correlation in Ancient sequences, but they do not make up the bulk of the typical deltaic stratigraphic sequence.

The development of radially arranged bar-finger sands of the modern Birdfoot delta can be considered an example of sediment supply grossly exceeding the reworking and transportation capabilities of the local marine environment. As we move on to look at other areas, we shall see that the sediment delivered to the mouth of a distributary is very often reworked and transported to build other deposits. In the Mississippi delta, the sediment supply is so great that the effective reworking of these deposits does not occur so long as the sediment is being supplied to the area. The sediment simply piles up at or near the distributary mouth.

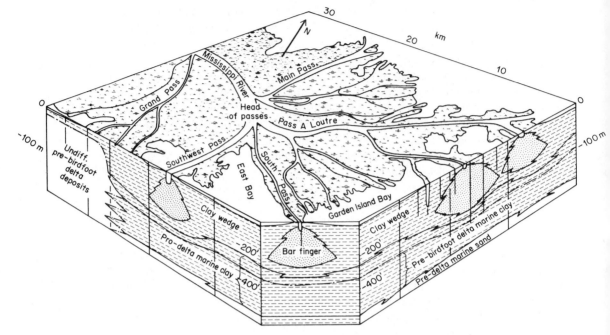

Figure 11.4 Block diagram indicating geometric arrangement of bar-finger sand, fine-grained sediments, and topset marsh deposits of the modern Birdfoot delta of the Mississippi River. (After Fisk *et al.*, 1954.)

Therefore, rapid progradation of this area is the general rule of deltaic sedimentation here.

Another outstanding attribute of Mississippi delta sedimentation is the varying position of delta sedimentation. As indicated in Figures 11.5 and 11.6, the focal point of delta sedimentation has changed radically at least seven times during the past 5000 years. Presumably, each of these subdeltas was built in the regular fashion, as was the modern Birdfoot delta, until the river was abruptly diverted from its channel during time of flood.

Thus, there is a strongly discontinuous and nonsystematic aspect to Mississippi delta sedimentation. The internal stratigraphy of any one delta lobe has no relationship to the internal stratigraphy of another delta lobe, no matter how close they are. Each delta lobe is capped with a marsh deposit, and each delta lobe contains bar-finger sands that have the coarsest material at the top and the fine downward. At that point, the systematics of sedimentation and stratigraphy end. As an entity, each delta lobe is unique.

The abrupt discontinuation of sediment supply to a delta lobe provides an opportunity for marine processes to rework the delta deposits. As long as sediment is being supplied to an active delta front, new sediment is added so rapidly that

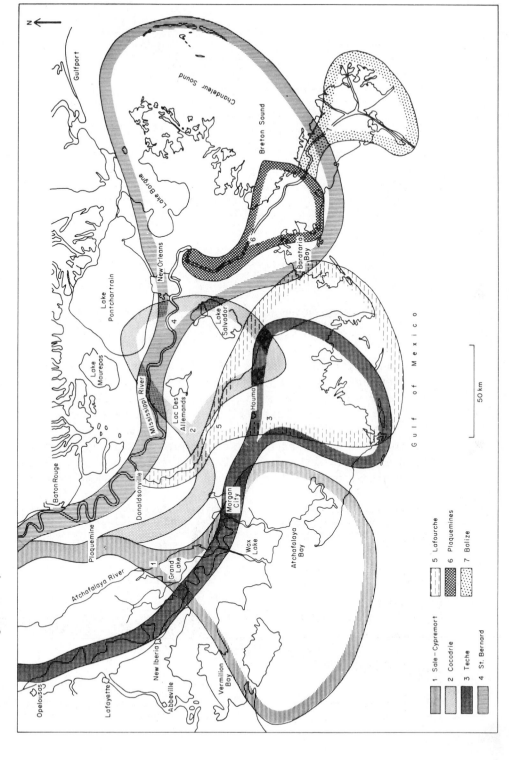

Figure 11.5 Subdeltas of the modern Mississippi delta complex during the last 5000 years. The site of major deltaic sedimentation has shifted numerous times. Chronology of these shifts are given in Figure 11.6. (After Kolb and Van Lopik, 1966.) Courtesy Houston Geological Society.

1 Sale – Cypremort
2 Cocodrie
3 Teche
4 St. Bernard
5 Lafourche
6 Plaquemines
7 Balize

Gulf of Mexico

50 km

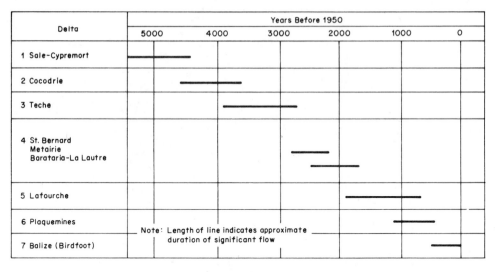

Delta	Years Before 1950					
	5000	4000	3000	2000	1000	0
1 Sale-Cypremort	▬▬▬▬					
2 Cocodrie		▬▬▬▬				
3 Teche			▬▬▬			
4 St. Bernard Metairie Barataria-La Lautre				▬▬ ▬▬		
5 Lafourche					▬▬▬	
6 Plaquemines	Note: Length of line indicates approximate duration of significant flow				▬▬	
7 Balize (Birdfoot)						▬▬

Figure 11.6 Chronology of the major delta lobes of the Mississippi during the past 5000 years. Position of each delta is indicated in Figure 11.5. (After Kolb and Van Lopik, 1966.) Courtesy Houston Geological Society.

redistribution by marine processes does little to change the basic radial arrangement of bar-finger sands. When the sediment supply is shifted to another site, a destructive phase of delta history may occur (Figure 11.7 and Scruton, 1960). For example, the Chandeleur Islands (Figure 11.2) mark the former seaward position of the St. Bernard delta. When the river shifted its site of major deposition, the St. Bernard delta was left exposed to marine processes without the protective cover of rapid addition of new sediments. Wave action winnowed the sediment, leaving relatively clean sand and silt and carrying away the fine-grained sediment. The net

Figure 11.7 Schematic cross section showing the vertical relationships between constructional and destructional deposits in imbricating deltas. Stratigraphic units of successive deltas merge and may appear correlative over wide areas if destructional deposits are not recognized. (After Scruton, 1960.)

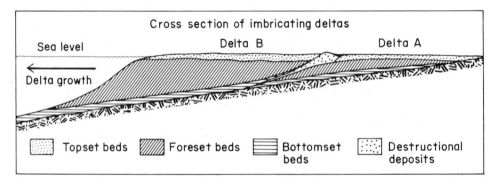

Cross section of imbricating deltas

Sea level

Delta growth

Delta B   Delta A

Topset beds   Foreset beds   Bottomset beds   Destructional deposits

result is twofold. First, deflation of the old St. Bernard delta has produced a subtidal area where there was once topset marsh deposits. Secondly, the destructive
phase has resulted in a relatively clean sand deposit of large areal extent. The
geometry of these sand deposits associated with the destructive phase of delta
sedimentation is in sharp contrast to the bar-finger sands of the constructive phase
of Mississippi delta sedimentation. Thus, Mississippi delta sedimentation results
in two distinctive types of sand-body geometry: (1) the constructive radial bar-
finger sands and (2) the laterally extensive blanket sands of the destructive phase.

Extensive coring in southern Louisiana has provided unparalleled information
concerning the relation of modern sediments to the underlying unconformity surface. Of particular interest is the isostatic and compactional deformation of the

Figure 11.8 Map depicting the relationship between Recent Mississippi delta sediment accumulation and the Pleistocene surface. Dashed contours indicate
present-day bathymetry of the Recent sediment surface. Solid contours indicate
the present-day topography on the Pleistocene erosional surface that underlies
Recent delta sediment. Pattern (2) indicates the present-day subaerial outcrop
of Pleistocene erosional surface. Pattern (1) indicates the deeply incised
Mississippi trench, through which the Mississippi River flowed at the time of
the Wisconsin low stand of the sea. Radiometric dating of materials obtained
from boreholes demonstrates that the former Pleistocene erosional surface has
been deformed into its present configuration by the loading of Recent
Mississippi delta sediment onto the area. (After Fish and McFarlan, 1955, with
modification.)

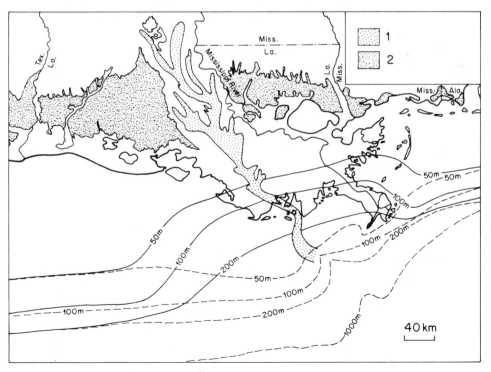

unconformity surface under the load of delta sediments. Figure 11.8 is a map of the present configuration of the Pleistocene subaerial surface beneath the modern Mississippi delta, and Figure 11.9 presents two generalized cross sections through the Recent delta. In cross section *A-A'*, note that the Pleistocene subaerial surface is lowest beneath the thickest accumulation of Recent delta material. To the west, as delta material thins gradually, the Pleistocene subaerial surface rises gradually. To the east, as delta thickness decreases abruptly, the Pleistocene subaerial surface rises abruptly. The pronounced embayment of contours in Figure 11.8 conveys the same information.

Radiometric dating of deposits that trangress this surface indicates that the present configuration of the subaerial surface is not paleogeographic but rather is the result of local subsidence brought about by the loading of sediment onto the area. Before the loading of sediment onto this area, the 100-meter contour, for example, was simply that line connecting 100-meter contours to the west and to the east of the modern delta. With the loading of sediment into this area, the paleogeographic 100-meter contour has been depressed by as much as 100 to 150 meters.

Figure 11.9 General facies relationships within the modern Mississippi delta and the relationship of these sediments to the former Pleistocene subaerial erosion surface. Sediment types are as follows: (1) deltaic plain facies, bar-finger sand, and marsh deposits; (2) prodelta facies, predominantly muds with marine fauna; (3) alluvial facies associated with Mississippi River sedimentation during the late Wisconsin low stand; (4) former Pleistocene subaerial surface.

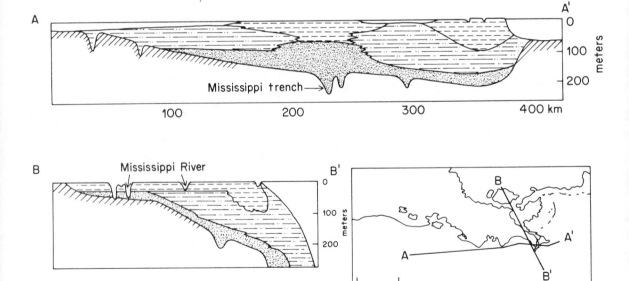

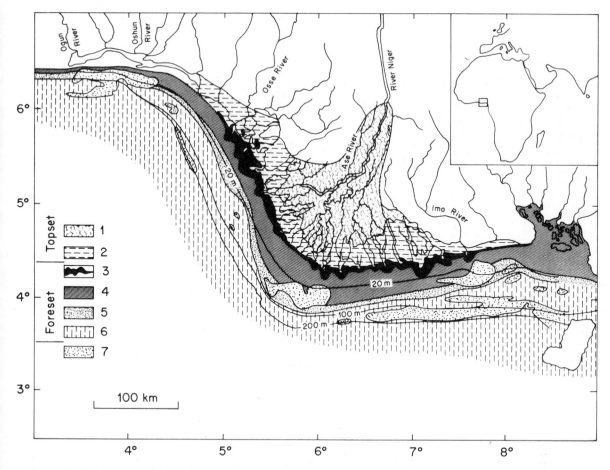

Figure 11.10 Major sedimentary facies of the modern Niger delta. Contemporaneous delta sedimentation is occurring around the entire perimeter of the Niger delta. Principal sedimentary facies are as follows: (1) alluvial flood plain, (2) mangrove swamp, (3) beaches and river-mouth bars, (4) very fine sand and coarse silt, (5) clayey silt, (6) silty clay, (7) nondepositional (older sediments). (After Allen, 1965, with modification.)

Thus, as predicted from general considerations in Chapter 4, a sea-level rise allows for the accumulation of sediment thicknesses at least twice as great as the amount of the sea-level rise. Compaction and isostatic adjustment associated with the loading of the unconformity surface allow for the accommodation of a thickness of marine sediments significantly in access of the actual amount of the sea-level rise.

The Niger delta. The Niger delta (Figure 11.10) provides an example in which marine processes of transportation and redeposition are in closer balance with sediment supply. In the Mississippi delta, the constructional phase involved such

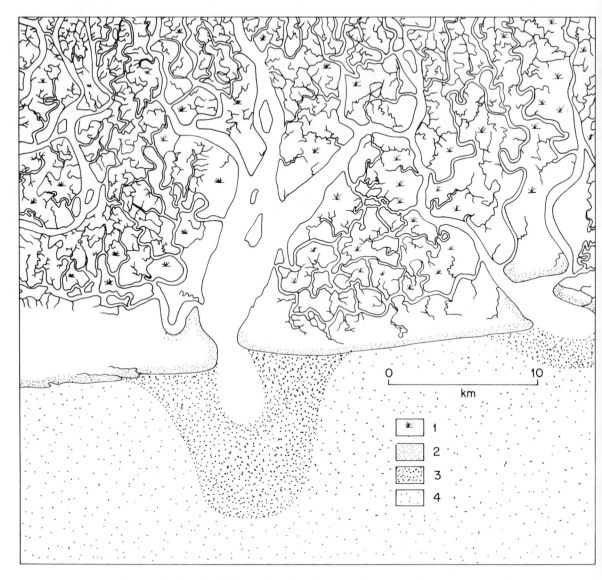

Figure 11.11 Map showing the physiography and sediment types near the mouth of the Brass River, Niger delta distributary. Note the intricate pattern of tidal distributaries within the mangrove swamp topset deposits. Sediment types are as follows: (1) mangrove swamp, (2) barrier beaches, (3) subtidal river-mouth bar, (4) offshore fine sand. Water depth throughout the offshore area is less than 10 meters. (After Allen, 1965, with modification.)

rapid sedimentation that bar-finger sands prograded rapidly into open marine waters. Redistribution of delta sediments occurred only after the subdelta entered the destructive phase following a major shift in river course. In contrast, in the Niger delta, constructional and destructional phases are essentially superimposed upon each other contemporaneously. Marine processes are capable of reworking newly deposited river-mouth bars almost as fast as the bars are being deposited. The net result, then, is concentric growth along the entire front of the Niger delta. Whereas the area between river-mouth bars in the Mississippi delta was filled in by fine-grained sediments, the area between the river-mouth bars in the Niger delta is filled in by the extensive development of beaches composed of sand derived from those bars (Figure 11.11). Consequently, there is an almost continuous body of sand-bars and beaches at or near sea level around the entire perimeter of the Niger delta. Instead of rapid progradation of single bar-finger sands, as in the Mississippi delta, the Niger delta barrier bar complex is prograding simultaneously over its entire front.

As in the Mississippi delta, grain size of the marine sediment decreases away from the delta front. Behind the barrier-bar complex, extensive mangrove swamps occupy the topset position.

The barrier island, Galveston, Texas. In the Mississippi delta, river-supplied sediments are deposited so fast that river-mouth bars prograde to form bar-finger sands. In the Niger delta, river-supplied sediments are locally reworked to form an extensive barrier-beach complex around the whole perimeter of the delta. Now let us look at an example of coastal sedimentation in which the sand supply is from the marine environment rather than from intimately associated river discharge. The marine environment may supply sediment for coastal deposition either by reworking material in front of the coastline or by longshore transport from deltas located elsewhere along the coast.

Galveston Island lies some 500 kilometers west of the modern Birdfoot region of the Mississippi delta (see Figure 11.1). The predominant longshore current throughout the northwestern Gulf of Mexico is westward. Because Galveston Island is far removed from the sediment source to the east, sedimentation here is a rather simple example of transportation and sedimentation by longshore currents. Therefore, we shall examine Galveston Island as a simple example of longshore sedimentation. Then we shall proceed eastward into the Chenier Plain of southwestern Louisiana to look at additional complications brought about by greater sediment supply.

In the Galveston area (see Figure 11.12), the post-Wisconsin transgression has trapped local sediment supplies at the head of estuaries. Both the San Jacinto River and the Trinity River supply sediment to Galveston Bay, but the sediment supply has not been sufficient to fill in the bay during Recent time. As a result, the sand supply that has accumulated to form Galveston Island and Bolivar

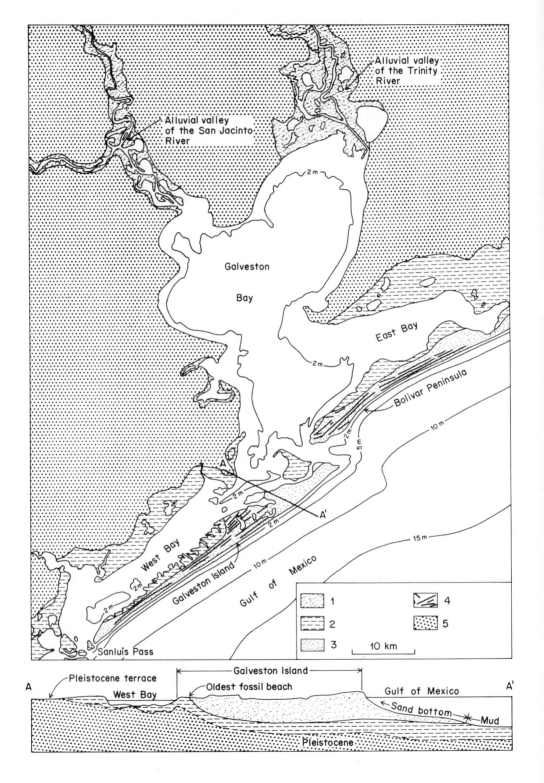

Alluvial valley
of the Trinity
River

Alluvial valley
of the San Jacinto
River

2 m

Galveston

Bay

East Bay

Bolivar Peninsula

10 m

2 m

5 m

A

A'

West Bay

2 m

2 m

Galveston Island

10 m

Gulf of Mexico

15 m

2 m

Sanluis Pass

		1			4
		2			5
		3	10 km		

Galveston Island

Pleistocene terrace

West Bay

Oldest fossil beach

Gulf of Mexico

Sand bottom

Mud

Pleistocene

A

A'

188

peninsula was brought to the area from the marine environment rather than from local river discharge.

Cross section *A-A'* in Figure 11.12 depicts the relationship of Galveston Island to the preexisting topography of the Pleistocene subaerial surface. The island consists of a series of accretionary beach ridges that have prograded seaward about 5 kilometers since the post-Wisconsin sea-level rise. The first accumulation of beach conditions appears to be associated with the rapid convergence of Pleistocene surface with sea level. Wave action delivers the coarsest sediment to the beach environment and wells out silt and mud that are deposited in the marine environment in front of the beach. With continued accumulation of sand on the beaches and mud in the offshore area, the beach has prograded seaward by some 5 kilometers.

Behind the barrier island, West Bay is now a relatively protected body of shallow water. Tidal marshes prograde toward the center of the bay from both the mainland and the barrier island. West Bay retains fairly good communication with the open Gulf of Mexico through the two channels, Bolivar Roads and San Luis Pass.

Topset deposits are essentially lacking on the barrier island complex. Sand accretion ridges form the very low-lying topography of the island. In the swales between ridges, relatively thin marsh deposits are accumulating. On the ridges themselves, there is only a thin cover of grass.

From the barrier island seaward, once again we see a basic pattern of accumulation of the coarsest material at or near sea level and a general fining of the sediment outward into the marine area.

The Chenier Plain of southwestern Louisiana. The Chenier Plain (see Figures 11.13 and 11.14) has a history quite similar to Galveston Island, but the internal stratigraphy is somewhat more complicated because of its closer proximity to the sediment supply, the Mississippi delta. Extensive radiometric dating of these deposits by oil company researchers makes them especially interesting to the sedimentologist-stratigrapher who is interested in the dynamics of how stratigraphic units develop.

In comparison to Galveston Island, the effects of increased sediment supply to the Chenier Plain are several. Note, for example, that progradation has moved the shoreline seaward by some 15 kilometers in the area south of Grand Lake.

Figure 11.12 (Left) Physiography and sediment types of Galveston Island, and surrounding area, Texas. Sediment types are as follows: (1) alluvial sediments, mostly backswamp; (2) marine to freshwater marsh deposits; (3) beach sand; (4) predominant lineations within the beach-bar complex; (5) older sediments, mostly Pleistocene. (After LeBlanc and Hodgson, 1959, with modification.)

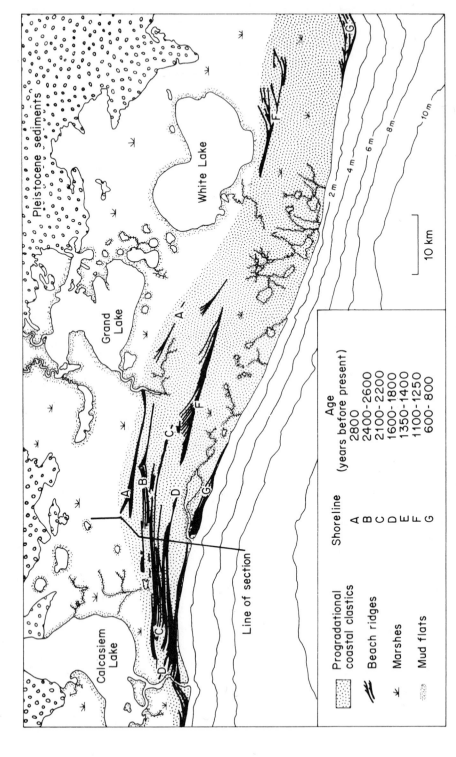

Figure 11.13 The Chenier Plain of southwest Louisiana. View in conjunction with Figure 11.14. Westerly longshore drift occasionally causes rapid progradation of tidal flats southward into the Gulf of Mexico. As the sediment supply slackens, reworking of the tidal flats by wave activity and longshore currents produce beach-sand ridges. (After Gould and McFarlan, 1959, with modification and simplification.)

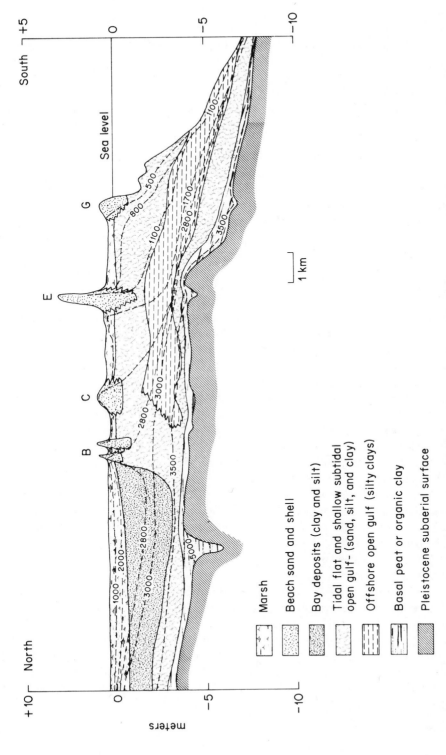

Figure 11.14 Cross section through the Chenier Plain of southwest Louisiana. Time lines are provided by numerous radiocarbon dates on peat and mollusc shells. (After Gould and McFarlan, 1959, with modification and simplification.)

Legend:

Marsh

Beach sand and shell

Bay deposits (clay and silt)

Tidal flat and shallow subtidal open gulf- (sand, silt, and clay)

Offshore open gulf (silty clays)

Basal peat or organic clay

Pleistocene subaerial surface

This distance contrasts to the progradation of about 5 kilometers in the Galveston area. In addition, Calcasieu, Grand, and White lakes are now freshwater lakes, because coastal sedimentation has effectively cut off large-scale communication of these water bodies with the open Gulf. The area between Calcasieu Lake and White Lake was once a shallow bay behind a barrier island, not unlike West Bay behind Galveston Island today. In the Chenier Plain, these bays have become filled in and are now supratidal marshes.

By far the most striking contrast between Galveston Island and the Chenier Plain is the wide spacing between sets of accretion beach ridges. Although accretion beach ridges in the Galveston area are each separated by no more than a 100-meter swale, sets of accretion beach ridges in the Chenier Plain are commonly separated by as much as 3 to 10 kilometers. On a map this separation appears as marsh deposits in between the sandy ridges rising above the marsh surface. In cross section (Figure 11.14), we see that the marsh is simply a thin veneer over tidal-flat and shallow, gulf-bottom, silty clay deposits. The development of Galveston Island represents a rather continuous process of sediment supply, reworking, and deposition; whereas the Chenier Plain represents a discontinuous process of rapid sedimentation followed by a destructional phase. The scheme is the same as the one that we saw in the Mississippi delta. Occasional superabundant sediment supply from the Mississippi delta results in rapid deposition of poorly sorted tidal-flat and shallow-marine silty clays. Sedimentation is so fast that reworking in the shallow-marine coastal environment does not have time to produce well-sorted beach sediments. As sediment supply slackens, winnowing processes can begin to develop well-sorted beach sands on the seaward margin of the newly deposited silty clay coastal sediments. As winnowing continues, the well-sorted chenier sand bodies become a protective armor, preventing the further erosion of the silty clay tidal-flat deposits from which they themselves were formed.

The net result of this two-phase process—sedimentation followed by winnowing—is the development of a far more complicated internal stratigraphy within the Chenier Plain as compared to the stratigraphy of Galveston Island. For example, the general rule of the coarsening-upward unit begins to break down in the Chenier Plain. If we have a core or other stratigraphic section that culminates in a chenier sand ridge, we shall observe the expected unit. If, however, we have a core through an area in between chenier ridges, our general rule will fail to apply.

Nevertheless, sedimentation models based on Galveston Island or on the Chenier Plain have certain similarities. In both cases, major accumulations of bay deposits occur only on the updip end of the transgression. With seaward progradation of foreset facies (beach and offshore marine), topset sedimentation is restricted to a very thin veneer of marsh deposits. In both cases, seaward progradation into the Gulf of Mexico results in a clastic wedge that systematically thickens seaward, its upper surface being governed by sea level and its lower surface being governed by topography on the Pleistocene subaerial erosion surface.

The continental shelf of the northwest Gulf of Mexico, offshore east Texas and western Louisiana. A discussion of sedimentation on Galveston Island and the Chenier Plain would be incomplete if we did not mention Recent sedimentation on the continental shelf offshore in these areas. In short, there is very little truly modern sedimentation occurring on that shelf. This fact makes the story of the post-Wisconsin transgression even that much more interesting.

The continental shelf south of Galveston and the Chenier Plain is a remarkably flat piece of subtidal real estate. The average slope of the bottom is about .5 meter per kilometer. The surficial sediment over all of this continental shelf is relict sediment. When geologists set out to study the Recent sedimentation in this area, they very rapidly realized that they were studying a thin veneer of abandoned beach, lagoonal, and marsh deposits that bear no relationship to the present-day bathymetry. Beneath this layer of shallow-water deposits is the Pleistocene sub-aerial surface. It is evident that the post-Wisconsin transgression passed over this area so fast that it left only a very thin and scattered record of its passing. Today, the bulk of sedimentation activity is taking place in the foreset position, from the beaches out to the mud deposits several kilometers offshore. For example, note in Figure 11.14 that Recent sediment above the Pleistocene unconformity surface is approximately 8 meters thick beneath the position of the modern beach; whereas Recent sediment is only about 1 meter thick, a scant 5 kilometers seaward from the modern beach in 8 meters of water. Shallow water over the continental shelf extends more than 150 kilometers seaward from this position. Sedimentation there today is essentially nil. The fact that a transgression such as the post-Wisconsin transgression can pass over a vast area and leave such an exceedingly scanty record will be an important part of any model concerning the interaction of clastic coastal sedimentation with events of submergence.

Tidal-flat sedimentation. Where longshore transport provides sand to build barrier-beach islands, extensive lagoons may be cut off between the barrier island and the mainland (see Figure 11.15). As the beach environments accept the brunt of wave energy from the open-marine environment, the lagoons behind the beach may be predominantly influenced by tidal currents. Tidal currents in semirestricted lagoons are generally asymmetrical; flood tides come in faster and ebb tides go out slower. As the net result, sediment is pumped from the open-marine environment into the lagoon and is accumulated under the influence of flowing tidal currents. Klein (1970) provides an excellent example of primary structures resulting from tidal action.

During each half of the tidal cycle, sediments may be transported and deposited by the flowing water. Thus, sedimentation in tidal channels and on tidal flats is to a large extent quite analogous to meandering-stream sedimentation: The channels are like the streams; the tidal flats, like the backswamps. The coarsest material accumulates as high-angle cross strata in the bottom of tidal channels;

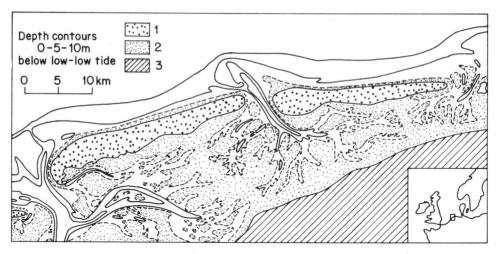

Figure 11.15 Tidal flats of the Wadden Sea, the Netherlands. Where barrier islands take the brunt of wave activity, extensive coastal environments may be influenced primarily by tidal action. Sediment types are as follows: (1) barrier islands; (2) tidal flats, which are sand deposits that are submerged at high tide, exposed at low tide; (3) older sediments. [After L. M. J. U. Van Straaten, "Sedimentation in Tidal Flat Areas," *Jour. Alberta Soc. Petrol. Geologists*, **9**, 203–226, (1961), with modification.]

the finest material accumulates on the tidal flat. With time, progradation occurs, producing a fining-upward unit (see Klein, 1971, for example). However, the presence of marine or brackish-water fauna and the presence of multidirectional cross-stratification distinguishes tidal-flat deposits from meandering-stream deposits.

A Basic Model for Clastic Sedimentation in Coastal Environments

In accord with our systematic approach to the development of models, let us first look at the general attributes of the sediment themselves, then examine the geometric relationships among sediment types, and finally consider the complications that can develop within the model under conditions of emergence or submergence.

The coarsening-upward cycle. Clearly, the most striking generality to come out of our review of coastal clastic sedimentation is the coarsening-upward cycle (see Figure 11.16). Whether we are studying the bar-finger sands of the Mississippi Birdfoot delta, the beach deposits of Galveston Island, or any complex in between, somewhere within the strata we expect to see a well-developed coarsening-upward cycle.

The *coarsening-upward cycle* is the cycle generated by foreset deposition. Foreset deposition may begin in sharp contact with an underlying erosion surface or

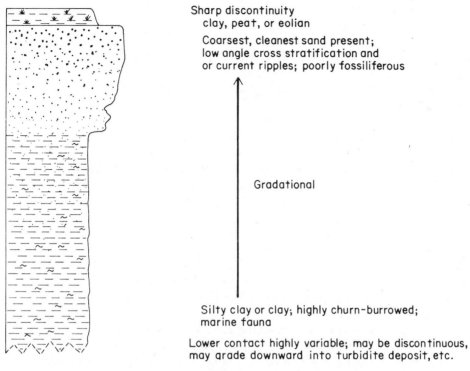

Sharp discontinuity
clay, peat, or eolian

Coarsest, cleanest sand present;
low angle cross stratification and
or current ripples; poorly fossiliferous

Gradational

Silty clay or clay; highly churn-burrowed;
marine fauna

Lower contact highly variable; may be discontinuous,
may grade downward into turbidite deposit, etc.

Figure 11.16 General model of a coastal clastic coarsening-upward cycle.

in gradational contact with fine-grained, open-marine, bottomset beds. The finest-grained sediment present in the cycle occurs at the bottom of the section. It must be emphasized that we cannot say precisely what this grain size will be. Grain size will be a function of size distribution supplied by the river and also a function of the currents in the marine environment at the site of deposition. These fine-grained sediment present in the cycle occurs at the bottom of the section. It must be will be the predominant primary structure. Toward the top of the section, grain size gradually increases, burrow-mottling becomes less predominant, fossils generally become less abundant and more typically disarticulated and abraded. In addition, low-angle cross-stratification (on a beach) or high-angle cross-stratification (in a river-mouth bar or tidal channel) become the predominant sedimentary structures. The foreset coarsening-upward unit may be abruptly overlain by topset beds consisting of natural levee, marsh, or eolian deposits.

Tidal-flat deposits. Because of the rapid rise in post-Wisconsin sea level, clastic tidal-flat deposits are not aerially extensive today. Stratigraphic evidence suggests that such deposits were much more abundant at various times in the geologic record. It is therefore appropriate that we study their identification.

Tidal-flat deposition does not obey the general rule of the coarsening-upward unit that applies to most other coastal clastic sediments. Transportation and deposition within the tidal flat are much more akin to those of a meandering stream. The generalized cycle is a fining-upward one, but it has important attributes to set it apart from the fluvial meandering streams.

Marine or brackish-water fossils are likely to be present in tidal-flat deposits. Furthermore, cross-stratification tends to indicate multiple directions of sediment transport in the tidal-flat environment, whereas similar features are more or less unidirectional in the fluvial meandering-stream facies.

Systematic seaward-thickening of coastal clastic sedimentation units. The thickness of a coastal clastic sedimentation unit is governed by the plane of sea level above and the preexisting topographic (bathymetric) surface below. For a great many paleogeographies that will interest us, reexisting topographic surfaces slope generally seaward. Thus, as progradation occurs, by whatever style it may be occurring, we would expect the development of a gross relationship in which overall thickness of the sedimentation unit increases in the seaward direction, (for example, see Figures 11.9, 11.12, and 11.14). Henceforth, we may find it convenient to summarize this relationship by referring to the *coastal clastic wedge*.

Geometric arrangement of coarsening-upward units within the coastal clastic wedge. Sedimentation in the modern Birdfoot delta of the Mississippi River and longshore transport and deposition of sand in the Galveston Island area provide us with two end members in a spectrum depicting geometric configurations of coarsening-upward units. The Niger delta and the Chenier Plain of southwestern Louisiana illustrate two situations intermediate between the end members. All four examples have already been described, so they require only a brief summary in this discussion of a general model.

At one end of the spectrum, the radial arrangement of bar finger sands generates an extremely complicated stratigraphy of sand bodies. We cannot predict either the number or the direction in which the bar-finger sands will develop in any one subdelta. Moreover, relationships within one subdelta are totally independent of relationships within any other subdelta. Deflationary sheet-sand deposits of the destructive phase may provide good lithologic correlation over large areas within any one subdelta. Bear in mind, however, that deflationary sheet sands of any one subdelta are almost certainly unsystematically diachronous with the deflationary sheet sands of other subdeltas. But note that a number of deflationary sheet sands were associated with essentially the same period of high sea level. Thus, although they are diachronous in any strict sense, they may represent the closest approach to a time line that is available within the section. The search for a good time line is open-ended. We take the best we have and keep on looking for something better.

At the other end of the spectrum, a more or less continuous supply of sand to a barrier beach results in an extremely simple, widespread development of the coarsening-upward cycle (Galveston Island, for example). Virtually, the entire coastal clastic wedge is deposited as a coarsening-upward unit. Wherever we take a core or wherever we see an outcrop, this phenomenon will be apparent. In this situation, the top of the coarsening-upward unit will probably be a very good surface for regional lithologic correlation. You must realize, however, that this surface is systematically diachronous; the oldest is toward the land and youngest toward the sea.

The Niger delta provides an example of something in between the two preceding extremes. River-mouth bars do, in part, prograde seaward; but they also supply sand for the development of well-sorted beaches in between river-mouth bars. Therefore, the coarsening-upward unit is being developed throughout the broad concentric front of the delta. In some places, the upper portion of the cycle is river-mouth bar sedimentation; in other places, the upper portion of the cycle is beach sedimentation. The whole complex is prograding seaward and is leaving behind the coarsening-upward cycle around the entire perimeter of the delta.

Discontinuous sedimentation and beach deflation in the Chenier Plain produce a stratigraphy of coarsening-upward unit arranged subparallel to the general strike of the coastal clastic wedge. Thus, individual beach sands are excellent time-stratigraphic markers along the strike, but lithostratigraphic correlations of these markers downdip is systematically diachronous.

Associated lagoons and topset deposits. Having recognized coarsening-upward units and the general geometry of the coastal clastic wedge, what do we expect to find overlying the top of the foreset unit (on a beach or a river-mouth bar)? First and foremost, we have noted topset deposits in all of our examples from the Recent epoch. These are typically marsh, swamp, or eolian sand. Secondly, our examples of longshore transportation and deposition commonly have a subtidal lagoon updip from the coarsening-upward unit. Even though we did not see it in the Recent relics, perhaps that lagoon could migrate seaward as the coarsening-upward unit progrades into slowly rising sea level. Finally, in the delta areas, there are always fluvial environments upstream that could presumably prograde out onto the delta as the delta itself progrades out into the sea. Let us examine each of these situations in a bit more detail.

Wherever we have examined marsh and swamp deposits in the topset position (Figures 11.4, 11.12, 11.14), we have found them to be rather thin. Limiting their thickness from underneath is the fact that progradation has built a surface at or near sea level before the marsh or swamp takes over. Limiting their thickness above is the fact that these environments must be very wet to keep going. If they begin to accumulate too thick a deposit, the top of the deposit becomes too high above sea level and conditions become too dry for the marsh or swamp to continue to thrive. The process, therefore, is self-limiting and closely related to a water table controlled by sea level.

At first glance, eolian sand dunes appear to offer a considerable opportunity for the development of thick topset deposits over the progradational foreset deposits. However, here again we find that what we see in the environment today is not necessarily what we would expect to see incorporated into the stratigraphic record. Whereas stabilization of eolian sand is the first fundamental step in its incorporation into the geologic record, the large dunes that we see today in south Texas, for example (see Figure 11.1), are transient, migrating features.

An effective way to stabilize eolian sand is to permanently wet it near the water table. Dry sand at the top of the dune continues to blow and migrate, but wet sand near the water table tends to remain in place so long as it remains slightly wet. Thus, it is not the eolian dunes that we would expect to see incorporated into the stratigraphic record as topset deposits. Instead, the planar deflation surfaces associated with the water table offer good opportunity for the lower portion of migrating eolian dunes to become incorporated into the geologic record. Once again, therefore, we see that the thickness of topset deposits over progradational foreset deposits is effectively limited by a water table directly related to sea level. So long as the sea level remains constant, we would not expect to see development of thick topset marsh, swamp, or eolian deposits, no matter how far the foreset beds may prograde.

Similar arguments apply to the possible development of a subtidal lagoon over foreset deposits during a time of progradation into constant sea level. There is simply no room between sediment and sea level. If there were room, swamp and marsh deposits would likely expand quickly to fill that space.

Thus, the basic model based on the Recent epoch would suggest that the best way to get lagoonal deposits progradational over foreset coarsening-upward deposits would be to prograde into rising sea level. We shall explore this fascinating possibility shortly.

Finally, we might expect to find fluvial meandering-stream deposits prograding into the topset position over a delta as the foreset of the delta prograde seaward into constant sea level. Such a model is in keeping with our concept of the equilibrium profile in meandering-stream deposition. As the ultimate seaward end of the equilibrium profile (the river-mouth bar) prograde seaward, the equilibrium profile will be raised somewhat, thereby accommodating meandering-stream deposition over earlier deltaic deposits. However, it is axiomatic that this relationship can develop only where rivers join the coastal clastic wedge. Progradational foreset deposits resulting from longshore transportation and deposition need not necessarily have fluvial deposits updip and therefore may not be prograded by meandering-stream facies.

These considerations have two important consequences for us when we look at the stratigraphic record. First, classical stratigraphers often attached formation significance to such descriptions as "alternating sand and shale." Alternating sand

and shale describes both the coastal clastic wedge and the fluvial meandering-stream facies. Thus, puzzling thickness relationships and puzzling lithostratigraphic correlations may result from the fact that one section of "alternating sand and shale" involves only coastal clastic wedge and meandering-stream deposits. Secondly, progradation of meandering-stream facies onto the delta generates greater topographic relief than will be present in adjacent coastal clastic environments resulting from longshore transportation. Differences in geometry of the depositional surface may grossly affect the development of facies with the next transgression.

Foreset progradation. As we talk of progradation in Recent sedimentary environments, we commonly speak of changes in map view: "the beach has prograded 3 kilometers seaward," and so forth. We must bear in mind, however, that progradation is not a function of area or distance but rather a function of sediment volume. For a beach to prograde seaward, the area in front of the beach must first be filled in. Thus, the rate of progradation of foreset deposits is a function of sediment supply and preexisting foreshore bathymetry. However, the situation can become more complicated.

We must consider the question of sediment *deposition in* an area as opposed to sediment *transportation through* an area. Why do some sediments come to their final resting place on that particular stretch of beach, whereas other sediments are transported past that spot and deposited further down the beach? Clearly, there exists some balance between sediment supply and the energy conditions of that particular stretch of coast. The more sediment brought to the area, the better chance there is that some of it will come to its final rest in that area. The more wave and current action on the beach, the more sediment may be carried on past that particular area.

Size distribution of the sediment supply and the concept of effective wave base may further complicate foreset progradation. When ocean waves approach a coast line, they begin to exert a significant force on the bottom as water depth shallows from 50 to 75 meters. Whether or not this force will resuspend a sediment is a function of the grain size of the bottom sediment. If abundant fine-grained sediment is present, then it may be mobilized by incoming waves and redistributed to form an equilibrium profile in which wave energy is dissipated evenly from deep water into shallow water. Thus, an incoming ocean wave will first expend some energy in mobilizing and transporting fine-grained sediment in 100 meters of water, utilize still more energy mobilizing and transporting silt and fine sand in less than 50 meters of water, and finally utilize its last energy mobilizing and transporting coarse sand in the beach environment. With continuing sediment supply, all sediment types, from fine-grained mud in deep water to coarse-grained sand on the beach, will be transported and deposited. With continued sorting, transportation, and deposition, the whole coastal clastic wedge prograces seaward. Note, however, that if fine-grained material is lacking in the offshore area, inconsistencies may enter this scheme.

Consider a coastal area to which a given mix of sand, silt, and clay are transported by longshore currents. At first, this mix of particle sizes is adequate for initiating progradation. The sand accumulates, let us say, from the beach out to 10 meters of water; the silt, in water depth from 10 to 30 meters; and clay-sized material, in water depths of 30 meters and greater. As progradation continues, the amount of sand and silt required to produce one areal unit of seaward progradation (that is, to prograde the beach 1 kilometer seaward, for example) remains constant. However, the amount of clay-sized material required for fill in water depth of 30 meters or greater in front of the prograding beach becomes greater and greater as the clastic wedge progrades further and further out onto the continental shelf. Sand and silt may not be suitable substitutes for the clay. The weak wave agitation and low-velocity currents acting in water depths of 30 meters or greater may not be capable of sand or silt transportation.

So what happens? Beach sand and offshore silt go on accumulating for a while as though they were continuing to prograde. This process results in a steepening of the forebeach depositional topography, which puts the beach in closer proximity to incoming ocean waves and allows less distance for ocean wave energy to be dissipated before reaching the beach. The net result is increased agitation and stronger currents in the beach and offshore silt environments. Increased agitation and stronger currents result in greater transportation of sediment *through* the environment and less accumulation of sediment *in* the environment. Progradation, therefore, may become self-limiting in the absence of sufficient fine-grained sediment to produce equilibrium profiles in deeper foreshore waters.

Basin filling by vertical accumulation of bottomset beds. It is the natural fate of geologic basins to become filled with sediment. Progradation by coastal clastic wedges is a common method of basin filling. It is also the method involved in the origin of a great many sedimentary units that are of economic importance for their petroleum accumulations. For these reasons, we have spent a lot of time on foreset progradation and related phenomena of coastal clastic wedges. Let us note, at least in passing, that essentially vertical sedimentation of clays and pelagic organisms can also contribute to basin filling. Where clastic sediments are abundant, basin-margin sedimentation rates (coastal clastic wedges) will generally exceed basin-center sedimentation rates. When we study carbonate sedimentation, we shall find sharp contrast between carbonate and clastic sedimentation with reference to this generality.

Isostatic considerations. As a final point in the consideration of our constant sea-level model for coastal clastic wedges, we must note the profound importance of isostatic subsidence during the loading of basin margins. Consider, for example, a basin such as the Gulf of Mexico. For the sake of discussion, assume that we are going to replace 4000 meters of water with sediment of the coastal clastic wedge environment. Four thousand meters of water with a density of 1.0 is being replaced by sediment with a density of approximately 2.5. Isostatic considerations suggest that the floor beneath the basin margin subsides from 3000 to 6000 meters to accommodate the new load in isostatic equilibrium. Indeed, seismic investigations indicate that some 15,000 meters of Tertiary sediment have accumulated in southern Louisiana.

Similarly, each eustatic transgression will accommodate sediment accumulation to at least twice the amount of the eustatic sea-level change (see Figures 11.8 and 11.9, for example).

Response of coastal clastic wedges to emergence. As we begin to consider the response of coastal clastic wedges to conditions of changing relative sea level, we must keep in mind two general questions. First and foremost, what happens to sediment supply? Secondly, how are existing deposits affected by the change?

With a condition of relative sea-level lowering (emergence), the answer to both of these questions is rather simple. Meandering streams continue to supply sediment from continental areas. In fact, probably a little more sediment may come down to the coastal area as the meandering-stream facies incises into older deposits to approach the equilibrium profile dictated by the new lower sea level.

Old marshes, bays, and lakes sit high and dry, ceasing to accumulate sediment and perhaps undergoing considerable surficial modification and erosion as terrestrial soil zones develop. Actively prograding river-mouth bars and beaches go right on prograding; their rate of seaward progradation is somewhat speeded up by the fact that decreasing water depth allows a given volume of sediment to prograde farther seaward.

The systematic seaward thickening of coastal clastic wedges accumulated under constant sea-level conditions clearly does not apply to coastal clastics accumulating under conditions of emergence.

Progradation under conditions of constant sea level produces an extremely flat upper surface to the coastal clastic wedge. The subsequent transgression will flood a large area of essentially uniform bathymetry. But coastal clastic sediments deposited under conditions of emergence result in an upper surface that may have considerable seaward slope to the top of the regressive unit. The subsequent transgression will therefore have a more or less continually rising topography on which to transgress.

Response of coastal clastic wedges to submergence. Rising sea level may seriously diminish sediment supply to coastal clastic environments. Initially, rising sea level raises the lower end of the equilibrium profile of the meandering stream that is supplying sediment to the coastal area. The meandering-stream environment may temporarily become the deposition site of clastic sediments that otherwise would have been carried downstream to the coastal area (recall Figures 10.32, 10.33, and 10.34 for example). In addition, rivers commonly occupy valleys that are incised into preexisting topography. With rising water, the site of new deltaic sedimentation may be pushed well up into the head of the newly formed estuary as the river valley becomes flooded. Given this situation, all clastic sediments supplied by the river must go into building a new delta that will fill the estuary before any sediment can again be supplied to longshore currents. In Galveston Bay, for example (see Figure 11.12), sediment supplied by the Trinity River has barely begun to fill the bay from the northeast end. It will be many years before the Trinity River supplies a significant amount of sediment to the barrier beaches of the Texas Gulf Coast.

Thus, we have two generalities concerning sediment supply to coastal clastic wedges during a time of submergence: (1) The supply is decreased by the rise of equilibrium profile in the meandering stream; and (2) those clastics that are supplied to the coastal environments must build new deltas out of the estuaries before longshore transport of sediment can be resumed.

It is difficult to compile data concerning the effect that the rate of sea-level rise has on sediment supply. If we want an answer from field studies, accessibility and radiometric dating techniques dictate that we look only at the effects of the post-Wisconsin sea-level rise. Even so, quality studies of this problem are presently limited to the Mississippi delta and areas westward to Galveston Island.

Data concerning the post-Wisconsin transgression suggest that sea-level rises of 8 meters/1000 years (the approximate average for post-Wisconsin sea-level rise) are sufficiently rapid to generate grossly discontinuous responses of coastal environments to the rising water. Figure 11.6 shows that the first subdelta of the Recent Mississippi delta formed approximately 5000 years ago. As indicated in Figures 11.13 and 11.14, the Chenier Plain developed did not really get into full swing until approximately 3000 years ago, some 2000 years after the sea had effectively flooded this area. This time lag is presumably related to a lack of sediment supply from the infant Recent Mississippi delta. At any rate, a sea-level rise of 8 meters/1000 years is sufficient to wreck the continuity of sediment accumulation in coastal clastic environments.

Clearly, there is some slow rate of sea-level rise that is slow enough that deltaic sedimentation and longshore transport and sedimentation will go on essentially unchanged. In the delta area, the rate of vertical accumulation of swamp, marsh, and natural levee deposits must be sufficient to confine the river to its channel if deltaic sedimentation is to continue to occur along the outer periphery of the pre-existing delta (see Figure 11.17). Given this situation, longshore transportation of sediment away from the delta should also go on as usual, even though the sea level is rising slowly.

Turning now to the prograding barrier-bar environment, we notice some things that have considerable stratigraphic significance. The beach environment of the barrier bar ignores the rising sea level. From day to day, this beach environment simply goes on accumulating sediments and prograding seaward. Topset deposits and lagoonal deposits, on the other hand, take advantage of the rising water and expand with it. In our static sea-level model, marsh and swamp deposits behind

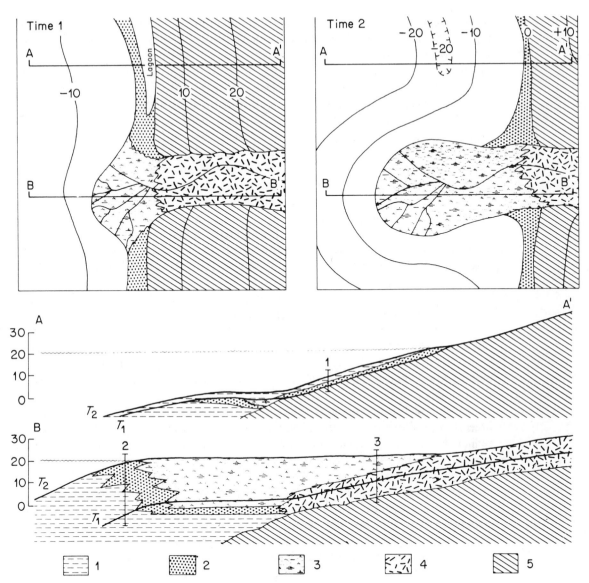

Figure 11.17 Theoretical model of stratigraphic relations generated by the interaction between coastal clastic sedimentation and rising sea level. In map view,
time (1) shows the geographic distribution of major sediment types before
submergence. Time (2) indicates one potential geographic distribution of
major sediment types following an event of "moderately rapid" submergence.
The delta has more or less maintained its position, whereas the beach environments to the north and south have migrated landward with rising water. Cross
sections *AA'* and *BB'* demonstrate the vertical stratigraphies that may be
generated by such an event. Note the "transgressive" sequences (1) and (3)
are properly correlated with the "regressive" sequence (2). Observe also the
thick accumulation of topset deposits in the delta area.

barrier bars were thin because thick accumulation would have put them up too high above sea level and growth would have ceased; the process was self-limiting, governed by a water table closely tied to sea level. Because of similar limitations, lagoonal deposits could only accumulate to some trivial thickness in our constant sea-level model. Now that we are developing a model that involves a rising sea level, we observe that these deposits are given room to expand, both in a vertical sense and in a horizontal sense. Thus, continued seaward progradation of a barrier beach during rising sea level not only generates a "regressive sequence" during the time of rising water but also allows thick and widespread marginal-marine topset beds to accumulate. These beds will undergo lateral expansion both in the seaward direction (regressive) and in the landward direction (transgressive). Note the confusion that can arise from loose use of the words "transgression," "regression," and "progradation."

Given a different balance between sediment supply and rate of sea-level rise, we could argue for the progressive transgression of beach facies onto the land surface as sea level rises. This relationship is the classical one that we used extensively in Chapter 2.

So long as there is a gently sloping topography along which the beach can extend, this model is perfectly reasonable. With rising water, beach facies transgress onto the land surface and former beach facies are overlain by fine-grained offshore sediments. As indicated in Figure 11.17, the model developed here suggests that it is perfectly reasonable to correlate a "transgressive" sequence in an area of low-sediment supply with a "regressive" sequence in an area of high-sediment supply.

In all models for continuous sedimentation during conditions of submergence, the preexisting sediments will be covered up with new sediment. The more rapidly a sediment becomes covered with new sediment, the better the chance is that that portion of sedimentation history will become incorporated into the permanent stratigraphic record. There is, therefore, a strong preservational bias for stratigraphic sequences deposited under conditions of submergence, as opposed to stratigraphic sequences deposited under conditions of emergence. Note that this statement does not use the terms "transgressive" and "regressive" sequences, which appear in classical stratigraphic terminology. Both transgressive and regressive sequences may be deposited under conditions of submergence.

An Ancient Example: Bottomset to Meandering-Stream Transition; Reedsville Shale-Oswego Sandstone, Upper Ordovician of Central Pennsylvania

The work of Horowitz (1966) provides a simple, yet instructive, test of our general model. First, let us discuss the section in classical stratigraphic terms.

Regional setting and earlier stratigraphic work. As indicated in Figures 5.3, 5.4, and 5.5, it has long been recognized that throughout the sedimentary Appalachians earlier Paleozoic limestone sedimentation gives way with time to marine clastic sedimentation, which in turn gives way gradually to nonmarine clastic sedimentation. The timing of these events varies along the Appalachian front. In south central Pennsylvania, the transition from Reedsville shale to Oswego sandstone was recognized rather early as one of these transitions from marine clastics to continental clastic sedimentation.

The Reedsville shale is gray and silty, with occasional fine-grained sand beds, and generally contains marine fossils. Typically, the Oswego sandstone is green and red, high-angle, cross-stratified, and interbedded with shaly siltstone. The formation is generally unfossiliferous except for occasional bituminous fragments. Clearly, the distinction between Reedsville and Oswego lithologies provides the basis for the definition of separate and distinct formations. But where should we place the contact? Inasmuch as the Reedsville is predominantly shale and silt and the Oswego is predominantly sandstone, there is a certain logic to placing the formation contact so that the predominantly silt and shale lithology is below the predominantly sand lithology. Note that the formation boundary thus defined may have little paleogeographic significance.

Figure 11.18 Generalized stratigraphic section of the contact zone between Reedsville shale and Oswego sandstone. (After Horowitz, 1966, with modification.)

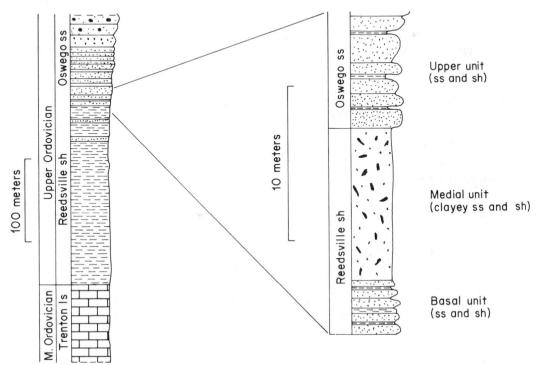

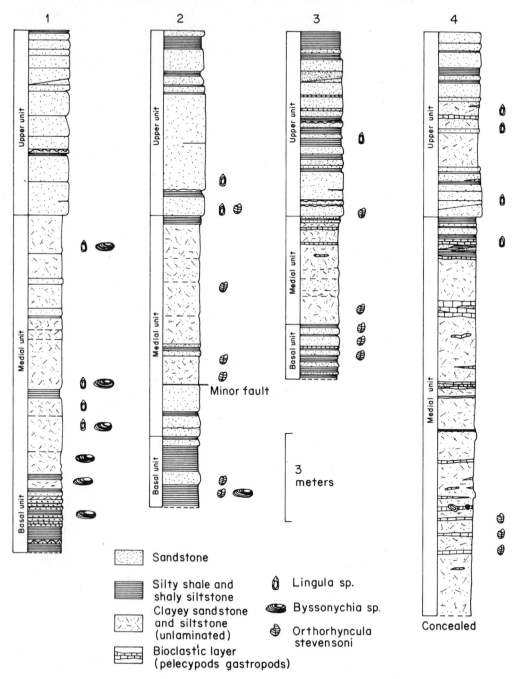

Figure 11.19 Four measured sections through the contact between Reedsville shale and Oswego sandstone. (After Horowitz, 1966, with modification.)

Minor fault

3 meters

Sandstone

Silty shale and shaly siltstone

Clayey sandstone and siltstone (unlaminated)

Bioclastic layer (pelecypods gastropods)

Lingula sp.

Byssonychia sp.

Orthorhyncula stevensoni

Concealed

Sedimentary observations within the Reedsville-Oswego contact interval. Horowitz (1966) studied the contact between the Reedsville and Oswego formations in considerable detail (see Figure 11.18). Four of his measured sections are shown in Figure 11.19. Horowitz subdivides the transition from Readsville to Oswego into three units that he considers have environmental significance. The units of the transition zone are designated as *basal, medial,* and *upper;* their lithologic and paleontologic attributes are summarized in Table 11.1. In traditional stratigraphic terminology, the basal and medial units belong to the Reedsville shale. The upper unit, being predominantly sand, is designated as Oswego sandstone. A brief lithologic description of typical Oswego sandstone is also included in Table 11.1 for comparison.

Observe particularly the transition from medial unit to upper unit. Massively bedded, burrow-mottled, fossiliferous, clayey siltstones and very fine-grained sandstones give way upward to interbeds of evenly laminated very fine-grained sandstone and thick, cross-bedded, very fine-grained sandstone. These sandstones are poorly fossiliferous; those fossils that are present are poorly preserved. Note further that

Table 11.1. Summary of Lithologic and Faunal Characteristics of Units of Deltaic Sequence

Unit	Fossils	Lithologies
Oswego (excludes basal 20 feet)	Unfossiliferous. Small bituminous coal-like fragments scattered in a few shaly beds.	Cross-bedded, very fine-grained to fine-grained sandstone grades upward into shaly siltstone. Beds were *red* before diagenetic reduction.
Upper	Fossils rare and fragmental; mostly *Lingula* and poorly preserved calcareous forms.	Evenly laminated, very fine-grained sandstone; thick, cross-bedded, fine-grained sandstone; unlaminated, clayey siltstone and very fine-grained sandstone; silty shale; fine-grained quartzite and other less common rock types.
Medial	Small pelecypods and gastropods concentrated in thin beds. *Lingula* occurs near top; brachiopod *Orthorhyncula stevensoni* and pelecypod *Byssonychia* sp. near base.	Mostly massive or thick-bedded, poorly sorted, clayey siltstone and very fine-grained sandstone. Internal structures suggest that sediment was churned or disturbed. Thin wavy fossiliferous beds are locally conspicuous features. Chert "balls" of probable slump origin are present.
Basal	Brachiopod *Orthorhyncula stevensoni* and pelecypod *Byssonychia* sp. locally abundant in shaly layers. Cephalopods, bryozoans, and small brachiopods rare.	Graded, very fine-grained to fine-grained sandstone interbedded with silty shale. Fossil fragments occur at base of sandstones.

After Horowitz, 1966.

the upper unit is overlain by a sequence of unfossiliferous fining-upward units in which high-angle cross-stratified sandstone grades upward into shaly siltstone. The sequence from medial transition beds of the Reedsville into the fining-upward units of the Oswego presents us with three suites of lithologies that are quite familiar to us from our foregoing discussions of Recent sedimentation.

Sedimentological interpretation. The basal unit of the transition zone presumably represents slump or storm-deposited material in essentially the marine bottomset position. Accumulation of fossil fragments at the base of sandstone layers and the generally graded texture of the sandstones suggest mass transport and resedimentation of earlier marine deposits. This point will be discussed further in a later chapter. The medial unit has many of the attributes of the marine foreset deposits that we discussed in some detail with respect to the Mississippi delta. The upper unit is the river-mouth bar-beach complex of the upper portion of the foreset beds. The fining-upward units, which constitute the bulk of the Oswego, are meandering-stream deposits that have prograded over the coastal clastic environments.

The environmental interpretation of the Reedsville-Oswego contact gives us a completely different unifying concept for the designation of formations. In particular, the upper unit of the transition zone has previously been considered part of the Oswego sandstone on the basis of the idea that "shale below is Reedsville and sandstone above is Oswego." If we could redesignate the formation boundary, we would probably prefer to place it at the point where coastal clastic sediments lie below and fluvial meandering-stream sediments lie above.

An Ancient Example: Coastal Clastic Sediments in the Subsurface Lower Oligocene of Southeast Texas

The work of Horowitz described earlier relies heavily on detailed examination of outcrops and hand specimens of the rock under consideration. Upon moving into subsurface problems, the geologist does not have these kinds of sample materials to study. Where core is available, detailed examination of the lithology is possible. However, coring is an expensive way to drill a hole. The geologist is quite often left to carry out his investigations on the basis of electrical logs and well cuttings.

In the Gulf Coast, cuttings are of no value for sandstone petrology because the sands are so poorly cemented that only loose sand is recovered. Thus, all information concerning primary structures and systematic grain-size variation is lost in this drilling process. Cuttings from shale intervals, however, are valuable for the stratigraphic control provided by the foraminifera contained in the shales. The geologist studying the subsurface in the Gulf Coast, therefore, generally uses (1) electrical logs, which provide his only lithology information, and (2) cuttings from the shale intervals, in which foraminifera provide paleontological control on his stratigraphy. The work of Gregory (1966) provides a good example of how a geologist proceeds to draw important paleogeographic conclusions from the study of these limited types of information.

Regional setting and earlier stratigraphic work. As noted in Chapter 5, the major stratigraphic subdivisions of Gulf Coast Tertiary stratigraphy are based on transgressive tongues of limestone or shale that give way vertically to regressive (progradational) deposits of marginal-marine and finally continental clastic sediments.

Figure 11.20 indicates how these major sedimentation units are further subdivided on the basis of benthonic foraminiferal zones. Recall that a formation is simply a mappable unit. If we choose to define our mappable unit on the basis of paleontology rather than lithology, that definition is perfectly acceptable.

Sedimentological observations in the Vicksburg formation. Figure 11.21 presents a stratigraphic cross section of the Vicksburg formation oriented perpendicular to regional strike. There are several things in this cross section that should start a petroleum geologist thinking. To begin with, the Vicksburg formation abruptly thickens between wells (6) and (7). Abrupt thickenings of the section downdip are a general rule in Gulf Coast stratigraphy. Oil and gas accumulations are commonly associated with these *flexure zones*. Usually a large number of sands interlayered with the shale occur just downdip from the flexure zone. In the cross section we are looking at, the entire section downdip from the flexure zone is shale. This fact is disappointing, but at least we have one place on the map through which the flexure zone passes. Perhaps at some other place, we shall find sand downdip to the flexure zone.

Turning our attention to the Vicksburg formation updip from the flexure zone, we note particularly the seemingly continuous sandy zone developed within the Middle Vicksburg between wells (2) and (5). Sands such as these, in among a lot of shale, always interest the petroleum geologist. The sand can provide the reservoir for oil and gas; the shale can provide the seal on the trap. Thus, a little structure or some stratigraphic peculiarity can make these sands petroleum reservoirs. Observe that the sand body does not extend southward to the Vicksburg flexure zone. Well (6) indicates that a relatively thin Vicksburg section is all shale before the section thickens across the flexure zone. Note that the sand body also shales out on the updip end of the cross section [well (1)].

Making use of our first piece of paleoecological information provided by examination of the foraminifera content of the shales, we observe that these updip shales do not contain the marine benthonic foraminifera that are traced by biostratigraphers for stratigraphic control. Presumably, the absence of these marine benthonic foraminifera implies that this updip shale zone, stratigraphically equiva-

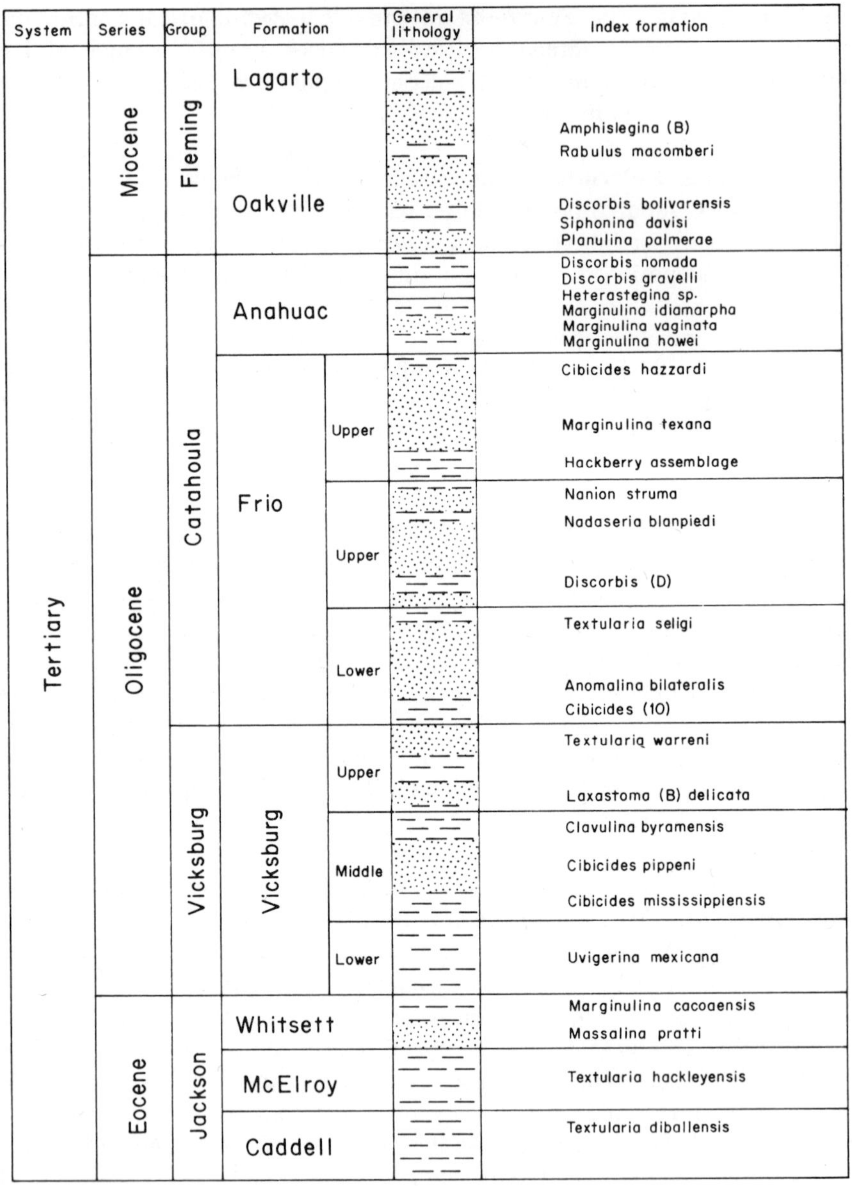

System	Series	Group	Formation		General lithology	Index formation
Tertiary	Miocene	Fleming	Lagarto			
						Amphislegina (B)
						Rabulus macomberi
			Oakville			Discorbis bolivarensis
						Siphonina davisi
						Planulina palmerae
	Oligocene	Catahoula	Anahuac			Discorbis nomada
						Discorbis gravelli
						Heterastegina sp.
						Marginulina idiamorpha
						Marginulina vaginata
						Marginulina howei
			Frio	Upper		Cibicides hazzardi
						Marginulina texana
						Hackberry assemblage
				Upper		Nanion struma
						Nadaseria blanpiedi
						Discorbis (D)
				Lower		Textularia seligi
						Anomalina bilateralis
						Cibicides (10)
		Vicksburg	Vicksburg	Upper		Textularia warreni
						Laxastoma (B) delicata
				Middle		Clavulina byramensis
						Cibicides pippeni
						Cibicides mississippiensis
				Lower		Uvigerina mexicana
	Eocene	Jackson	Whitsett			Marginulina cacoaensis
						Massalina pratti
			McElroy			Textularia hackleyensis
			Caddell			Textularia diballensis

Figure 11.20 Generalized stratigraphic column, Vicksburg and adjacent formations, subsurface of southeastern Texas. (After Gregory, 1966.) Courtesy Houston Geological Society.

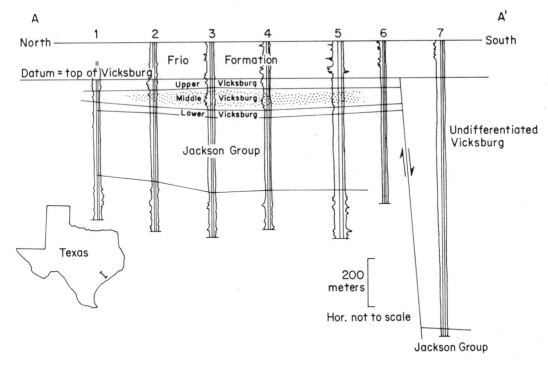

Figure 11.21 Stratigraphic cross section indicating the distribution of Middle Vicksburg sand development. (After Gregory, 1966.) Courtesy Houston Geological Society.

lent to Middle Vicksburg sand, is not a marine shale but rather a freshwater or brackish-water shale.

Figure 11.22 is a map of the net thickness of sand contained in the Middle Vicksburg stratigraphic interval. Note that there are essentially two axes of thickening within the areal distribution of Middle Vicksburg sand. The thickening axis running from southwest to northeast more or less parallels regional strike; the thickening axis extending off to the northwest is almost perpendicular to regional strike. Note further the pronounced downdip bulge of net sand where these two axes of thickening come together. Surely this contour is beginning to look familiar to us.

Now we would like to relate the net sand map to the paleogeography in which it formed. Because of postdepositional tectonic deformation, we cannot construct a simple contour map of the Middle Vicksburg top that will reflect its initial sedimentary topography. Postdepositional tectonic activity has grossly distorted the depositional surface. A map showing the top of the Middle Vicksburg would be a map of structural deformation rather than of depositional topography.

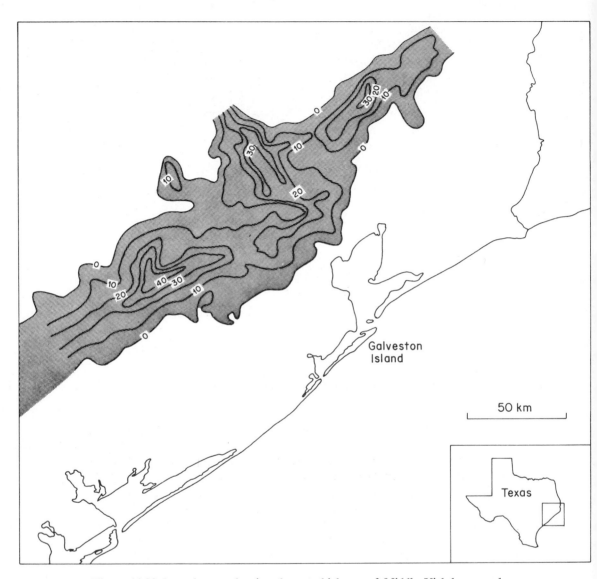

Figure 11.22 Isopach map showing the net thickness of Middle Vicksburg sand. (After Gregory, 1966, with modification.) Courtesy Houston Geological Society.)

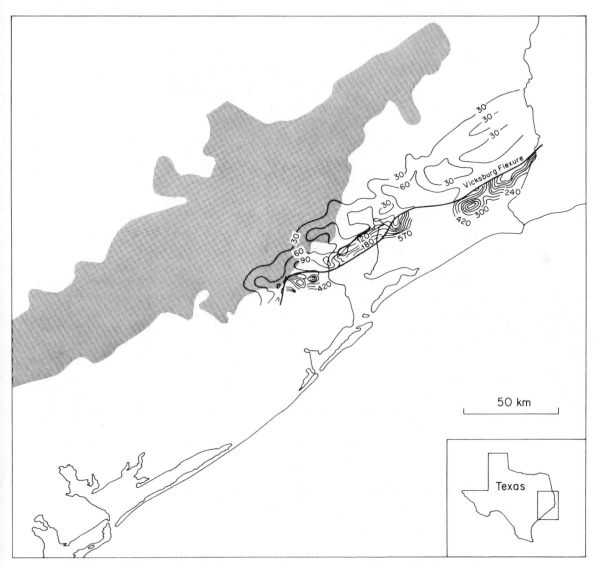

Figure 11.23 Isopach map depicting the total thickness of the Upper Vicksburg stratigraphic interval. Stippled area indicates distribution of Middle Vicksburg sand (Figure 11.22). Note that Middle Vicksburg sand accumulation lies tens of kilometers landward of the Upper Vicksburg flexure. (After Gregory, 1966, with modification.) Courtesy Houston Geological Society.

A common way to get around this problem is to construct an isopach map of the sedimentary unit directly above the topographic surface that interests us. Clastic sediments do not generally pile up in big mounds; they tend to fill in the low spots. Thus, a map depicting the thickness of Upper Vicksburg sediments may be a crude reflection of the topography upon the Middle Vicksburg depositional surface. Where the Middle Vicksburg depositional surface was low, thicker accumulation of Upper Vicksburg material would be expected. Granted, this scheme is imperfect, but it is a step in the right direction and is probably the best that can be done on the basis of data generally available to the practicing petroleum geologist.

Figure 11.23 is an isopach map of Upper Vicksburg sediment within the study area. The tendency to abrupt thickening at or near the Vicksburg flexure is quite apparent. Observe that major sand accumulation in the Middle Vicksburg is considerably updip from the zone of abrupt thickening in Upper Vicksburg sediments. In short, Middle Vicksburg sands were deposited on a shallow-marine shelf marginal to deeper water.

Sedimentological interpretation. On the basis of sand geometry, reconstructed paleobathymetry based on the thickness of overlying deposits, and paleontological data from associated shale, Gregory (1966) put together the environmental interpretation for the deposition of Middle Vicksburg sands, as indicated in Figure 11.24. If we briefly review his lines of evidence, we can recognize marine shales from electric log and micropaleontological data. Sands are shown to be sands from the electric log and no other evidence. Marshy, back-bay, mud-flat deposits and tidal-flat deposits are known from the electric log, which indicates shale, and from micropaleontological data, which indicate that they do not contain marine benthonic foraminifera. The rest of the interpretation is based on the way in which these three general rock types fit together and fit the paleogeography provided by the Upper Vicksburg isopach map.

An Ancient Example: Stratigraphic Oil Trap in Coastal Clastic Sands of the Subsurface Upper Cretaceous of New Mexico

In both of the earlier Ancient examples, we have found a certain degree of reliability in the Ancient record as we have recognized sediment types and geometries that appeared in our study of Recent sedimentation. As we move on to look at Bisti field, New Mexico (Fig. 11.25), we shall begin to test our dynamic model of coastal clastic sedimentation. We shall be looking at a section that undoubtedly contains a lot of emergence, submergence, and progradation, if we can only sort them out.

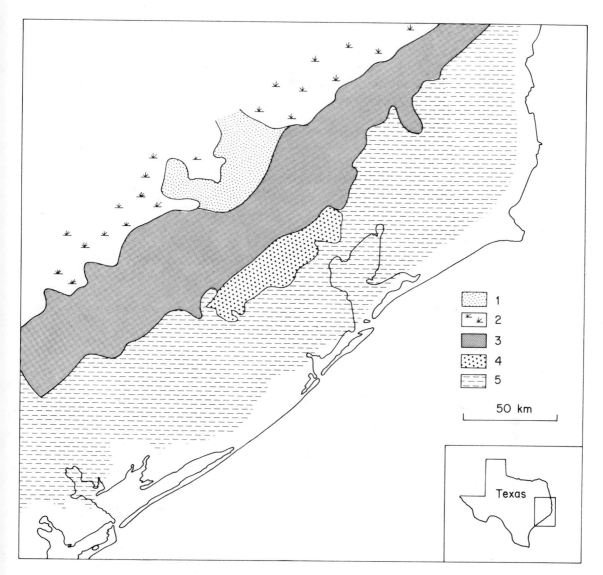

Figure 11.24 Environmental interpretation of Middle Vicksburg sediments. Sediment types are as follows: (1) meandering-stream deposits; (2) brackish-water to freshwater bay deposits, mudflats, and marshes; (3) beach and near-shore coastal clastics; (4) subtidal prodelta sands; (5) marine shales. (After Gregory, 1966, with modification.) Courtesy Houston Geological Society.

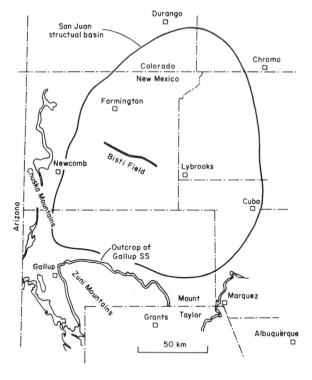

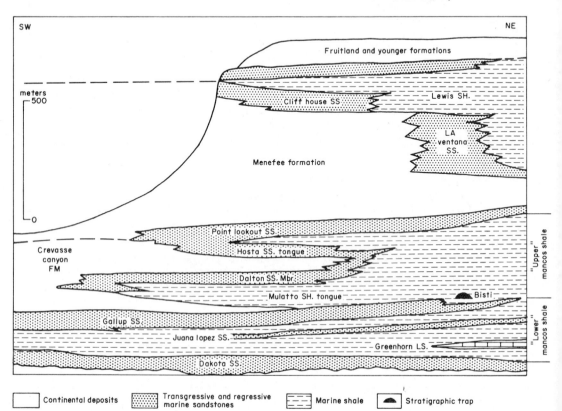

Figure 11.25 (Left) Index map of the San Juan basin, New Mexico, showing Bisti field and outcrops of the Gallup sandstones. (After Sabins, 1963, with modification.)

Figure 11.26 (Below) Upper Cretaceous stratigraphic diagram for the San Juan basin. No horizontal scale. (After Sabins, 1963, with modification.)

Regional setting and earlier stratigraphic work. The Bisti oil field is located near the center of the San Juan structural basin of northwestern New Mexico (see Figure 11.25). Upper Cretaceous stratigraphy of the area begins with the transgression of the Dakota sandstone over a major unconformity surface (Figure 11.26). From that time onward, the depositional history of the San Juan basin records a delicate balance between marine transgression from the northeast and continental regression from the southwest. Formation and member names are typically given to laterally extensive continental deposits, marginal-marine sandstones, and marine shales. The obvious stratigraphy of the area is decidedly lithostratigraphic.

Figure 11.27 is a structural contour map of a horizon just above the producing zone in Bisti field. The map shows only regional dip to the northeast. A structure map of a horizon below the producing zone would similarly show only northeast dip. Bisti field, therefore, is decidedly not a structurally controlled accumulation of petroleum. The oil occurs in that particular sand unit for some sedimentological-stratigraphic reason. Sabins (1963) attempts to explain the controlling factors on this stratigraphic trap.

Figure 11.27 Structural contour map of the Bisti field area. Note that this occurrence of petroleum is not associated with structural closure. (After Sabins, 1963, with modification.)

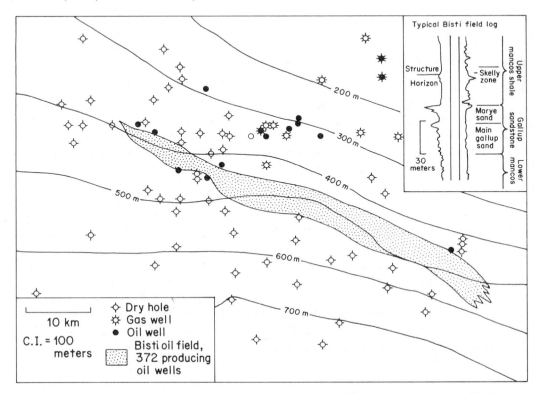

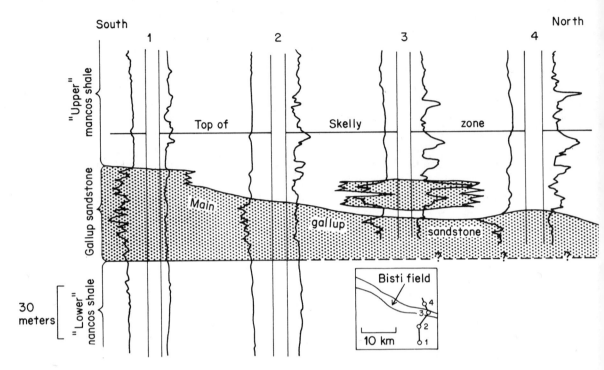

Figure 11.28 Stratigraphic cross section through Bisti field. Bisti field produces oil from sand bodies that are physically separated from the updip main Gallup sandstone by the occurrence of low-permeability material such as in well (2). (After Sabins, 1963, with modification.)

Sedimentological observations. Figure 11.28 presents Sabins' correlation of four wells crossing Bisti field from updip (south) to downdip (north). In brief, detailed electric log correlation has led Sabins to suspect that the sand bodies of Bisti field are petroleum reservoirs because they are physically separated from permeable Gallup sandstone updip by an impermeable zone such as occurs in well (2).

Cores are generally available within this field, so Sabins was able to make a detailed petrographic examination to test this theory. Consider first the main Gallup sandstone southwest of Bisti field (landward with respect to the paleogeography).

The basal contact of the main Gallup sandstone is gradational with Lower Mancos shale below. The lower portion of the Gallup sandstone is very fine-grained and contains abundant clay matrix. It tends to become coarsest toward the top, and/or clay matrix becomes less important. The upper 8 to 30 meters of the main Gallup sandstone are typically medium to coarse-grained clean sands. Thus, it is reasonable to summarize this section southwest of Bisti field as a coarsening-upward cycle from marine Lower Mancos shale gradationally up into the Gallup sandstone and culminating in unfossiliferous medium to coarse-grained clean sand. The upper contact of this cycle with the Upper Mancos shale is sharp.

Next, let us consider a horizontal traverse of those facies that appear to correlate lithostratigraphically with the upper 20 meters of the main Gallup sandstone (see Figure 11.28). In the paleogeographic position represented by well (2), sediments of this lithostratigraphic interval are sandy and silty shales, highly burrowed, with benthonic foraminifera common. Moving on to well (3) (the producing horizon of Bisti field), we see that this field is occupied by relatively clean, medium to fine sand, with abundant glauconite grains and a few pyrite-filled foraminifera but no other elements of a marine fauna. In well (4) (seaward from the producing zone in Bisti field), the sediments of this lithostratigraphic interval are sandy and silty shales, which contain an abundant marine fauna dominated by the pelecypod *Inoceramus,* by fish bones and teeth, and by planktonic foraminifera. Similar sediment types predominate in the Upper Mancos shale, which overlies the Gallup sandstone and related sandbars.

Sedimentological interpretation. Throughout the preceding lithologic descriptions, we avoided the use of genetic terms such as beach, offshore shales, and coastal sandbars. Now let us go back and see whether some of those words might fit. To begin with, we almost said "beach deposits" when we looked at the coarsening-upward cycle in well (1) (Figure 11.28). Sabins' lithologic description of a transition from Lower Mancos shale to the culmination of the Gallup sandstone certainly appears like the kind of sequence that would be generated by prograding, coastal clastic, longshore sedimentation. Indeed, when we look back at the regional stratigraphy (Figure 11.26), we see the Green Horn limestone as the maximum transgressive facies in this section. The constant thickness of the main Gallup sandstone and the gradual thickening of the upper portion of the Lower Mancos shale are certainly compatible with progradation of a Gallup sandstone beach building outward into the originally deeper portions of this shallow sea.

Clearly, the Gallup sandstone progradation was ended rather abruptly by submergence. That the submergence was rapid rather than gradual is suggested by the Mulatto shale tongue overlying the lower portion of the continental Crevasse Canyon formation with no intervening marine sandstone member. According to our generalized model for coastal clastic sedimentation under conditions of submergence, rapid submergence may be expected to temporarily trap the sediment supply in alluvial deposits and in newly constructed deltas, thus allowing water to rise over the former land area without the deposition of associated coastal coarse-grained clastics.

Bisti field, therefore, is the sandbar complex along the front edge of a major progradational sand beach. The reason we see it preserved today is that progradation was abruptly terminated by renewed submergence. As indicated in Figure 11.29, Sabins concludes that the lead edge of the Gallup sandstone progradational

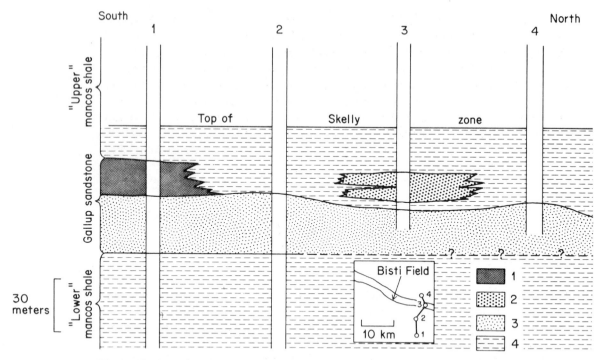

Figure 11.29 Environmental interpretation of the stratigraphic cross section presented in Figure 11.28. Sediment types are as follows: (1) beach sand; (2) forebeach sandbars; (3) offshore marine sands and silts; (4) low-permeability marine deposits, shales, and shally siltstones. (After Sabins, 1963, with modification.)

surface held its position for a time during the initiation of submergent conditions. Beach sediments accumulated to thicknesses of 8 to 30 meters in the area southwest of Bisti field. In the area of Bisti field itself, at least 2 generations of sandbar development took place before submergence finally dominated the area, driving coarse clastic sedimentation far to the southwest and replacing it with Upper Mancos shale sedimentation over the Bisti field.

Note that a stratigraphy of coastal clastic sediments such as the one indicated in Figure 11.26 affords magnificent opportunity for the development of (1) numerous other examples showing the delicate balance between progradation and submergence and (2) other stratigraphic traps like Bisti field.

Citations and Selected References

ALLEN, J. R. L. 1965. Late Quaternary Niger delta, and adjacent areas: sedimentary environments and lithofacies. Bull. Amer. Assoc. Petrol. Geol. 49: 547–600.

ASQUITH, D. O. 1970. Depositional topography and major marine environments, Late Cretaceous, Wyoming. Bull. Amer. Assoc. Petrol. Geol. 54: 1184–1224.
Detailed discussion of a complicated stratigraphy of shelf, slope, and basin clastic sediments.

BEALL, A. O., JR. 1968. Sedimentary processes operative along the western Louisiana shoreline. J. Sed. Petrology 38: 869–877.

CURRAY, J. R. 1960. Sediments and history of Holocene transgression, continental shelf, northwest Gulf of Mexico, p. 221–266. *In* F. P. Shepard *et al.* (eds.), Recent sediments, northwest Gulf of Mexico.

EVANS, G. 1965. Intertidal flat sediments and their environments of deposition in the wash. Geol. Soc. London Quart. J. 121: 209–245.

————. 1970. Coastal and nearshore sedimentation: a comparison of clastic and carbonate deposition. Geol. Assoc. Canada Proc. 81: 493–508.

FISHER, W. L., and K. H. McGOWEN. 1969. Depositional systems in Wilcox Group (Eocene) of Texas and their relation to occurrence of oil and gas. Bull. Amer. Assoc. Petrol. Geol. 53: 30–54.
Recognition of deltaic systems in the subsurface and their importance to petroleum exploration.

FISK, H. N., E. McFARLEN, C. R. KOLB, and L. J. WILBERT, JR. 1954. Sedimentary framework of the modern Mississippi delta. J. Sed. Petrology 24: 76–99.
Basic sediment geometries of the Mississippi delta.

FISK, H. N., and E. McFARLAN, JR. 1955. Late Quaternary deltaic deposits of the Mississippi River, p. 279–302. *In* A. Poldervart (ed.), Crust of the earth. Geol. Soc. Amer. Spec. Paper 62.

FRAZIER, D. E., and A. OSANIK. 1969. Recent peat deposits—Louisiana coastal plain, p. 63–86. *In* E. C. Dapples and M. E. Hopkins (eds.), Environments of coal deposition. Geol. Soc. Amer. Spec. Pub. 14.
Note especially the detailed stratigraphy of the St. Bernard delta complex.

GOULD, H. R., and E. McFARLAN, JR. 1959. Geologic history of the Chenier Plain, southwest Louisiana. Gulf Coast Assoc. Geol. Soc., Trans. 9: 261–270.

————. 1970. The Mississippi delta complex, p. 3–30. *In* J. P. Morgan (ed.), Deltaic sedimentation, modern and ancient. Soc. Econ. Paleont. Mineral. Spec. Pub. 15.
Good general summary.

GREGORY, J. L. 1966. A Lower Oligocene delta in the submarine of southeastern Texas, p. 213–228. *In* M. L. Shirley and J. A. Ragsdale (eds.), Deltas in their geologic framework. Houston Geol. Society, Houston, Tex.

HALBOUTY, M. T. 1969. Hidden treasures and subtle traps in the Gulf Coast. Bull. Amer. Assoc. Petrol. Geol. 53: 3–29.
A successful petroleum geologist discusses sedimentation.

HOROWITZ, B. H. 1966. Evidence for deltaic origin of an Upper Ordovician sequence in the Central Appalachians, p. 159–170. *In* M. L. Shirley and J. A. Ragsdale (eds.), Deltas in their geologic framework. Houston Geol. Society, Houston, Tex.

JOHNSON, K. G., and G. M. FRIEDMAN. 1969. The Tully clastic correlatives (Upper Devonian) of New York State: a model for recognition of alluvial, dune (?), tidal, nearshore (bar and lagoon), and offshore sedimentary environments in a tectonic delta complex. J. Sed. Petrology 39: 451–485.

KLEIN, G. DEV. 1970. Depositional and dispersal dynamics of intertidal sand bars. J. Sed. Petrology 40: 1095–1127.
Well-studied spectacular examples of intertidal primary structures in clastic sediments of Nova Scotia.

———. 1971. A sedimentary model for determining paleotidal range. Bull. Geol. Soc. Amer. 82: 2585–2592.

KOLB, C. R., and J. R. VAN LOPIK. 1966. Depositional environments of the Mississippi River deltaic plain—southeastern Louisiana, p. 17–62. *In* M. L. Shirley and J. A. Ragsdale (eds.), Deltas in their geologic framework. Houston Geol. Society, Houston, Tex.

LANE, D. W. 1963. Sedimentary environments in Cretaceous Dakota sandstone in northwestern Colorado. Bull. Amer. Assoc. Petrol. Geol. 47: 229–256.
Primary structures suggest a transgressive sequence within the formation.

LEBLANC, R. J., and W. D. HODGSON. 1959. Origin and development of the Texas shoreline. Gulf Coast Assoc. Geol. Soc., Trans. 9: 197–220.

MCCUBBIN, D. G. 1969. Cretaceous strike-valley sandstone reservoirs, northwestern New Mexico. Bull. Amer. Assoc. Petrol. Geol. 53: 2114–2140.
Stratigraphic analysis in an area approximately 30 miles northwest of Bisti field produces an interpretation of linear sand bodies that is grossly different from the interpretation of Sabins, 1963. It is interesting to consider whether or not the two papers are talking about the same type of sand body.

MCGREGOR, A. A., and C. A. BIGGS. 1968. Belle Creek field, Montana: a rich stratigraphic trap. Bull. Amer. Assoc. Petrol. Geol. 52: 1869–1887.
Paleogeography and facies relationships were important in the discovery of this 200-million-barrel oil field.

MORGAN, J. P., and R. H. SHAVER (eds.). 1970. Deltaic sedimentation, Modern and Ancient. Soc. Econ. Paleont. Mineral. Spec. Pub. 15. 312 p.
Symposium volume.

REINECK, H. E. 1967. Layered sediments of tidal flats, beaches and shelf bottoms, p. 191–206. *In* G. H. Lauff (ed.), Estuaries. Amer. Assoc. Adv. Sci. Pub. 83.

RICH, J. L. 1951. Three critical environments of deposition and criteria for recognition of rocks deposited in each of them. Bull. Geol. Soc. Amer. 62: 1–20.
A landmark paper concerning the conceptual value of recognizing bottomset, foreset, and topset sediments in the stratigraphic record.

SABINS, F. F., JR. 1963. Anatomy of stratigraphic trap, Bisti field, New Mexico. Bull. Amer. Assoc. Petrol. Geol. 47: 193–228.

SCRUTON, P. C. 1960. Delta building and the deltaic sequence, p. 82–102. *In* F. P. Shepard, F. B. Phleger, and T. H. van Andel (eds.), Recent sediments, northwest Gulf of Mexico.

SHEPARD, F. P., F. B. PHLEGER, and T. H. VAN ANDEL (eds.). 1960a. Recent sediments, northwest Gulf of Mexico. Amer. Assoc. Petrol. Geol., Tulsa, Okla. 394 p. Summary of extensive Recent sediment studies carried out between 1951 and 1958.

————. 1960b, Mississippi Delta: Marginal environments, sediments and growth, p. 56–81. *In* F. P. Shepard *et al.* (eds.), Recent sediments, northwest Gulf of Mexico.

————. 1960c, Rise of sea level along northwest Gulf of Mexico, p. 338–344. *In* F. P. Shepard *et al.* (eds.), Recent sediments, northwest Gulf of Mexico.

SHIRLEY, M. L., and J. A. RAGSDALE (eds.). 1966. Deltas in their geologic framework. Houston Geol. Society, Houston, Texas. 251 p. Eleven papers dealing with Recent and Ancient deltaic sedimentation. Also contains vital statistics on 24 modern deltas.

VAN ANDEL, T. H. 1967. The Oronoco Delta. J. Sed. Petrology 37: 297–310.

VAN STRAATEN, L. M. J. U. 1961. Sedimentation in tidal flat areas. J. Alberta Soc. Petrol. Geol. 9: 203–226. Summary article with extensive bibliography.

WITHROW, P. C. 1968. Depositional environments of Pennsylvanian red fork sandstone in northeastern Anadarko Basin, Oklahoma. Bull. Amer. Assoc. Petrol. Geol. 52: 1638–1654. Oil exploration based on paleogeography and the depositional history of alluvial and coastal clastic sand bodies.

12

Carbonates of the Shelf Margin
and Subtidal Shelf Interior

In the previous chapter, we dealt with clastic sedimentation from the coastline out onto the continental shelf. In the next two chapters, our discussion of chemical and biochemical sedimentation will deal with this same geographic area: the relatively shallow waters adjacent to land masses. The fundamental difference will be that *in situ* origin allows a greater variety of chemical and biochemical sediments than does clastic sedimentation, which depends on a source of sediment supply. Whereas clastic sediments are by and large inert objects *brought to* the depositional environment, chemical and biochemical sediments are generally the *product of* the depositional environment.

When comparing carbonate sedimentation to clastic sedimentation, we may find it convenient to think of the carbonate sediments as being, in a very real sense, "alive." Clastic sediments are inanimate objects that must be acted upon by external forces; but chemical and biochemical sediments are self-generating, thereby needing no external energy source to deliver them to the sedimentary environment. Although external forces are still an important agent in carbonate sedimentation, local production of the sediment upon which these forces act is more important.

Perhaps the most striking example of the contrast between chemical and clastic sediments occurs at the outer margin of the continental shelf. Clastic sediments require a source of sediment and external forces to deliver that sediment to the depositional site. For clastics in general, therefore, the continental shelf margin is a long way from the modern coastline, and the water depth over the major physiographic break from continental shelf to continental slope is at least 100 meters. Thus, the shelf break is far from the source of clastic material. Clearly, the shelf break is occupied by relict sediments, that is, sediments deposited during the Wisconsin low stand of the sea (from 10,000 to 20,000 years ago).

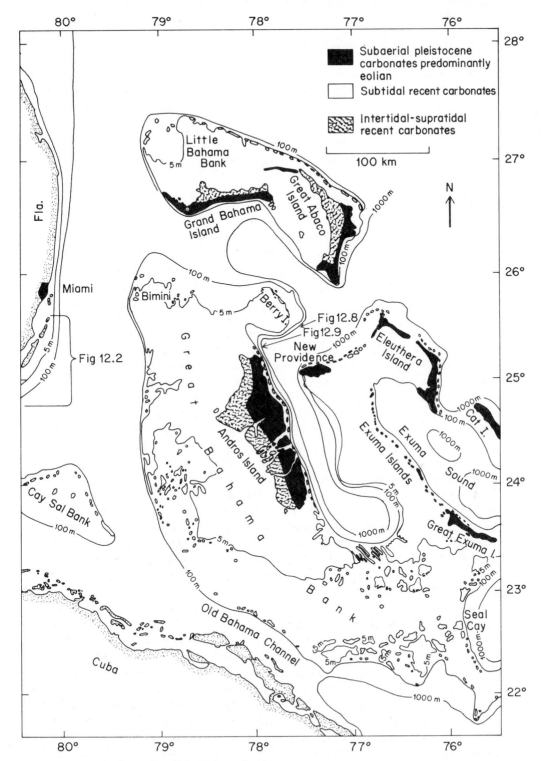

Figure 12.1 Physiography of the Bahama banks.

In contrast, carbonate sediments in this same physiographic position are self-generating. Coral reefs and oolite shoals commonly occupy the physiographic break from continental shelf to ocean basin. Because of this self-generating character of carbonate deposits, our study of carbonate sedimentation is in large part set free from considerations of sediment source and external forces. The water itself is a sufficient "source." Nor is there any need for physical forces to transport the material, because very large particles can accumulate essentially *in situ*.

Recent Carbonate Sedimentation

As in previous chapters, this discussion will be organized around the development of general principles rather than a description of regional sedimentation.

Because carbonate sediments form the major petroleum reservoirs of the world and because many complications have been encountered in attempting to locate new carbonate petroleum reservoirs, Recent carbonate sediments have been studied in great detail. By way of introduction, therefore, several of the principal areas of study of Recent carbonate sedimentation do deserve brief regional description.

The Bahama platform is a vast area of shallow subtidal carbonate sedimentation (Figure 12.1). Platforms rise abruptly out of several hundred meters of water and are constructed by the accumulation of carbonate and evaporite sediments at or near sea level during the continuing subsidence of this area since Cretaceous time. Oolite sedimentation predominates at the shelf margin, where daily tides cause rapid flow of water onto and off of the bank. The subtidal interior of the platform is the site of extensive carbonate mud sedimentation. Portions of many of the larger islands, western Andros Island for example, are supratidal accumulations of Recent carbonate sediments.

South Florida is another area where carbonate sedimentation has been studied extensively (see Figures 5.11 and 12.2). The shelf margin facing the Florida Strait is the site of discontinuous coral-reef development. Fifteen to thirty kilometers interior from, and parallel to, the platform margin, outcrops of Pleistocene limestone form the Florida keys. The Florida keys separate the south Florida reef tract from Florida Bay and appear to exert a pronounced effect on carbonate sedimentation in Florida Bay.

Along the Caribbean coast between Mexico and Honduras in Central America, the coral reefs of British Honduras (Figure 12.3) are the most extensive development of barrier reefs and normal salinity backreef lagoons to be found anywhere in the Western Hemisphere. Well-developed barrier reefs exist on the eastern edge of a shallow shelf marginal to the deep water of the Caribbean. Behind the barrier reefs is an extensive development of *in situ*, shallow-water, carbonate-skeletal sands referred to as the Barrier Platform. Interior to the shallow waters of the Barrier Platform, the Main Channel Lagoon is the site of moderately deep-water lime-mud and clay-mud sedimentation. The patch reefs rising out of 30 to 60 meters of water in the southern portion of the Main Channel Lagoon are unique in the Western Hemisphere and have been carefully studied.

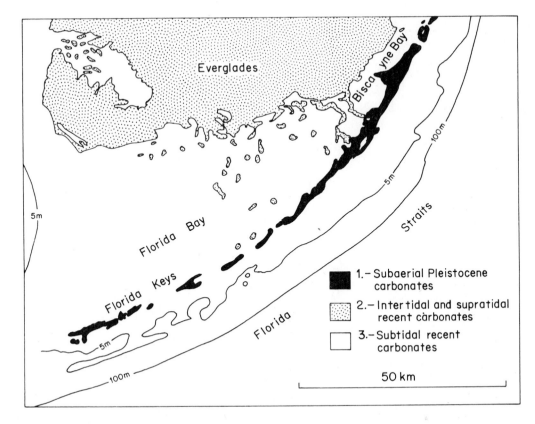

Figure 12.2 Physiography of environments with Recent carbonate sedimentation in south Florida. (After Ginsburg, 1964.)

The Great Barrier Reef of eastern Australia is the largest such feature in the world. The Barrier Reef Province extends for 1200 miles along the eastern coast of the Australian continent, and its backreef platform is as much as 200 miles wide. Much of the area is virtually uncharted. Because of its immense size, sedimentation studies in this area have been predominantly reconnaissance in nature. (See Maxwell and Swinchatt, 1970, for further discussion.)

In contrast, coral reefs off the north coast of Jamaica are of special interest because they have been examined in extreme detail. This region is where Tom Goreau pursued his pioneering studies on coral-reef communities and physiology (Goreau, 1959; Goreau and Land, 1972).

The Persian Gulf (Figure 12.4) interests us primarily because of its shallow water and increased salinities. In all of the areas outlined above, shallow-water carbonate sedimentation begins abruptly at a shelf margin that is adjacent to deep water. In the Persian Gulf, the transition from moderately deep water to shallow water is exceedingly gradational. Here we see a geometric condition that has undoubtedly occurred numerous times in the Ancient record but has not been dupli-

227

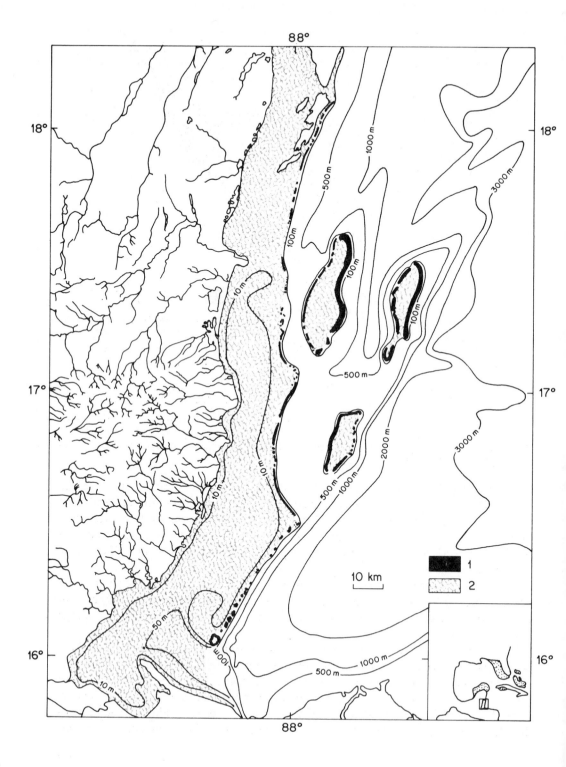

Figure 12.3 (Left) Physiography of regions with Recent carbonate sedimentation in British Honduras. Sediment types are as follows: (1) barrier reefs and associated skeletal sands; (2) subtidal shelf-interior sediments, predominantly skeletal carbonates, with significant clastic input along the mainland coast throughout the area shown in the southern two-thirds of the map.

cated in any of the well-studied areas just outlined. In addition, high evaporation rates and the virtual absence of rainfall make the entire Persian Gulf an area of increased salinities. This fact is a sharp contrast to high-rainfall areas such as South Florida.

The southern portion of the island of Bonaire, off the northcentral coast of Venezuela, has been studied in considerable detail as an example of a restricted lagoonal environment where evaporite sedimentation may lead to refluent dolomitization, a subject we shall discuss in Chapter 13.

Shark Bay, Western Australia, also provides a wonderful natural laboratory for studying the effect of increasing salinity on carbonate sedimentation. Here a series of large "pools" or bays offer a wide range of salinities, depending on the relative ease of communication between the pool and open water (Logan *et al.,* 1970).

Figure 12.4 Physiography of the Persian Gulf.

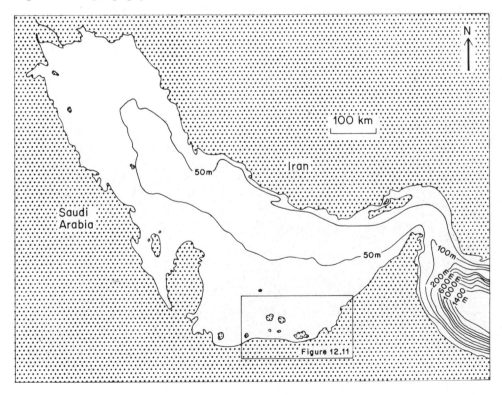

Figure 12.5 Oblique aerial photograph of the British Honduras barrier reefs. Deep water of the Caribbean is to the right and the shallow carbonate environments of the barrier platform are to the left.

Figure 12.6 Oblique aerial photograph showing spur and groove structures in the forereef of the British Honduras barrier reef.

With this brief and general introduction to several areas where carbonate sedimentation has been studied, let us take a more detailed look at the various components that we shall try to fit together into a general model.

The shelf margin. Our discussion of coastal clastic sediments logically began at the delta, where clastic sediments are brought to the marine environment. Our discussion of carbonate sedimentation logically begins at the shelf margin, where the water from which the carbonates are to be precipitated first comes onto the shelf.

Two sediment types will predominate in the bank-margin environment: coral reefs and oolite sands. We can gain important input for our general model from southern British Honduras, northern Jamaica, south Florida, the Bahamas, and the Persian Gulf.

The shelf margin of southern British Honduras strikingly shows how coral-reef development dominates a shelf margin (Figure 12.5). Let us begin at the reef-crest community and study first the forereef environments and then the backreef environments.

Goreau (1959) has an authoritative discussion of forereef zonation. Reconnaissance suggests that his descriptions generally apply to British Honduras reefs as well. From surf zone to depths of approximately 5 meters, modern coral reefs are dominated by *Acropora palmata,* the moosehorn coral. From 5 to 15 meters, reef-front coral developments are predominantly *Acropora cervicornis,* the staghorn coral, or *Montastrea annularis,* a large head coral. Within this zone, it is common for thick coral growth to alternate with channels through which reef-derived debris is transported to deep-water forereef environments. Such an arrangement is shown in Figure 12.6. The elongate patches of reef development are topographically high and are referred to as *spurs.* The channels, down which reef-derived carbonate-skeletal debris is transported to the deep forereef environments, are called *grooves.* It is difficult to estimate how much reef-derived skeletal detritus is transported down the grooves to the forereef environment. The mere presence of the sand-floored grooves indicates a rather active process. If transportation of sediment down these grooves were not significant, coral growth would colonize the channel floors. At a water depth of approximately 15 meters, there is a transition from the *A. cervicornis* and/or *M. annularis* buttress zone to the deep-water, coral-head zone. Whereas shallower zones of the reef are dominated by single species or few species, the deep-water, coral-head zone accommodates a great variety of corals.

Turning our attention now to the area behind the reef crest, we note abundant evidence of reef-derived material that has been transported into the immediate backreef area. In Figure 12.5, the predominantly white area nearest the bank margin is the reef flat. This zone is composed almost entirely of cobbles and boulders of *Acropora palmata,* ripped loose from the reef-crest zone and thrown back by wave action.

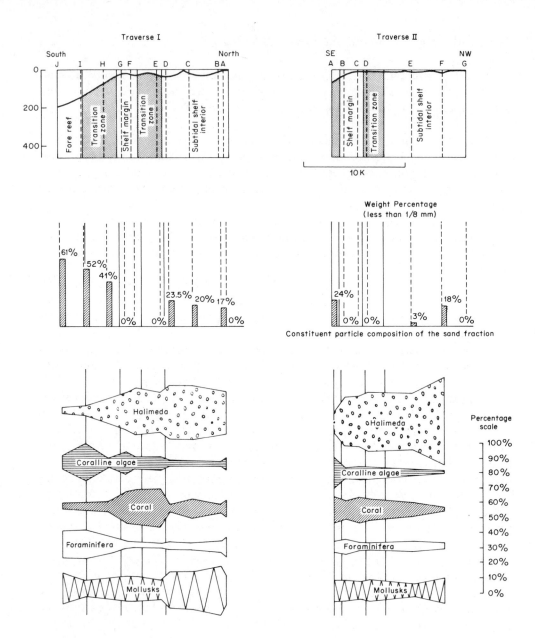

Immediately behind the reef flat is the sand apron, which consists almost entirely of reef-derived skeletal sand that is transported only a slight distance back onto the barrier platform. The sand bottom in this environment is constantly shifting. Thus, no marine grasses or other attached forms can become established.

The shelf margin in southern Florida interests us because it is somewhat different from the British Honduras example and because available data allow us to say more about the forereef environment.

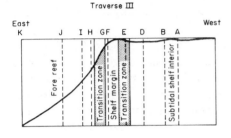

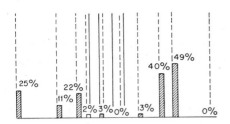

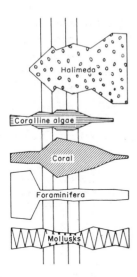

Figure 12.7 Composition by grain size and constituent particles of carbonate sediments in the Florida reef tract. (After Ginsburg, 1956.)

Coral reefs occupy the shelf-margin position in southern Florida, but they are by no means as continuous as the barrier reefs of southern British Honduras. Let us look first at forereef sedimentation.

In south Florida, limited data are available concerning sediment in the forereef. There is a pronounced tendency for south Florida forereef samples to become muddy with increasing depth (see Figure 12.7). Indeed, the absence of carbonate mud appears to be a striking characteristic of the shelf-margin environment. In front of the shelf margin, forereef sediments contain mud. Behind the shelf-margin environment, shelf-interior sediments tend to contain considerable mud. The south Florida reefs are not such an obvious source of sediment supply for the backreef area as the reefs are in British Honduras. At Key Largo Dry Rocks, for example, a *Acropora cervicornis* thicket grows profusely immediately behind the *A. palmata* reef-crest community. In British Honduras, we saw this geographic position occupied by the reef flat, a debris pile derived from the *A. palmata* community. Thus, in the absence of a large accumulation of reef-derived detritus, quiet-water coral communities may predominate in the area immediately behind the reef-crest community.

Similarly, reef-associated sands show no obvious geometric relationship to shelf-margin reefs in south Florida. Sand composition indicates a significant portion of the skeletal material probably originated on the reef (Figure 12.7). Yet sediment supply is not great, and there is ample time for reworking of the sand bodies into their own discrete geometries.

Most of the Great Bahama Bank (Figure 12.1) exemplifies a completely different type of shelf-margin sedimentation. In many areas, coral reefs are substantially absent from the shelf margin. Usually, there are only occasional coral heads and coralline algae on the foreslopes of the shelf.

The predominant sediment type of this shelf-margin environment is the oolite shoal (Figure 12.8). The daily tides have currents that flow rapidly onto and off of the shelf margin. The oolite sand is locally transported by these currents. As a result, the oolite-shoal environment has two characteristics. (1) Continual resuspension of sand grains provides opportunity for additional oolitic laminae to grow around each grain by chemical or biochemical precipitation of aragonite. (2) The constant movement of sand results in such bottom instability that benthonic organisms cannot flourish. Sessile forms find no place to attach themselves; bur-

Figure 12.8 Cross section from shelf margin to shelf interior, Frazer's Hog Cay area, Great Bahama Bank. Sediment types are as follows: (1) Pleistocene and older carbonate rocks, (2) coralgal facies, (3) oolite sandbars, (4) grapestone facies, (5) pellet mud facies. (After Purdy and Imbrie, 1964, and Buchanan, 1970, with modification.)

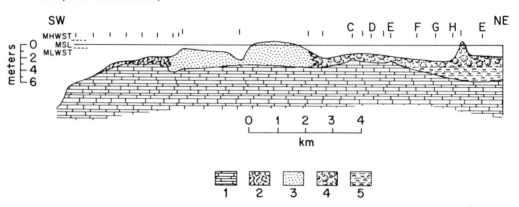

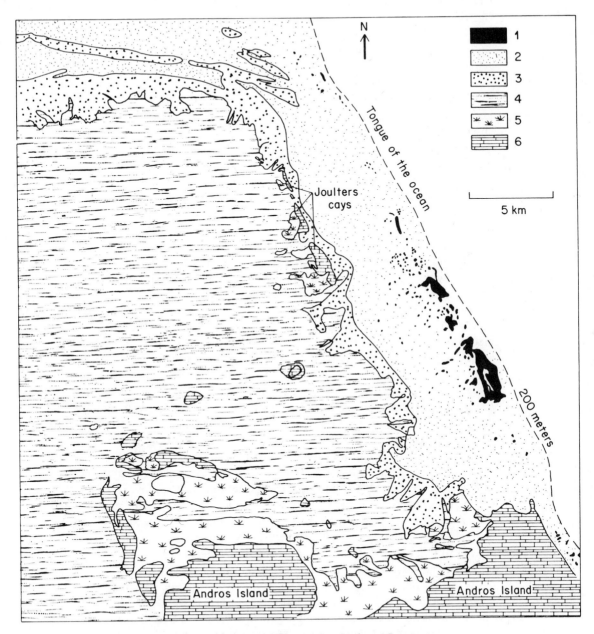

Figure 12.9 Map showing Recent sedimentary facies in the Joulters Cay area, Great Bahama Bank. Sediment types are as follows: (1) coral reefs, nearly awash at low tide; (2) thin veneer of coralgal sand over Pleistocene rock; (3) oolite shoals, awash at low tide; (4) subtidal platform-interior deposits, mostly carbonate mud; (5) supratidal mudflats; (6) land, Recent or Pleistocene in age. (After Purdy and Imbrie, 1964, with modification.)

rowing organisms are continually uprooted by the erosion of sand from around their burrows. Thus, it is very difficult for any sedimentological succession to occur in the oolite-shoal environment. So long as new fauna and flora cannot establish themselves, the environment remains a mobile sand belt. So long as the environment remains a mobile sand belt, new fauna and flora cannot establish themselves. This system will not change unless something changes in front of the mobile sand belt.

Observations in the Jolters Cay area of the Bahamas indicate that changes can occur in front of such sand belts (Figure 12.9). Specifically, if there is enough shallow platform in front of the oolite mobile sand belt, coral reefs may establish themselves seaward from the oolites and dissipate some of the wave and tidal energies that formerly went into maintaining the mobile sand belt.

Mobile oolite sand belts and shelf-margin coral reefs have a different set of requirements for their origin. As we generally understand the oolite sand belts, shallow-water conditions must be maintained throughout their development. Five meters of water is probably too much; tidal currents would not have sufficient velocity to initiate sand-bottom mobility. On the other hand, the absence of bottom mobility is a prime requisite for the establishment of coral reefs; the larvae must attach themselves to a firm substrate that will remain stationary throughout the first years of coral growth.

Furthermore, reef-building corals have a greater depth tolerance than do oolite shoals. Apparently, coral communities can establish themselves in as much as 15 meters of water and grow upward from this base to sea level. This growth appears to be happening in the Jolters Cay area today, for example. The oolite shoals are extensive and are located in the very shallow water where the shelf-margin topography closely approaches modern sea level. The coral reefs out in front seem to be just getting started; there is no reef-flat debris accumulation nor obvious abundance of reef-associated sand. With continued reef development, the oolite-shoal environment will become a relatively quiet-water environment, thus decreasing bottom mobility and ceasing oolite production.

In all of the above examples, the post-Wisconsin sea-level rise has flooded onto a shallow carbonate shelf; the shallow shelf margin is juxtaposed to relatively deep water. Presumably, the existence of well-developed topographic prominence at the shelf margin is itself related to shelf-margin sedimentation associated with previous high stands of the sea. The Campeche Bank of southeastern Mexico (Figure 12.10) provides an interesting example of a shelf margin on which carbonate sedimentation has not kept pace with high stands of the sea.

Here the major topographic break from shelf to slope lies at about 70 to 100 meters below present sea level. With the post-Wisconsin eustatic sea-level rise, only a few isolated coral banks managed to keep pace with the rising water. Over the majority of the shelf, shallow-water carbonate sedimentation could not develop as sea level rose. Consequently, the bottom layer over most of the bank today is a thin veneer of relict shallow-water carbonate sediments combined with deep-water carbonate sediments of the modern environment. The result is some rather unusual

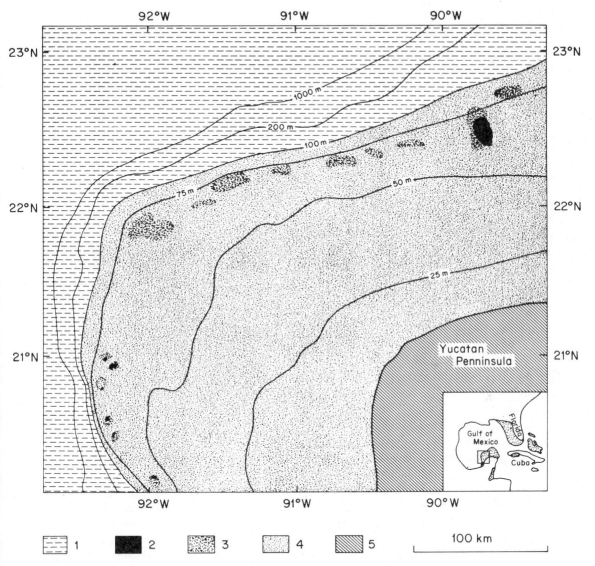

Figure 12.10 General facies and physiography of the western portion of Campeche Bank. Sediment types are as follows: (1) pelagic ooze; (2) shallow-water coral reefs; (3) deeper-water hardbank communities; (4) thin veneer of Recent carbonate sediment, predominantly skeletal sands; (5) land. (After Logan *et al.*, 1969.)

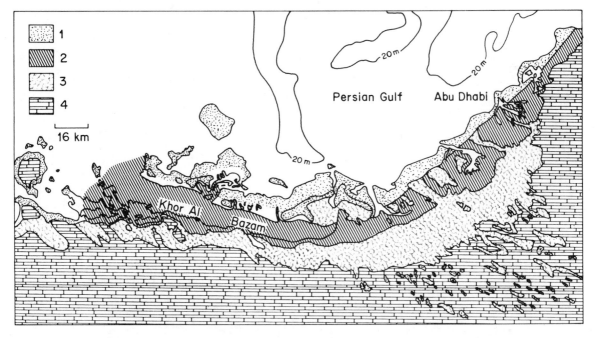

Figure 12.11 Physiography and sedimentary facies of a portion of the southern Persian Gulf. Sediment types are as follows: (1) high-energy shelf-margin facies, includes coral reefs, oolites, and eolian sands; (2) subtidal shelf-interior, predominantly pelletal, lime mud with some nonskeletal carbonate sands; (3) intertidal-supratidal deposits, from algal stromatolites to sabkha evaporites; (4) older rock. (After Kendall and Skipwith, 1969, with modification. Courtesy Geological Society of America.)

sediment types. Unless we have studied an area like Campeche Bank, we may find it rather difficult to conjure up an explanation for an oo-micrite with abundant pelagic foraminifera, for example.

To round out our discussion of the shelf-margin environment, let us now look at the southern shore of the Persian Gulf (Figures 12.4 and 12.11). First, note that the shelf-margin geometric relationships are completely different from anything we have looked at earlier. In our previous examples, shelf-margin facies are very close to deep water. In the Persian Gulf, the shelf margin is just a slight wrinkle in a bathymetric surface where shallow water extends basinward for tens of kilometers. In short, the steep geometries of our previous forereef environments have been replaced by the very low-relief, shallow, subtidal, open-marine environment. Such a forereef geometry must have existed numerous times in the stratigraphic record however. There are a lot of places in the continental United States where there simply could not have been enough space available for the development of the precipitous forereefs that characterize most carbonate shelves today.

In the Persian Gulf, reefs and subtidal oolite shoals blend together to form a shelf-margin complex. Reefs predominate in front of islands, and oolite shoals predominate in the tidal channels between islands. Eolian dunes on many of the islands are composed of Recent oolite with a marine origin. The dunes are an interesting example of a topset facies in close association with shelf-margin sediments.

The subtidal shelf interior. The shelf-margin environment marks the transition from relatively deep water to the relatively shallow water of the carbonate shelf. Sedimentation on the shelf interior varies greatly as a function of water circulation over the shelf. The very fact that the shelf is generally shallow impedes circulation to some extent. Furthermore, the shallow shelf-interior water may be altered considerably by either freshwater input or by net evaporation.

Four examples from the Recent epoch will help us understand the variability of subtidal sedimentation of the shelf interior. In southern British Honduras, normal Caribbean seawater passes over the barrier platform and produces a net flow to the south in the main channel lagoon. Reduced salinities are restricted to coastal areas; the majority of the main channel lagoon and all of the barrier platform are left with normal salinity environments. In south Florida, the existence of the Florida keys restricts circulation both in the inner reef tract and in Florida Bay. Furthermore, net evaporation alternates with the input of fresh water from the Everglades and so causes large fluctuations in Florida Bay salinity. In the Bahamas, shelf-interior circulation is somewhat impeded by both the sheer size of the platform and the "shadow effect" of Andros Island. Salinities that are 20% or 30% above normal marine are common here. The shallow lagoons along the southwest coast of the Persian Gulf are the site of gross net evaporation. Salinities two or three times normal marine are common in the subtidal environments. Let us examine each of these examples in more detail.

Figure 12.12 presents a map and cross section of a portion of the southern British Honduras shelf interior. The shelf interior of British Honduras exhibits a large amount of topographic relief. In particular, the deep water of the main channel lagoon is a feature that will be lacking in examples of more restricted shelf interiors.

Undoubtedly, the existence of the main channel lagoon plays an important role in maintaining normal marine salinities throughout the southern British Honduras shelf interior. This lagoon serves as a conduit for net southward flow of water brought onto the shelf during flood tide. Furthermore, net southward flow in the main channel lagoon serves to hold fresh water discharge from the mainland close to the shore. Only in the extreme southern portions of the British Honduras reef tracts do freshwater wedges begin to impinge on the reef environments. Finally, the main channel lagoon serves as a settling basin to separate carbonate sedimentation of the barrier platform from clastic sedimentation of the coastal areas. Without this settling basin, clastic sediments would undoubtedly exert a harmful influence on barrier-platform carbonate sedimentation.

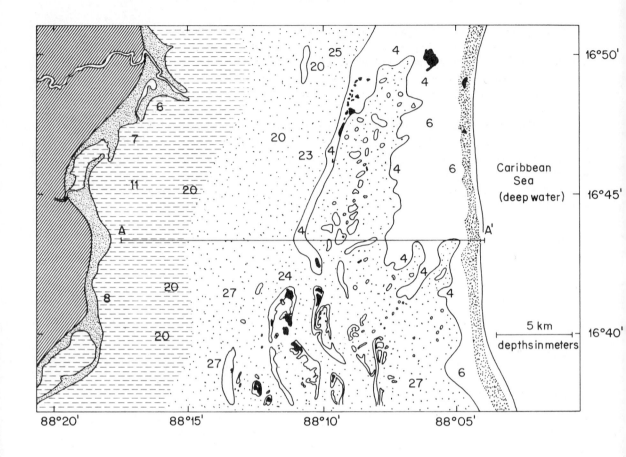

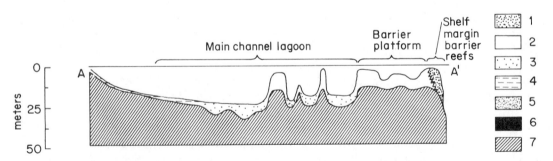

Figure 12.12 Physiography and facies of central British Honduras. Sediment types are as follows: (1) shelf-margin coral reefs and clean reef-derived sands; (2) shelf-interior, *in situ,* skeletal sands; (3) shelf-interior, deep-water, lime muds; (4) terrigenous mud; (5) nearshore to subaerial coastal clastic sediments; (6) cays, predominantly mangrove swamp; (7) older rock.

   Distribution of sediment types on the southern British Honduras shelf interior is related to bathymetry, proximity to carbonate shoals, and proximity to the mainland coast. The barrier platform sediments are predominantly clean skeletal sands composed in large part of the green algae *Halimeda,* molluscs, porcellaneous foraminifera, hyaline foraminifera, and lesser amounts of coral and coralline algae. This sediment is apparently forming *in situ* on the barrier platform and contains only minimal contributions of transported skeletal detritus from the barrier-reef shelf-margin environment. Significant accumulation of skeletal detritus from the barrier reef seems to be limited to the reef flat and sand apron (discussed earlier).

   Sediments of the main channel lagoon are predominantly muds: carbonate mud near the barrier platform, sandy clay mud near the mainland coast, and a spectrum of carbonate-clay mud in between. The main channel lagoon has its own distinctive fauna, consisting primarily of thin-shelled molluscs and hyaline benthonic foraminifera. By and large, the water depth precludes a benthonic flora.

Figure 12.13 Generalized facies of south Florida. Sediment types are as follows:
   (1) shelf-margin coral reefs and clean reef-derived sands; (2) shelf-interior muddy skeletal sands; (3) Pleistocene key rock; (4) grass-covered, carbonate mudbanks, awash at low tide; (5) thin veneer of skeletal sand; (6) intertidal-supratidal deposits, commonly mangrove swamp.

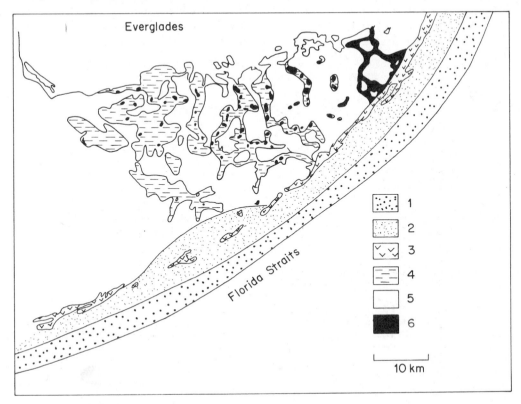

The bulk of the lime mud is apparently generated by skeletal comminution in the barrier-reef and barrier-platform environments and is transported westward to its final resting place in the deeper water settling basin that is the main channel lagoon (Matthews, 1966).

Platform-interior sediments of south Florida offer an interesting contrast to those of southern British Honduras (see Figure 12.13). Physiographic differences between the two areas undoubtedly sets the stage for the sedimentological differences. To begin with, the south Florida shelf possesses no main channel lagoon, nor does it have any built-in mechanism to control the net flow of water across the shelf. Replenishment of the very shallow water of the shelf interior under normal circumstances must be accomplished by oscillatory tidal flow from the shelf margin. The sheer size of the shelf may place serious limitations on this process.

Communication between the shelf interior and the Florida Strait is further inhibited by the existence of the Florida keys. For instance, faunas of the inner reef tract are found in Florida Bay only at the various passes through the Florida keys (Figure 12.13 and 12.14). If the keys were not there, inner reef tract faunas and Florida Bay faunas would seek their own equilibrium positions, which would undoubtedly be different from the positions now governed by the existence of the keys.

It is convenient to discuss southern Florida sedimentation in terms of three major environments. (1) The outer reef tract is the shelf-margin environment just discussed. (2) The inner reef tract is that belt of sediment between the shelf-

Figure 12.14 Generalized variations in sediment grain size and constituent composition from the south Florida reef tracts into Florida Bay. (After Ginsburg, 1956.)

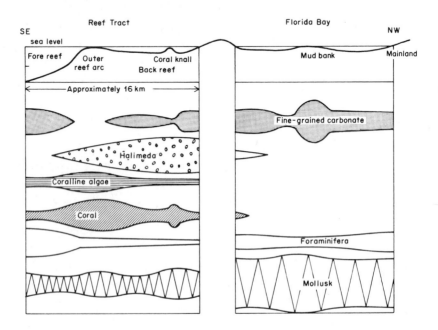

margin environment and the Florida keys. (3) Florida Bay is that triangular area of shallow-water sedimentation between the Florida keys on the southeast, the Gulf of Mexico on the west, and the Florida Everglades on the north.

The south Florida inner reef tract is more or less analogous to the barrier-platform environment of southern British Honduras. The major distinction is the tendency of lime mud to build up in the shallow waters of the inner reef tract; similar environments in southern British Honduras are predominantly clean skeletal sands. The accumulation of mud within the inner reef tract results simply from the fact that there is no place for the mud to go. In the absence of a well-developed current system on the shelf, mud transported to this area by flood tides simply accumulates here.

Some consideration should also be given to the role of marine grasses (*Thallasia*) in the stabilization of mud in shallow water. Scientists have observed that sediment samples from patches of *Thallasia* generally have more mud than samples from sediment-floored areas nearby. It can be argued that the blades of *Thallasia* tend to create a less agitated environment near the sediment-water interface and thereby encourage the accumulation of fine-grained sediments on the *Thallasia*. Note, however, that the inevitable question arises: Is the mud there because the *Thallasia* came first, or is the *Thallasia* there because it prefers to grow on a muddy bottom?

An alternate proposal for the sedimentology of carbonate-mud accumulation simply considers pelleted mud to be a soft sand particle (Ginsburg, 1969). The stuff is as gooey as we could want a mud to be, yet the hydrologic behavior of the individual soft pellets is that of a sand-sized particle. In this fashion, we can argue that a shallow-water mudbank accumulating under agitated conditions is simply an accumulation of coarse-grained "soft sand."

Crossing over the Florida keys and entering Florida Bay, we come into a new world (Figures 12.13 and 12.14). This region is a restricted shallow-water shelf interior with a large potential source of fresh water along its northern boundary. The combination of a restricted shallow-water shelf interior and a potential fresh-water input makes Florida Bay a highly variable environment. As indicated in Figure 12.15, salinity within the bay may vary from hypersaline to brackish. Salinity variations within the bay render this environment hospitable only to certain molluscs, foraminifera, incrusting red algae and soft-bodied bryozoa. Water depths are generally 2 meters or less and an intricate pattern of *Thallasia* mudbanks rise essentially to sea level. Intervening areas between mudbanks are flooded by approximately 15 centimeters of mollusc-foraminifera skeletal sand.

The origin of carbonate mud in shelf-interior environments has been a subject of considerable controversy. On the Florida shelf, it has been rather conclusively demonstrated that a single genus of calcified green algae, *Penicillus,* can produce the vast majority of the carbonate mud observed in these environments (Stockman *et al., 1967*). By gathering data on the weight of aragonite needles in single specimens of *Penicillus,* the population density of *Penicillus,* and the turnover rate (life

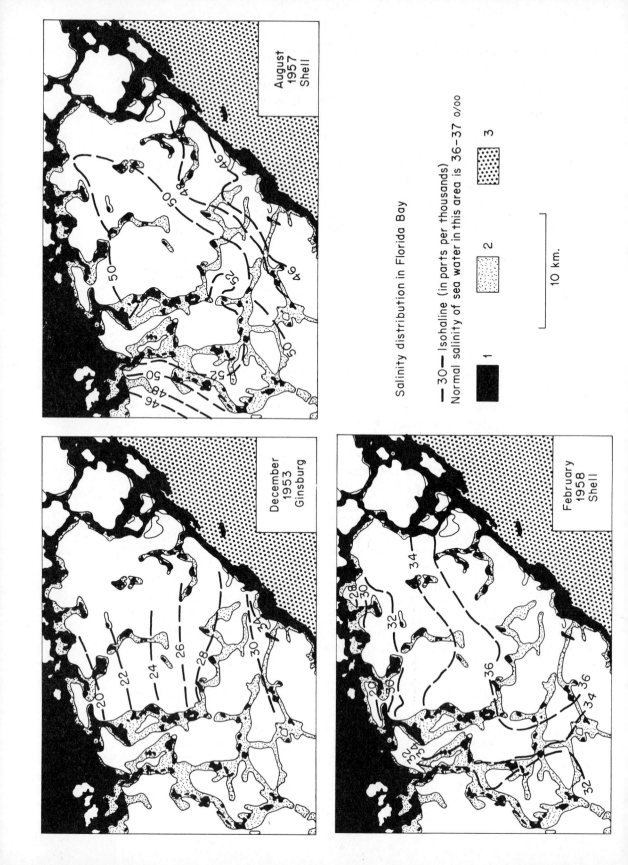

Salinity distribution in Florida Bay

—30— Isohaline (in parts per thousands)
Normal salinity of sea water in this area is 36-37 o/oo

1 2 3

10 km.

August 1957 Shell

December 1953 Ginsburg

February 1958 Shell

cycle) of *Penicillus,* we can estimate the rate of production of aragonite-needle carbonate mud. This rate, applied uniformly since the time when this area became flooded by the post-Wisconsin sea-level rise, is sufficient to account for the vast majority of mud accumulated in the shelf-interior environments of southern Florida.

In the Great Bahama Bank, we can examine shelf-interior sedimentation in a physiographic setting quite similar to south Florida but with two important differences: (1) Shelf-margin oolite shoals and Pleistocene outcrops (like Andros Island) provide greater restriction of the platform interior; and (2) there is no large source of fresh water, such as the Florida Everglades. Consequently, there is a tendency for the shelf-interior environment to have increased salinity.

Recent sediment thickness over the shelf interior is thin, averaging around 1 meter (Figure 12.16). The sediment of the skeletal-mud facies is similar to the muddy skeletal sands of the inner reef tract of south Florida. However, sediment types that predominate in the shelf interior are rather different from anything discussed thus far. In the pellet-mud facies, a large percentage of the sand fraction consists of lithified pelleted mud. These pellets are presumably similar in origin to the "soft sand" discussed earlier. Lithification resulted from some combination of increased salinity and a relatively long residence time at the sediment-water interface.

The grapestone facies is a relatively clean sand in which grapestone grains are predominantly sand-sized particles. Grapestone grains are aggregates of preexisting carbonate grains held together by chemical or biochemical carbonate cement. The original sedimentary particles are most commonly recrystallized oolite grains or indurated pellets. The cementation process by which they are united into grapestone grains is a combination of biochemical cementation by foraminifera or blue-green algae followed by physical-chemical recrystallization and precipitation of additional cementing material. The abundance of these sediment types in the shelf interior of the Bahamas can be attributed to some combination of increased salinity and long residence time at the sediment-water interface. Indeed, it has been proposed that the grapestone facies represents a surface of nondeposition on which no new grains are being produced; just old grains are being cemented together (Winland, 1971). Thus, the grapestone facies may represent a modern, shallow, subtidal environment of essentially zero sedimentation rate.

In the Persian Gulf, the extremely low relief profile of coastal sedimentary environments has resulted in substantially reducing the areal importance of subtidal shelf-interior sediments. The large area that was once shallow, subtidal

Figure 12.15 (Left) Salinity variations in Florida Bay. Sediment types are as follows: (1) Pleistocene rock and supratidal swamp areas; (2) shallow, subtidal, grass-covered mud banks; (3) subtidal platform-interior, Atlantic province. Open areas are Pleistocene rock surface overlain by less than 6 inches of mollusc foram sand. (After Ginsburg, 1964, with modification.)

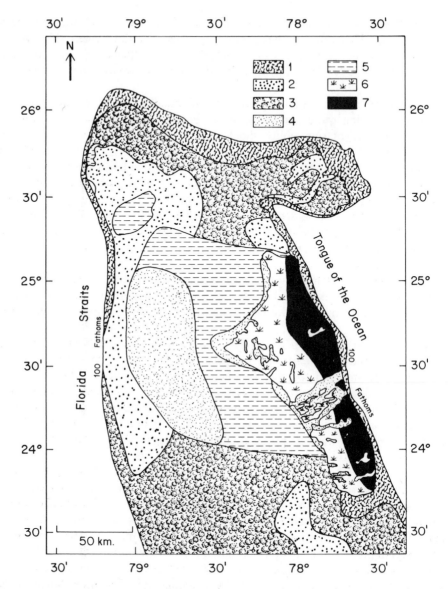

Figure 12.16 Generalized facies map of the Great Bahama Bank. Sediment types are as follows: (1) coralgal sand facies; (2) oolitic facies, mostly oolite tidal bars; (3) grapestone sand facies; (4) skeletal-mud facies; (5) pellet-mud facies; (6) supratidal carbonates; (7) lithified Pleistocene eolian sands. (After Imbrie and Purdy, 1962, with modification.)

shelf interior is now occupied by supratidal sediments of the sabkha environment. Subtidal shelf-interior sedimentation in this area is today restricted to relatively small, shallow lagoons. Salinity of these lagoons is typically two or three times normal marine values. In some lagoons, aragonite-needle lime mud and pelleted lime mud are predominant sediment types. In other lagoons, skeletal sands have become well lithified in the submarine environment, presumably by a process some-what akin to the formation of grapestone in the Bahamas. Here, the process appears to have gone to ultimate completion: the formation of laterally continuous, lithified sand layers.

Topographic control of Recent carbonate sedimentation. We have already noted a large-scale topographic control of carbonate sedimentation, the shelf-margin environment. Let us now look closer at some of the second-order effects of topography on carbonate-sediment accumulation.

In Biscayne Bay, Florida, scientists have demonstrated that preexisting topography on the underlying Pleistocene carbonate rock serves to modify patterns of tidal flow and thereby accounts for the geometric configuration of Recent carbonate-bank deposits (Wanless, 1970). Where Pleistocene topographic highs form islands, strong tidal jets exist between them. Elongate carbonate banks are thus generated perpendicular to the shelf margin. Elsewhere, the Pleistocene surface forms a continuous subtidal lip at the eastern margin of the bay. Here tidal velocities are increased over the shallow water of the lip and abruptly decreased in the deeper water immediately behind the lip. Sediment dumping associated with this decrease in velocity has resulted in the accumulation of a sediment bar running more or less parallel to the Pleistocene topographic prominence.

In southern British Honduras, the distribution of carbonate shoals within the main channel lagoon is controlled by preexisting topography. As indicated in Figure 12.12, there is a good correlation between carbonate-shoal areas in the lagoon and preexisting topography. Figures 12.17 and 12.18 show that much of this preexisting topography is directly related to underlying structural features that have been subsequently modified by subaerial geomorphologic processes.

It is interesting to note that preexisting topographic highs and carbonate-shoal sedimentation apparently become interrelated during times of sea-level fluctuation, such as exemplified by the Pleistocene and Recent epochs. During a period of high-stand sedimentation, the carbonate shoals generate more sediment than the surrounding deep-water environment. With subaerial exposure during a low stand, the carbonate-bank deposits of the topographic highs are permeable and thus allow rainfall to pass through them freely. Cavernous porosity may develop, but the carbonate-bank facies will nevertheless withstand the subaerial exposure and become a well-cemented cavernous limestone. With the next high stand of the sea, the hard cavernous limestones of the bank facies remain as topographic highs for rejuvenation of bank facies sedimentation (Purdy, 1972; Bloom, 1972).

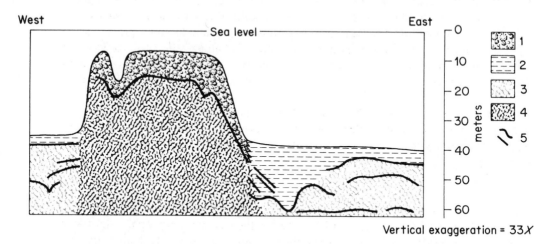

West   Sea level   East

0
10
20
30
40
50
60

meters

1
2
3
4
5

Vertical exaggeration = 33X

Figure 12.17 Reflection seismic profiles across a shoal area in main channel lagoon, southern British Honduras shelf. The upper profile records relatively shallow information, and the lower profile records deeper information in approximately the same area. Sediment types are as follows: (1) shelf-interior skeletal sands and patch reefs; (2) shelf-interior, deep-water, lime muds; (3) Pleistocene clastic sediments; (4) Pleistocene limestone; (5) prominent reflecting horizons. In the upper profile, note that the preexisting topography determines Recent distribution of sand facies and mud facies. Data in the lower profile suggest that the preexisting topography was in turn the result of previous differential sedimentation ultimately initiated on fault blocks. (After Purdy and Matthews, 1964, and Purdy, 1972, with modification.)

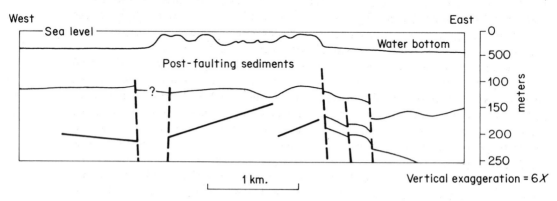

West   East

Sea level   Water bottom

Post-faulting sediments

?

0
500
100
150
200
250

meters

1 km.

Vertical exaggeration = 6X

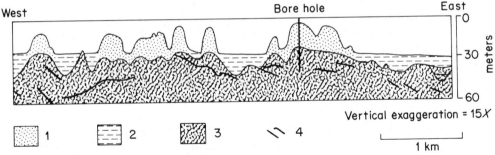

West   Bore hole   East

Vertical exaggeration = 15X

1   2   3   4   1 km

Figure 12.18 Additional example of topographic control of Recent carbonate sedimentation, main channel lagoon, southern British Honduras. Sediment types are as follows: (1) shelf-interior skeletal sands and patch reefs; (2) shelf-interior, deep-water, lime mud; (3) Pleistocene or older clastic sediments; (4) prominent reflecting horizons. In this case, carbonate shoals have developed on the topography related to a breached anticlinal fold. The borehole recovered soft, orange, terrestrial clay from approximately 40 meters. Other seismic data in the area confirm the existence of this broad anticlinal feature. (After Purdy and Matthews, 1964, and Purdy, 1972, with modification.)

Recent carbonate sedimentation rates. Estimations of Recent carbonate sedimentation rates can be made from two general sources: (1) from the study of living organisms in Recent environments or (2) from the thickness of Recent sediment overlying the Pleistocene surface, wherever the contact between Recent and Pleistocene can be well defined.

Studies of lime-mud production by the green algae *Penicillus* provide the most complete example of directly determining sedimentation rates from living organisms (Stockman *et al.*, 1967). *Penicillus* colonies stand 5 to 7.5 cm tall and are easily visible to the skin diver surveying the bottom of a carbonate environment. Measurement of sediment production rates for *Penicillus* consists of three steps. First, you establish the weight of aragonite sediment produced by a single specimen. Simply collect a large number of specimens and determine the average weight of aragonite per specimen. Next, you estimate the population density of *Penicillus* within the environment. At any one time, how many plants are growing per square meter of sea floor? Finally, you establish the life cycle of *Penicillus*. How long does it take an average specimen to grow to full size and die, thus releasing its aragonite to the sediment? Having these numbers, you can calculate the weight of aragonite sediment produced by *Penicillus* per square meter per year. Assumptions concerning porosity of the sediment allow translation of these weight figures into sediment volume figures. Resultant estimates of net lime-mud sedimentation rate range from .3 to 3.0 centimeters/1000 years in the southern Florida area.

A similar calculation should be possible for sediment production by corals. Calcification rates are fairly well known and indicate a potential for extremely rapid sediment accumulation (see Hoffmeister and Multer, 1964, for example).

However, the population density of corals and the life cycle of individual colonies are not known in sufficient detail to complete the calculation.

An alternative approach to carbonate sedimentation rates is to obtain a sample of buried sediment and determine its average age by carbon-14 analysis. For example, Goreau and Land (1972) report that numerous attempts to quarry into the face of forereef sediments on the north coast of Jamaica consistently yield ages of approximately 3000 years B. P. about 1 meter below the present sediment-water interface. Thus, it appears that the forereef slope is building outward at a rate of approximately 30 centimeters/1000 years.

Some boreholes through carbonate sediments have been dated radiometrically (see Thurber et al., 1965, for example). Such data has indicated net sedimentation rates of 2 to 3 meters/1000 years for borings on atolls.

Alternatively, the contact between Recent sediments and the underlying Pleistocene surface can be mapped by reflection seismic techniques or by probing through the soft sediment with a steel rod. These methods give the thickness of Recent sediment accumulation over the Pleistocene surface and the depth to the Pleistocene surface. This data can be combined with estimates of the post-Wisconsin sea-level rise to give an indication of the amount of sediment that has accumulated since the post-Wisconsin sea-level rise first flooded this area. Such estimates for Florida Bay and the inner reef tract indicate net sedimentation rates of 20 to 35 centimeters/1000 years (Stockman et al., 1967). Similar techniques in the Bahamas yield sedimentation rates from 45 centimeters/1000 years for oolite shoals down to 10 centimeters/1000 years for subtidal grapestone and mud.

In southern British Honduras (Figures 12.12, 12.17, 12.18), reflection seismic data indicate that the Pleistocene surface is deeper in the southern part of the study area than it is in the northern part of the study area. In the northern portion of Ambergris Cay, a Pleistocene reef facies outcrops on the island. In southern British Honduras, this horizon is as much as 30 meters below sea level. Presumably the lowering of this surface reflects postdepositional tectonic downwarping into the Cayman Trench. If we assume that this Pleistocene topographic surface is 120,000 years old, as is most of the Pleistocene topography beneath Recent shallow-water carbonates around the world, an average rate of tectonic downwarping of 90 centimeters or less per 1000 years is indicated. Thus, the British Honduras reef tract may record the results of an experiment that nature has run for us concerning sediment accumulation capabilities of these Recent environments.

As sea level passed on to the Pleistocene topographic highs, sedimentation was initiated. Has sedimentation kept pace with the rising water? In central British Honduras (just north of the area pictured in Figure 12.12), approximately 10 to 13 meters of Recent sediment overlie the Pleistocene surface that is situated 13 meters below present sea level, indicating sedimentation rates of approximately 150 centimeters/1000 years. The shelf-margin facies and barrier-platform clean sands are of approximately equal thickness, even though these two facies do not have a common sediment source. Apparently both facies have sediment accumulation rates sufficient to maintain shallow-water depths during rising sea level.

In contrast, to the south of the area depicted in Figure 12.12, where the Pleistocene surface lies 25 to 30 meters below sea level, the shelf-margin facies is discontinuous, and the barrier-platform, clean-sand facies does not completely occupy the preexisting topographic platform. Only occasional shoals rise out of deep water. The shelf-margin facies is apparently accumulated at a sedimentation rate of about 275 centimeters/1000 years. At this sediment accumulation rate, no truly continuous barrier reef has been maintained. Barrier-platform, clean-sand sedimentation apparently has been unable to keep pace with rising water in this area.

Note that sedimentation rates for southern British Honduras are considerably greater than sedimentation rates for related environments in southern Florida and the Bahamas. This point will be of considerable interest to us in the formulation of general models.

A Basic Model for Self-Margin and Subtidal Shelf-Interior Sedimentation

With the preceding examples of Recent environments serving as our guide, let us make some generalizations about what we would expect to see if we encountered similar deposits in the stratigraphic record.

Areal distribution of contemporaneous sedimentary facies. Generalities are summarized in Figure 12.19. The foreslope of the shelf margin is usually the site of accumulated skeletal sand and carbonate mud. Because access is difficult, this environment in the Recent epoch is not well studied.

The shelf-margin environment is the site of coral reef growth. In addition, current-transported skeletal sands or oolitic sands can accumulate here. Modern coral reefs of the shelf-margin environment have a pronounced zonation of coral communities; we would expect Ancient reefs to also show ecological zonation. The reef-crest community, therefore, may form an environmental datum upon which we can reconstruct the environment and the history of sea-level fluctuation. For example, the modern *A. palmata* community occurs from reef crest to water depths of 5 meters. Thus, if we encounter a vertical section of *A. palmata* facies 20 meters thick, we would suspect that the reef-crest community grew upward with the rising sea level.

Sands of the shelf-margin environment should show evidence of current transportation (Imbrie and Buchanan, 1965; Ball, 1967). Skeletal sands may predominate around shelf-margin coral reefs. In the absence of sediment supply from coral reefs, oolitic sands may typify the shelf-margin environment.

A generalization concerning the sediments of the shelf interior is more difficult than a generalization concerning shelf-margin sedimentation. Indeed, the farther

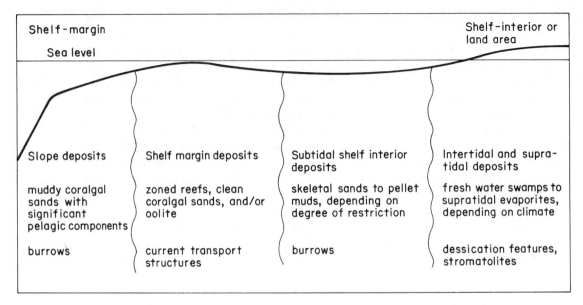

Shelf-margin			Shelf-interior or land area
Sea level			
Slope deposits	Shelf margin deposits	Subtidal shelf interior deposits	Intertidal and supratidal deposits
muddy coralgal sands with significant pelagic components	zoned reefs, clean coralgal sands, and/or oolite	skeletal sands to pellet muds, depending on degree of restriction	fresh water swamps to supratidal evaporites, depending on climate
burrows	current transport structures	burrows	dessication features, stromatolites

Figure 12.19 Generalized model for carbonate facies distribution from the shelf margin inward.

we get from the open-marine, sea-water conditions of the shelf margin, the more chances there are for one variable or another to predominate in the environment and change sediment type. In the shelf interior, circulation patterns, evaporation rates, and the presence or absence of freshwater input from nearby land masses become very important attributes of the paleogeography. Given open circulation, such as over the barrier platform of southern British Honduras, the shelf interior may be the site of widespread production of skeletal sands and carbonate mud. Sands may accumulate *in situ,* whereas muds may be transported by the thorough-going circulation pattern.

Given a somewhat more restricted circulation pattern, such as in the inner reef Tract of south Florida, both skeletal sands and muds may accumulate *in situ* in the shelf-interior, shallow-water environment.

The Bahamas apparently represent the next step in the restriction of the shelf interior. Water flow is limited, thus allowing the accumulation of carbonate mud in shallow-water environments. In addition, increased salinities restrict the fauna and therefore the skeletal constituents of the shelf-interior sediments. Lithification of mud pellets and the formation of grapestone probably reflect some combination of increased salinity and reduced sedimentation rate.

The concept of bottomset, foreset, and topset beds in carbonate sedimentation. In our discussion of coastal clastic environments, we found it convenient to use the top of the foreset (on a beach or a river-mouth bar) as our single, best environmental datum in the coastal clastic model.

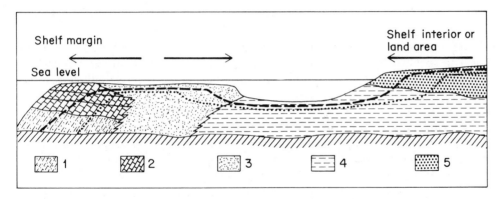

Figure 12.20 Schematic cross section indicating the various possibilities for progradation within carbonate environments. Sediment types are as follows: (1) forereef detritus; (2) shelf-margin reef or similar facies, with reef zonation indicated by solid lines; (3) shelf-interior skeletal sands; (4) shelf-interior subtidal muds; (5) intertidal-supratidal deposits, typically algal stromatolites, supratidal dolomites, and evaporites, or alternatively, coastal clastics. Dotted line and dashed line are time lines. Solid arrows indicate directions in which progradation may be expected.

In carbonate sedimentation, we shall continue to use the concept of foreset sedimentation, but with at least two important modifications. First, in any carbonate paleogeography, there may be numerous simultaneously active foresets: the shelf margin and the subtidal-supratidal transition, for example (see Figure 12.20). Secondly, the concept of systematic decrease in sedimentation rate from upper foreset to bottomset deposits seldom applies to carbonates because of *in situ* production of sediment within the local environment.

Vertical accumulation in carbonate sediments. In our discussion of coastal clastic sedimentation, we observed that virtually all of the Recent sediment accumulates within the foreset slope environment. Only very fine-grained clays achieve sufficiently wide distribution from the source area to be regarded as truly bottomset. *In situ* production, and therefore vertical accumulation of carbonate sediments, require that we reconsider this generality and the facies relationships that may be predicted by it.

In a very real sense, the Barrier Platform, carbonate-skeletal sands of southern British Honduras are bottomset deposits to the prograding coastal clastics of the mainland (see Figure 12.12). Note, however, that the barrier platform sands have such a high sedimentation rate that the application of the term "bottomset" is a purely academic exercise. Indeed, the barrier platform is itself a source of sediment supply westward to the main channel lagoon. Thus, the concept of a foreset to bottomset transition has equal application to sedimentation from both margins of the main channel lagoon.

Because of *in situ* production of sediment, it is quite possible for a carbonate basin to be filled by vertical accumulation. According to our former terminology, this process means that bottomset beds may be directly overlain by topset beds with no intervening foreset deposits.

Progradation in carbonate sedimentation. In coastal clastic sedimentation, the ability of a shoreline to prograde was tied directly to sediment supply. Once again, *in situ* production from carbonate sediment opens several new possibilities. In southern British Honduras, for example (see Figure 12.12), progradation is occurring in three different geographic positions, each having their own separate sediment supply. Coastal clastics are prograding eastward into the lagoon; the barrier platform is prograding westward into the lagoon; and the barrier reefs of the shelf margin are prograding eastward into the Caribbean. Simultaneous progradations from independent sediment sources must be expected to produce complicated stratigraphies.

Seaward progradation of the shelf-margin facies is fundamentally important in carbonate sedimentation. The establishment of a well-developed shelf-margin facies creates the other carbonate environments, just as the development of a clastic barrier island creates the bay environments. Continued maintenance and progradation of the bank-margin facies enlarges the shelf-interior environment. Indeed, magnetic data indicate that the carbonate platform of Florida was initiated on discrete volcanic cones separated from each other by as much as 100 kilometers. Maintenance and progradation of the shelf-margin facies eventually led to coalescence and development of a single Florida platform.

If we look at a more detailed level, we can expect that progradation of reef facies during times of constant sea level will result in sheet-like deposits with a vertical zonation of coral similar to the depth zonation of the forereef at any one instant in time (see Figure 12.20).

Furthermore, we can see that the rate of seaward progradation of the shelf-margin facies may be governed in large part by the preexisting topography of the foreshelf environment (Figure 12.21). Later, we shall study the relationships between progradation capabilities and sea-level fluctuations.

Progradation of supratidal deposits over platform-interior deposits can also be expected to produce laterally extensive sheet deposits in which the vertical stratigraphy mimics map-view relationships seen at any one instant in time. These deposits are the subject of the next chapter.

Continued sedimentation at constant sea level. In looking at various examples of Recent sedimentation, we have noted varying degrees of restriction on the shelf interior. In Recent sedimentation, these distinctions are primarily a function of local geometry, tectonic activity, and climate. For example, it is almost certainly no accident that the carbonate shelf exhibiting the best circulation (southern British Honduras) is also the carbonate shelf that is actively subsiding.

Clearly, continuing sedimentation will lead to progressively greater and greater restriction of water circulation. Because the Recent epoch provides only a brief glimpse into this process, we must now combine several of our Recent models into a temporal succession of sedimentary facies that we would expect to find if a carbonate shelf continued to accumulate sediment over an extended period of constant sea level.

Shelf-interior sedimentation of the British Honduras barrier platform and main channel lagoon, the inner reef tract of south Florida, and the interior of the Great Bahama Bank offer us glimpses of three stages that might be superimposed one upon the other.

With open circulation, sediments are predominantly skeletal and there is good separation of *in situ* skeletal sands and transported carbonate muds (British Honduras). With increasing restriction, as sediments begin to fill the space between the shelf and the level of the sea, there will likely come a time when water chemistry is still satisfactory for the production of skeletal sands but currents are sufficiently weak that lime mud accumulates with the sand (south Florida inner reef tract). As restriction continues, water chemistry and nutrient supply become unfavorable for normal marine benthonic organisms. Thus, fauna is limited, and pelleted lime-mud grapestone predominate (Great Bahama Bank). Finally, we might expect supratidal or eolian deposits to prograde out over much of the shelf interior.

Note that the vertical profile generated in this discussion is not duplicated in any single example of Recent carbonate sedimentation. In all cases, the model has either not run long enough, or the interaction between preexisting topography and the Recent sea-level rise has set the model running in an already constricted configuration.

Response of carbonate sedimentation to conditions of emergence. As the sea level recedes, supratidal deposits, subtidal platform-interior deposits, and shallow-water bank-margin environments all come under stress. Supratidal flats may regress over former subtidal shelf-interior sediments as the water retreats. At the shelf margin, however, this lateral shift of facies must come to an end. With the emergence of the shelf interior, shelf-margin oolite production will probably cease because there is no longer extensive tidal flow onto and off of the shelf interior.

Shelf-margin coral reefs are also faced with serious problems as the sea level falls. There will be a tendency for shallow-water coral faunas to migrate outward over deep-water coral faunas. But note that such regression demands a firm substrate in the forereef area. So long as the coral reef is producing sufficient fore-reef detritus, regression can occur (see Figure 12.21). Otherwise, the reef will tend to build outward as an overhanging structure, supported only by the strength of its bonding to the older reef rock. Such structures will ultimately collapse and become large blocks of forereef detritus themselves. This phenomenon can result in a near vertical wall just seaward of the high-stand bank-margin environment.

Figure 12.21 Geometries important in the seaward progradation of coral reefs. The upper diagram summarizes the three angles important to the maintenance of a coral reef during time of emergence. The reef has developed on some preexisting bathymetric surface that slopes seaward (angle *A*). If the living reef is to prograde seaward and remain unharmed by emergence, the reef crest equilibrium profile (angle *B*) must be maintained. Maintenance of angle *B* during seaward progradation will require volumes of forereef sediment determined by preexisting forereef bathymetry. (After Matthews, 1969.)

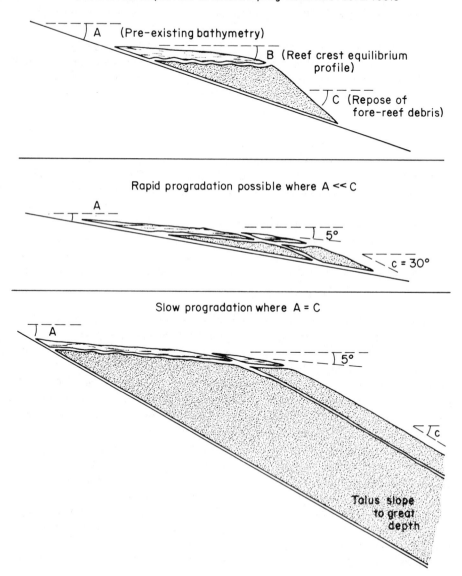

Vertical walls are an inhospitable place for reef-crest coral communities. Thus, only biogenic shelf-margin sedimentation has any chance of continuing during conditions of significant emergence. This sedimentation can take place only if there is adequate low-lying forereef topography or sufficient production of forereef detritus. The cessation of carbonate sedimentation by conditions of emergence is perhaps one of the most fascinating contrasts between clastic and carbonate sedimentation. No matter whether the sea level rises or falls, clastic sediments continue to come down the rivers and must be deposited somewhere. In contrast, when the sea level recedes from carbonate shelves, sedimentation ceases.

Response of carbonate sedimentation to conditions of submergence. In the discussion of coastal clastic sedimentation, we observed that rising water may tend to trap clastic sediment supply at the heads of estuaries, therefore forcing coastal areas into a condition of nondeposition. In carbonate sedimentation, *in situ* production of sediment relieves us from this consideration. Rising water will simply create more living space between the bottom and surface of the water and will tend to improve circulation of normal seawater to the shelf interior. At the same time, any problems created by clastic influx to the carbonate environment will be alleviated as clastic sediments are trapped in the estuaries and alluvial environments. Whereas rising water has a strong tendency to disrupt clastic sedimentation, it may actually improve conditions in carbonate environments (Figure 12.22).

Given new living space and improved circulation, each carbonate facies may tend to accumulate upward at a rate equal to or less than its own upward growth capability. If the rate of sea-level rise is sufficiently slow, the sediment-water interface will maintain its relationship to sea level and the character of the sediment being deposited will not change. If the sea-level rise is too fast, the sediment water interface will be left behind in deeper and deeper water and the carbonate facies will change accordingly.

For example, life on a shelf-margin coral reef will not change simply because sea level is rising at a rate of 1 meter/1000 years. The day-to-day processes of calcification and biological erosion go on as usual. Yet, 1000 years later, the net result of continuing reef sedimentation during a time of sea-level rise will be the accumulation of an additional meter of *in situ* reef framework above the original substrate. Indeed, if the production of forereef detritus has been sufficient, the reef may even prograde seaward during a time of rising water. On the other hand, if the sea-level rise exceeds the maximum upward-growth capability of the coral reef, the reef-crest community finds itself struggling to survive in deeper and deeper water. Eventually it will give way to the deeper-water communities.

Shelf-margin oolite sand belts are likewise controlled largely by the water surface. If water depth increases slowly enough, more and more oolite accumulates. The oolite shoals and channels near the water surface maintain their ever-changing patterns of migration and sedimentation; yet each time a channel migrates, a little more oolite is left below in its final resting place.

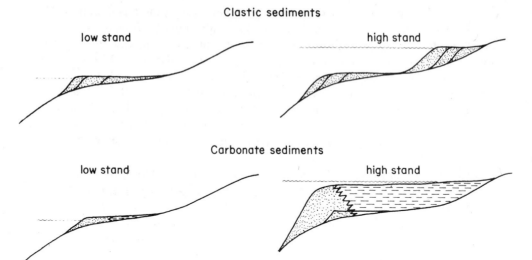

Figure 12.22 Comparison of clastic sedimentation with carbonate sedimentation during an event of submergence. Rising water tends to trap clastic sediments on the alluvial plain and in the newly drowned river-valley estuaries. Thus, sedimentation is disrupted by conditions of submergence and will be re-established at the new shore line during the new high stand. In contrast, carbonate sedimentation is invigorated by slowly rising water. First, rising water provides living space above the reef-crest community and thereby allows for vertical accumulation of reef-crest facies. Secondly, rising water tends to freshen marine circulation to the shelf interior, thereby increasing sedimentation rates. As a net result of this contrast between clastic sedimentation and carbonate sedimentation, Recent coastal clastic sediments are commonly found well back onto the modern continental shelf, whereas Recent carbonate sediments are typified by a shallow-water shelf-margin adjacent to deep water. Similar contrast would be expected in the stratigraphic record.

We have also seen that facies of the shelf interior have their own sedimentation rate. Note further, however, that the sedimentation rate is greatest in environments having the best water circulation over the shelf. Skeletal sands of the British Honduras barrier platform, for example, have accumulated at the rate of 150 centimeters/1000 years, whereas the pelleted muds of the more restricted Bahama shelf interior have accumulated at the rate of only 10 centimeters/1000 years.

Shelf interiors, therefore, present an additional complication during conditions of rising water. First and foremost, facies of the shelf interior do have a sedimentation rate. Furthermore, freshening of shelf-interior water-circulation patterns may produce changes in the sediment-producing benthonic communities, thus changing the facies and increasing the sedimentation rate in that geographic position. The freshening of shelf-interior circulation is undoubtedly dependent upon complicated interaction between rising sea level and responses of the shelf-margin facies.

Taking a closer look at the shelf-margin and shelf-interior environments, we must once again observe an important consequence of *in situ* production of carbonate sediment. Specifically, adjacent sedimentary facies may have their own self-contained sediment supply. For example, the sedimentation rate of the skeletal sand apron of British Honduras barrier reefs (Figure 12.5) is totally unrelated to the sedimentation rate of barrier-platform, *in situ,* skeletal sands. Similarly, each of the hard-bottom communities, sand-bottom communities, and mud-bottom communities within the shelf-interior possesses its own potential sedimentation rate.

During times of constant sea level or slightly rising sea level, these distinctions may be minimized. Sedimentation tends to smooth out the low spots and a mosaic pattern of coalescing facies typically predominates. However, with more rapid sea-level rise, one facies may keep pace with rising water while nearby facies do not. Topographic relief ensues. Inasmuch as light intensity is an important factor in carbonate-sediment production rates, the situation tends to become even more accentuated. Facies that maintain themselves at or near sea level will continue to receive sufficient light for rapid carbonate fixation. Facies that do not accumulate sufficiently fast to keep up with rising water will generally receive less light as water deepens and will therefore accumulate even more slowly.

The barrier-platform of southern British Honduras appears to exemplify this principle. Where the Pleistocene preexisting topography lies within 15 meters of the present sea level, the barrier platform is a mosaic of coalescing small patch reefs, clean sand bottom, and *Thallasia*-covered sand bottom. To the south, the same geomorphic features exist within the Pleistocene preexisting topography, but here they are as much as 25 meters below the present sea level. The pattern of coalescing facies noted to the north has broken down into isolated patch reefs standing high above the thin veneer of sediment in surrounding areas. The system is now locked in. The shoal areas receive sufficient light, so they can continue to calcify at rapid rates and thereby continue to grow upward toward sea level. The intervening areas have become too deep to support significant benthonic plant life; therefore, they continue to accumulate sediment at a slower rate than the shoal areas. Note that a long period of constant sea level will be required for this situation to rectify itself. The shoal areas can build upward only to sea level. Once they reach sea level, upward accumulation ceases. Then the intervening deep-water areas can begin to fill up, thus lessening the topographic distinction between shoals and surrounding areas.

An Ancient Example: Pleistocene Coral Reefs of Barbados, the West Indies

Perhaps your immediate reaction is to question the use of the Pleistocene as an "Ancient" example. Geologically speaking, the Pleistocene happened only yesterday. Most of the species present in the Pleistocene sediments of Barbados are present in modern carbonate environments of the Caribbean.

Yet, in a very real sense, the Pleistocene is as much a part of the geologic record as any Paleozoic sequence. Even though Pleistocene sediments are very

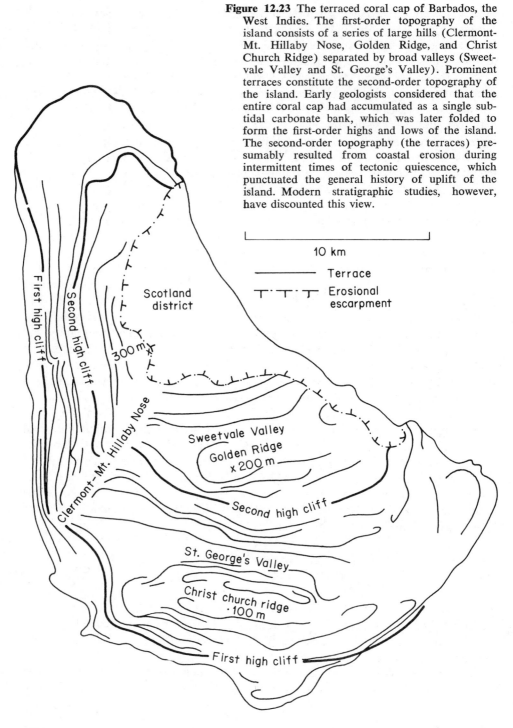

Figure 12.23 The terraced coral cap of Barbados, the West Indies. The first-order topography of the island consists of a series of large hills (Clermont-Mt. Hillaby Nose, Golden Ridge, and Christ Church Ridge) separated by broad valleys (Sweetvale Valley and St. George's Valley). Prominent terraces constitute the second-order topography of the island. Early geologists considered that the entire coral cap had accumulated as a single sub-tidal carbonate bank, which was later folded to form the first-order highs and lows of the island. The second-order topography (the terraces) presumably resulted from coastal erosion during intermittent times of tectonic quiescence, which punctuated the general history of uplift of the island. Modern stratigraphic studies, however, have discounted this view.

10 km

——— Terrace

T·T·T Erosional escarpment

First high cliff

Second high cliff

300 m

Scotland district

Clermont-Mt. Hillaby Nose

Sweetvale Valley

Golden Ridge
x 200 m

Second high cliff

St. George's Valley

Christ church ridge
·100 m

First high cliff

young, no one was there to see them deposited. By and large, in our approach to understanding these sediments, we must use the same sedimentological and stratigraphic inferences as we would apply to any Ancient sequence.

Regional setting and earlier stratigraphic work. Barbados is the easternmost island of the lesser Antilles. Unlike its volcanic neighbors to the west, Barbados is composed entirely of sedimentary rocks. Oligocene and Miocene pelagic sediments now stand as much as 300 meters above sea level, attesting to the general tectonic uplift of the sea. Over approximately 85% of the island, the folded and faulted Tertiary strata are overlain by shallow-water Pleistocene carbonate deposits that average about 80 meters in thickness (Figure 12.23).

The surface of these Pleistocene shallow-water carbonates is distinctly terraced; the origin of these terraces has been the subject of some controversy since the 1890's. From the 1930's until fairly recent times, the views of Trechmann (1933, 1937) prevailed. He claimed that the entire terraced coral cap of Barbados had been deposited as a single, large, shallow, subtidal bank at the time when the rising mass of folded and faulted Tertiary strata began to approach sea level. Subsequent tectonic uplift of the bank produced faults and broad gentle folds that are the first-order topography of the island today. As these structures rose out of the water, there was tectonic quiescence from time to time. During quiescent times, pronounced wave-cut terraces were gouged out of the carbonate rocks by the erosional action of the surf.

Thus, to summarize Trechmann's views, the Pleistocene coral cap of Barbados is isochronous. The terraces are of erosional origin, wave-cut during the intermittent uplifts that followed subtidal deposition of the coral cap.

Sedimentological observations. By the time Mesolella *et al.* (1969, 1970) renewed geological investigation of the Pleistocene carbonates of Barbados, considerable progress had been made in the study of Recent carbonate sedimentation. Particularly important were the developing ideas concerning coral zonation from reef crest into forereef slope. Before the days of scuba diving, the geologist's knowledge of coral reefs was confined primarily to a faunal list. On this basis, for example, Trechmann and workers before him recognized that the Pleistocene of Barbados did contain corals and was therefore a shallow-water deposit. With the advent of scuba diving, the coral-reef zonation from *Acropora palmata* reef crests to *Acropora cervicornis* foreslope to deep-water head zone has become well-documented in several areas.

Armed with this new knowledge of the Recent epoch, these workers have recognized that familiar corals occupy familiar positions (Figure 12.24). In particular, massive accumulations of *Acropora palmata* commonly occur at the forward edge of the terraces. On the foreslopes of terraces, *A. palmata* gives way downward to *A. cervicornis,* which in turn is replaced by a mixed coral assemblage predominated by large heads such as *Montastrea, Siderastrea,* and *Diploria sp.*

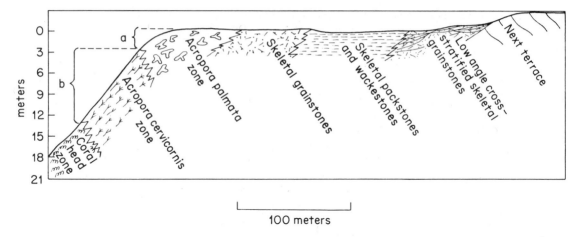

Figure 12.24 Sediment types encountered in a traverse through a typical Barbados terrace. Note that the facies contacts between *A. palmata* and *A. cervicornis* zones record two types of information. First, the depth *A* and (*A* + *B*) record the paleoecology of these coral zones. Secondly, the three-dimensional relationship between *A. palmata* and *A. cervicornis* should indicate whether reef growth occurred under conditions of submergence, constant sea level, or emergence. [After K. J. Mesolella, "Zonation of Uplifted Pleistocene Coral Reefs on Barbados West Indies," *Science*, **156**, 638–640, Table 1 (5 May 1967), with modification.] Copyright 1967 by the American Association for the Advancement of Science.

From the front edge back onto the terrace, *in situ Acropora palmata* usually passes into a rubble zone of transported *A. palmata* and skeletal sand. This sediment in turn gives way to clean skeletal sand with occasional coral heads or branching colonies. Further back on the terrace, there is skeletal sand with a lime-mud matrix. Finally, near the base of the next higher terrace, low-angle, cross-stratified, clean, skeletal sands are predominant.

On modern Caribbean reefs, *Acropora palmata* thrives in water depths ranging from zero to 6 meters. At some localities in the Pleistocene relicts of Barbados, *Acropora palmata* facies attain thicknesses of 30 meters. We must ask whether the ecological range of *Acropora palmata* has changed from Pleistocene to Recent, or whether this unusual thickness of *A. palmata* records the dynamic interaction between upward reef growth and conditions of submergence.

Data relevant to this question can be obtained by studying the contact between *A. palmata* and *A. cervicornis* zones. When we measure the position of this contact on the foreslope surface, relative to the general level of the terrace (see Figure 12.24), we determine a lower limit to the Pleistocene *A. palmata* facies. This limit agrees with similar data for Recent Caribbean reefs (Table 12.1). However, when we examine this contact between *A. palmata* and *A. cervicornis* in the third-dimension exposure provided by road cuts and quarries, the contact between the two zones typically has a strong vertical component.

Table 12.1. DEPTH RANGE OF CORAL ZONE BELOW REEF CREST, BARBADOS TERRACE DATA
(n.e., not exposed; n.d., not developed)

Section	A. palmata *zone* (meters)	Buttress zone (meters)	A. cervicornis *zone* (meters)	Coral-head zone (meters)
BD	0–6.8	n.d.	6.8–17.8 [1]	n.e.
CZ	0–4.0	n.d.	4.0–12.0 [1]	n.e.
BT	0–6.8	n.d.	6.8–18.2	18.2–27.7 [1]
BS	0–2.8	n.d.	2.8–10.2 [1]	n.e.
BE	0–3.7	n.d.	3.7–13.8	13.8–15.4 [1]
AEO	0–4.0	4.0–9.5	9.5 [2]–17.8 [1]	n.e.
DM	0–3.1	n.d.	3.1–13.5	13.5–20.3 [1]
BF	0–5.5	n.d.	5.5–13.5 [1]	n.e.
BY	0–4.0	n.d.	4.0–13.2	13.2–21.2 [1]
EH	0–9.2	n.d.	9.2–26.5	26.5–33.2 [1]
EW	0–5.2	n.d.	5.2–23.7 [1]	n.e.
AFM	0–6.2	n.d.	6.2–16.9	absent
D	0–4.6	4.6–12.3 [1]	n.e.	n.e.
IF	0–4.9	n.d.	4.9–15.1	15.1–23.7 [1]
AEU	0–3.7	n.d.	3.7–7.7 [1]	n.e.
E	0–2.2	n.d.	2.2–7.1	7.1–20.3 [1]
QU	0–4.6	4.6–10.2 [1]	n.e.	n.e.
HG	0–5.5	n.d.	5.5–10.2 [1]	n.e.
BV	0–3.7	n.d.	3.7–10.8	10.8–12.6 [1]
Average	*0–4.8*	*4.4–9.5*	*4.8–15.0*	*14.8*
Recent Jamaican reefs (Goreau) 1959	*0.5–6.0*	*1–10*	*7.0–15.0*	*15.0*

*(1) Minimum figure; base not exposed. (2) Lowered due to presence of buttress zone.
After Mesolella, 1967.*

Sedimentological interpretations. The data suggest that each Barbados terrace represents a discrete time of coral-reef development around an emerging island. Furthermore, observations concerning thick sections of *A. palmata* reef-crest community as well as the contact between the *A. palmata* community and the *A. cervicornis* foreslope community suggest that many of these reef-tract terraces were deposited under conditions of *submergence*. At first glance, this fact is rather surprising in view of the island's general history of *emergence* throughout the Pleistocene.

The key to reconciling this apparent contradiction comes from radiometric dating of the lower Barbados terraces by the thorium-230 growth method (Mesolella *et al.,* 1969). In particular, a 125,000-year-old terrace is well-known from widely separated parts of the world. It is commonly encountered 2 to 6 meters above the present sea level. On Barbados, this terrace is encountered at elevations ranging from 20 to 60 meters above present sea level.

Thus, the elevation of this terrace on Barbados not only confirms the general history of tectonic uplift of the island but also provides a basis for estimating the rate of tectonic uplift. On the average, Barbados is undergoing tectonic emergence at the rate of 30 centimeters/1000 years.

Immediately, we realize that 30 centimeters/1000 years is an extremely slow rate of *emergence* when compared to rates of *submergence* associated with Pleistocene sea-level fluctuations. For example, the post-Wisconsin transgression involved eustatic sea-level rise at the rate of 8 meters/1000 years (Figure 4.17).

Such considerations suggest that each Barbados coral-reef terrace records, not a tectonic event in the history of uplift of the island, but rather a eustatic high stand of the sea during the Pleistocene. Gradual tectonic uplift of the island has simply raised each terrace up out of the realm of the next eustatic high stand.

Thus, an understanding of Recent carbonate sedimentation and the dynamics of Pleistocene sea-level fluctuations has led to a major reevaluation of terraced coral caps. Prior to this work, for example, scientists interested in the history of Pleistocene sea-level fluctuations stayed away from tectonically active areas. How can we possibly recognize a *eustatic* terrace among all of those *tectonic* terraces?

The Barbados data cast serious doubt on the hypothesis of a strictly tectonic origin for terraced coral caps. Instead, it appears that tectonism is simply working for us as a natural strip-chart recorder. As the sea-level data are recorded by the formation of a coral-reef terrace, tectonic uplift moves that information along to where it can be preserved for us to read; it does not allow additional eustatic data to be superimposed in an unintelligible fashion. Indeed, the eustatic data contained in Pleistocene coral caps of tectonically emergent islands promise to be the foundation in formulating a proper understanding of Pleistocene history (Mesolella *et al.,* 1969; Veeh and Chappell, 1970), a point we shall return to in a later chapter.

Citations and Selected References

AKIN, R. H. JR., and R. W. GRAVES, JR. 1969. Reynolds oolite of southern Arkansas. Bull. Amer. Assoc. Petrol. Geol. 53: 1909–1922.
Subsurface study of the transition from shelf margin to subtidal shelf interior in the economically important Smackover formation Jurassic of the Gulf Coast.

BALL, M. M. 1967. Carbonate sand bodies of Florida and the Bahamas. J. Sed. Petrology 37: 556–591.
Dynamics and geometry of carbonate sand accumulation.

BATHURST, R. G. C. 1971. Carbonate sediment and their diagenesis. Development in Sedimentology, 12. Elsevier, Amsterdam. 620 p.
A thoughtful synthesis.

BLOOM, A. L. 1972. Geomorphology of reef complexes, *In* L. F. Laporte (ed.), Reef complexes in time and space: their physical, chemical, and biological parameters. Soc. Econ. Paleont. Mineral. Spec. Pub. No. 19.

BUCHANAN, H. 1970. Environmental stratigraphy of Holocene carbonate sediments near Frazer's Hog Cay, British West Indies. Ph.D. dissertation, Columbia University. 257 p.

CANN, R. S. 1962. Recent calcium carbonate facies of the north-central Campeche Bank, Yucatan, Mexico. Ph.D. dissertation, Columbia University. 225 p.

CLOUD, P. E. 1962. Environment of calcium carbonate deposition west of Andros Island, Bahamas. U. S. Geol. Survey Prof. Paper 350. 138 p.

DODD, J. R., and C. E. SIEMERS. 1971. Effect of Late Pleistocene karst topography on Holocene sedimentation and biota, lower Florida keys. Bull. Geol. Soc. Amer. 82: 211–218.

EVAMY, B. D., and D. J. SHEARMAN. 1965. The development of overgrowths from Echinoderm fragments. Sedimentology 5: 211–233.
Petrography of cementation in French Jurassic limestones.

FISHER, W. L., and P. U. RODDA. 1969. Edwards Formation (Lower Cretaceous), Texas: dolomitization in a carbonate platform system. Bull. Amer. Assoc. Petrol. Geol. 53: 55–72.
Paleogeography and sedimentology of an economically important carbonate unit.

FRIEDMAN, G. M. (ed.). 1969. Depositional environments in carbonate rocks. Soc. Econ. Paleont. Mineral. Spec. Pub. 14. 209 p.
Symposium volume.

GEBELEIN, C. D. 1972. Sedimentology and ecology of a recent carbonate facies mosaic, Cape Sable, Florida. Ph.D. dissertation, Brown University.

GINSBURG, R. N. 1956. Grain size of Florida carbonate sediments. Bull. Amer. Assoc. Petrol. Geol. 40: 2384–2427.
Grain size and constituent-particle composition of sediments from the Florida keys eastward to the shelf margin.

————. 1964. South Florida carbonate sediments. Guidebook for field trip no. 1. Geological Society of America annual convention, Miami Beach, Fla. 72 p.

————. 1969. Presidential address before the Society of Economic Paleontologists and and Mineralogists.

GOREAU, T. F. 1959. The ecology of Jamaican coral reefs. Part I: Species composition and zonation. Ecology 40: 67–90.

————, and L. LAND. 1972. Forereef slope ecology and depositional processes in Jamaica, *In* L. F. Laporte (ed.), Reef complexes in time and space: their physical, chemical, and biological parameters. Soc. Econ. Paleont. Mineral. Spec. Pub. No. 19.

HOFFMEISTER, J. E., and H. G. MULTER. 1964. Growth-rate estimates of a Pleistocene coral reef of Florida. Bull. Geol. Soc. Amer. 75: 353–358.

————, J. W. STOCKMAN, and H. G. MULTER. 1967. Miami limestone of Florida and its recent Bahamian counterpart. Bull. Geol. Soc. Amer. 78: 175–190.

HRISKEVICH, M. E. 1970. Middle Devonian reef production, Rainbow area, Alberta. Bull. Amer. Assoc. Petrol. Geol. 54: 2260–2281.

IMBRIE, J., and E. G. PURDY. 1962. Classification of modern Bahamian carbonate sediments, p. 253–272. *In* W. E. Hamm (ed.), Classification of carbonate rocks. Amer. Assoc. Petrol. Geol. Mem. 1.

————, and H. BUCHANAN. 1965. Sedimentary structures in modern carbonate sands of the Bahamas, p. 149–172. *In* J. V. Middleton (ed.), Primary sedimentary structures and their hydrodynamic interpretation. Soc. Econ. Paleont. Mineral. Spec. Pub. 12.

JONES, O., and R. ENDEAN (eds.). 1972. Biology and geology of coral reefs. Vol. I: Geology. Academic Press, New York.

KENDALL, C. G. ST. C., and P. A. D. SKIPWITH. 1969. Geomorphology of a recent shallow-water carbonate province: Khor Al Bazam, Trucial Coast, southwest Persian Gulf. Bull. Geol. Soc. Amer. 80: 865–892.

KLEIN, G. DEV. 1965. Dynamic significance of primary structures in the Middle Jurassic Great Oolite Series, southern England, p. 173–191. *In* G. V. Middleton (ed.), Primary sedimentary structures and their hydrodynamic interpretations. Soc. Econ. Paleont. Mineral. Spec. Pub. 12.

KONISHI, K., S. O. SCHLANGER, and A. OMURA. 1970. Neotectonic rates in the central Ryukyu islands derived from thorium-230 coral ages. Marine Geol. 9: 225–240.

LOGAN, B. W., J. L. HARDING, W. M. AHR, J. D. WILLIAMS, and R. G. SNEED. 1969. Carbonate sediments and reefs, Yucatan shelf, Mexico, p. 5–198. Amer. Assoc. Petrol. Geol. Mem. 1, Section 1.

————, G. R. DAVIES, J. F. READ, and D. E. CEBULSKI. 1970. Carbonate sedimentation and environments, Shark Bay, Western Australia. Amer. Assoc. Petrol. Geol. Mem. 13. 223 p.

MAIKLEM, W. R. 1968. The Capricorn reef complex, Great Barrier Reef, Australia. J. Sed. Petrology 38: 785–798.
Physiogeography and gross aspects of sedimentation from the southern portions of the Great Barrier Reef.

MATTHEWS, R. K. 1966. Genesis of Recent lime mud in southern British Honduras. J. Sed. Petrology 36: 428–454.
Proposes a predominantly skeletal origin for lime mud in southern British Honduras.

MAXWELL, W. G. H., and J. P. SWINCHATT. 1970. Great Barrier Reef: regional variation in a terrigenous-carbonate province. Bull. Geol. Soc. Amer. 81: 691–724.
A convenient reference to the general features of the Great Barrier Reef.

MESOLELLA, K. J. 1967. Zonation of uplifted Pleistocene coral reefs on Barbados, West Indies. Science 156: 638–640.

————, R. K. MATTHEWS, W. S. BROECKER, and D. L. THURBER. 1969. The astronomical theory of climatic change: Barbados data. J. Geology 77: 250–274.

————, H. A. SEALY, and R. K. MATTHEWS. 1970. Facies geometries within Pleistocene reefs of Barbados, West Indies. Bull. Amer. Assoc. Petrol. Geol. 54: 1899–1917.

PURDY, E. G. 1963. Recent calcium carbonate facies of the Great Bahama Bank. J. Geology 71: 334–355, 472–497.
The basic definition of Bahamian carbonate facies.

————. 1972. Reef configurations: some causes and effects, *In* L. F. Laporte (ed.), Reef complexes in time and space: their physical, chemical, and biological parameters. Soc. Econ. Paleont. Mineral. Spec. Pub. No. 19.

————, and J. IMBRIE. 1964. Carbonate sediments, Great Bahama Bank. Guidebook for field trip no. 2. Geological Society of America annual convention, Miami Beach, Fla. 66 p.

————, and R. K. MATTHEWS. 1964. Structural control of Recent calcium carbonate deposition in British Honduras [Abstract]. Geological Society of America annual convention, Program with abstracts: 157.

SHINN, E. A. 1969. Submarine lithification of Holocene carbonate sediments in the Persian Gulf. Sedimentology 12: 109–144.

STOCKMAN, K. W., R. N. GINSBURG, and E. A. SHINN. 1967. The production of lime mud by algae in south Florida. J. Sed. Petrology 37: 633–648.

Green algae appear to be capable of producing all the aragonite mud present in the Recent environment.

STODDART, D. R., and M. YONGE (eds.). 1971. Regional variation in Indian Ocean coral reefs. Zool. Soc. London, Symposia 28. Academic Press, New York. 608 p.

SWINCHATT, J. P. 1965. Significance of constituent composition, texture, and skeletal breakdown in some Recent carbonate sediments. J. Sed. Petrology 35: 71–70.

THURBER, D. L., W. S. BROECKER, H. A. POTRATZ, and R. L. BLANCHARD. 1965. Uranium series ages of Pacific atoll coral. Science 149: 55–58.

TRECHMANN, C. T. 1933. The uplift of Barbados. Geol. Mag. (Great Britain) (823): 19–47.

————. 1937. The base and top of the coral rock in Barbados. Geol. Mag. (Great Britain) 74 (878): 337–358.

VEEH, H. H., and J. CHAPPELL. 1970. Astronomical theory of climate change: support from New Guinea. Science 167: 862–865.

WANLESS, H. R. 1970. Influence of preexisting bedrock topography on bars of "lime" mud and sand, Biscayne Bay, Florida [abstract]. Bull. Amer. Assoc. Petrol. Geol. 54: 875.

WANTLAND, K. F., and W. C. PUSEY. 1971. The southern shelf of British Honduras. New Orleans Geological Society, New Orleans, La. 98 p. (plus appendices). Guidebook. Well-illustrated and containing much three-dimensional information.

WINLAND, H.D., 1971, Diagenesis of carbonate grains in marine and meteorite waters: PhD. dissertation, Brown University.

13

Carbonates and Evaporites

of the Intertidal

and Supratidal Shelf Interior

This chapter is the logical continuation of the subject matter in Chapter 12. In the previous chapter, we broke off our discussion simply to consolidate some of the observations concerning the shelf margin and subtidal shelf interior. Now we return to our consideration of the restricted environments of the shelf interior.

Recent Intertidal and Supratidal Environments

We have noted that subtidal shelf interiors tend to become restricted environments, dominated by conditions other than those of the normal marine environment. On the Great Bahama Bank and in the Persian Gulf, increased salinity resulting from poor circulation generally excludes many marine organisms from the subtidal shelf interior. In Florida Bay, a similar reduction in fauna has been caused by poor circulation and the additional complication of a variable freshwater input from the north.

As we move into the intertidal and supratidal environments, we see that the fauna must face the further hazard of intermittent exposure above sea level. The result is even further restriction of the fauna. This situation characterizes intertidal-supratidal environments.

Subtidal to supratidal transitions. A volumetrically significant amount of carbonate and evaporite sedimentation occurs in the supratidal environment. For example, much of the dolomite to be found in Recent environments occurs in association with supratidal sediments. Inasmuch as dolomite is an abundant carbonate mineral in Ancient sequences, Recent supratidal dolomite is of considerable interest to us.

All Recent sedimentation of evaporite minerals essentially occurs on supratidal flats marginal to subtidal-marine environments. Furthermore, there is reasonably good evidence, and strong geologic implication, that pore water from supratidal evaporite deposits may flow back through underlying carbonate sediments and dolomitize them.

In addition, the intertidal-supratidal environment appears to be a major site of stromatolite deposition and preservation. By recognizing these deposits in Ancient rocks, we may have an excellent environmental datum from which to make estimates on the water depth of associated environments.

Finally, the subtidal-supratidal couplet, in which supratidal sediments are closely superimposed upon subtidal sediments, is one of the primary tipoffs of progradation with carbonate sequences. We shall try to identify these transitions for precisely the same reasons that we sought to identify the coarsening-upward cycle in coastal clastic sedimentation.

Let us now look at the transition from subtidal to supratidal environment in six examples where the water of the supratidal environment ranges from fresh to very highly saline. In the transition from the subtidal Florida Bay into the Everglades, there really is no transition from subtidal-marine to supratidal sedimentation. Subtidal marine simply becomes subtidal freshwater as we examine the coastal swamps. In portions of the Bahamas, supratidal carbonate flats exhibit well-developed stromatolites and there is supratidal dolomitization without associated evaporite minerals. In the sabkha environment of the Persian Gulf, gypsum and anhydrite precipitation in the supratidal environment is added to the Bahamas model. On Bonaire, in the Netherlands Antilles, we can add certain embellishments to an evaporite-reflux dolomitization model. In Hamblen Pool, Shark Bay, Western Australia, high salinity in the subtidal-marine environment allows supratidal stromatolite sedimentation to extend well into the subtidal environment. The last example concerns the supratidal flats of Baja California, where halite accumulation is added to the gypsum and anhydrite sedimentation previously encountered.

In the northwest portion of Florida Bay, the transition from marine deposition in the bay to freshwater deposition in the Florida Everglades has been studied in considerable detail in connection with problems of the Holocene sea-level rise (Scholl and Stuiver, 1967). Only in the broadest sense does this transition go from subtidal-marine sedimentation to supratidal sedimentation. We would give a better description by saying that this is the area where fresh water and saline water meet. Water is everywhere, and the entire sediment-water interface is at or very near sea level. Just such features as natural levees on tidal channels are persistent topographic highs.

Traversing from Florida Bay into the Everglades, we find that Florida Bay lime mud gives way to a coastal mangrove swamp, which in turn gives way to the freshwater swamps of the Everglades. In vertical profile through the marine deposits, the sequence is repeated and records the way the Recent sea-level rise transgressed onto this very low-lying, carbonate erosion surface (Figure 13.1).

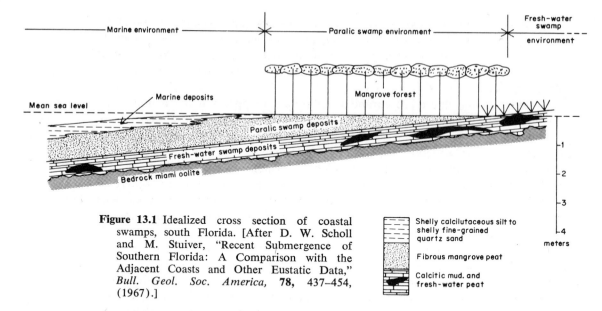

Figure 13.1 Idealized cross section of coastal swamps, south Florida. [After D. W. Scholl and M. Stuiver, "Recent Submergence of Southern Florida: A Comparison with the Adjacent Coasts and Other Eustatic Data," *Bull. Geol. Soc. America,* **78,** 437–454, (1967).]

Shelly calcilutaceous silt to shelly fine-grained quartz sand

Fibrous mangrove peat

Calcitic mud. and fresh-water peat

Progradation in this area has a very interesting style. Just as we did not find bay deposits prograding seaward over the Chenier Plain (Figures 11.13 and 11.14), we would not expect to find mangrove or freshwater swamp deposits prograding out over such coastal marine deposits as the Cape Sable beach sands. In both cases, there is a problem of space so long as sea level is constant. Note, however, that the marine mudbanks of Florida Bay (Figures 12.2, 12.13, and 12.15) tend to cut off isolated ponds among them. Similar features have been documented beneath the thin veneer of freshwater swamps east of Cape Sable (Gebelein, 1972). Here marine mudbanks cut off ponds that developed progressively fresher water as sedimentation continued. Thus, a complex pattern of marine mudbank deposits with intervening freshwater lake deposits is generated. In a very real sense, the net result of this process is progradation, that is, migration of the "shoreline" seaward by sedimentation processes. The major distinction between this situation and other examples of progradation is simply that "shoreline" must remain a fantastically ill-defined term where extremely shallow-marine water meets the freshwater swamp environment.

In the northwest portion of Andros Island, the Bahamas, the transition from subtidal to supratidal takes place (1) without the influence of large amounts of fresh water, such as occurs in southern Florida, and (2) with some fair amount of topographic relief (Figure 13.2). The topography in this area is apparently generated by subaerial accumulation of pelleted mud, presumably trapped on the palm hammocks by the dense vegetation during times of unusually high water (during storms, for instance). Subtidal sediments here are likewise pelleted carbonate muds. Between these two sediment types, laminated algal mats and supratidal dolomites record the intertidal to supratidal sedimentation today.

Filamentous algae thrive wherever the sediment surface is wet. They are ubiquitous in subtidal, intertidal, and low-lying supratidal environments. In all of these regions, filamentous algae are capable of binding sediments together (Gebelein, 1969). In the subtidal and intertidal environments, however, the browsing marine benthos (principally gastropods) crop the algae as fast as they grow; burrowing activity also tends to disrupt any algal structure that may be formed. Therefore, it is quite common in modern environments to find laminated algal structures preserved only in the supratidal environment where stress conditions place severe limitations on browsing faunas. (See Logan, Rezak, and Ginsburg, 1964; and Shinn, 1968, for details of lithology and petrography of stromatolites.)

The Bahamas is our first example of the formation of Recent dolomite under supratidal conditions. Scientists do not completely understand the details of this dolomite formation. However, two generalities have been made. First, most Recent supratidal dolomite appears to form by the replacement of aragonite rather than by direct precipitation of dolomite. Secondly, studies of pore-water chemistry indicate that the dolomitizing fluid has a magnesium-calcium ratio much greater than that of normal marine seawater.

Two mechanisms have been proposed for the generation of a high magnesium-calcium ratio in the pore water of supratidal sediments. If seawater is evaporated to the point where calcium sulfate is precipitated, a significant portion of seawater calcium is lost from the fluid phase, thus generating a higher magnesium-calcium

Figure 13.2 Recent occurrence of supratidal dolomite, Andros Island, the Bahamas. (After Shinn *et al.,* 1965.)

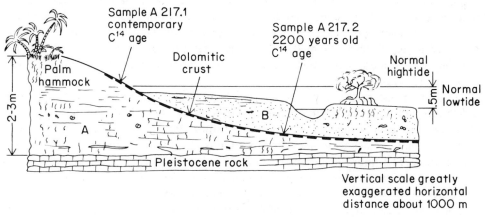

A Dark gray pelletal mud containing roots, land snails, desiccation laminations, and few forams and marine gastropods.

B Light tan pelletal sediment containing numerous marine gastropods, foraminifera, and red mangrove roots.

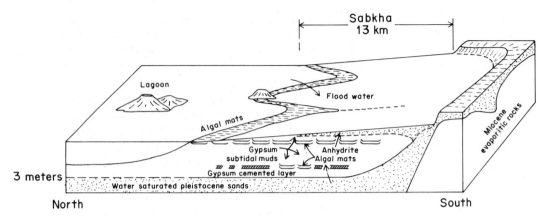

Figure 13.3 Schematic block diagram of facies relationships in the sabkha environment, southern Persian Gulf. (After Butler, 1969.)

ratio in the remaining brine (see Butler, 1969, for example). Alternatively, algae tend to concentrate magnesium in their vital fluids. Upon the death of the algae, these fluids are presumably released to the pore water and impart a high magnesium-calcium ratio (Gebelein and Hoffman, 1971). Clearly, the former mechanism requires evaporating conditions, whereas the latter mechanism does not. In the Bahaman example of supratidal dolomite, evidence of significant precipitation of calcium sulfate is lacking. It can be argued that calcium sulfate was never precipitated in this environment or, alternately, that calcium sulfate is precipitated during the dry season and dissolved away by rainwater during the wet season. This problem is important, so we shall dwell on it at some length as we attempt to formulate a general model for dolomite within stratigraphic sequences.

In the Persian Gulf, the Bahamas model of algal mats and supratidal dolomite is carried a step further. Here there is large-scale precipitation of calcium sulfate and thus the formation of extensive supratidal evaporite deposits (Figures 12.11, 13.3, and 13.4).

The southwest coastline of the Persian Gulf is an area of virtually no rainfall and exceedingly high evaporation. For these reasons, the salinity of the Gulf as a whole is considerably higher than normal marine. Pore fluids within the low-lying supratidal flats adjacent to the Persian Gulf are even more concentrated by evaporation. The supratidal flats (sabkhas) are deflation surfaces, controlled by the level of the water table, which is in turn controlled by sea level. Sediment in proximity to the water table remains wet and is not blown away during windstorms. On the other hand, sediment well above the water table may dry out completely and is highly subject to wind transportation.

Because the water table is close to the deflation surface, evaporation from the water table is significant. Water lost to the atmosphere by evaporation must be

272

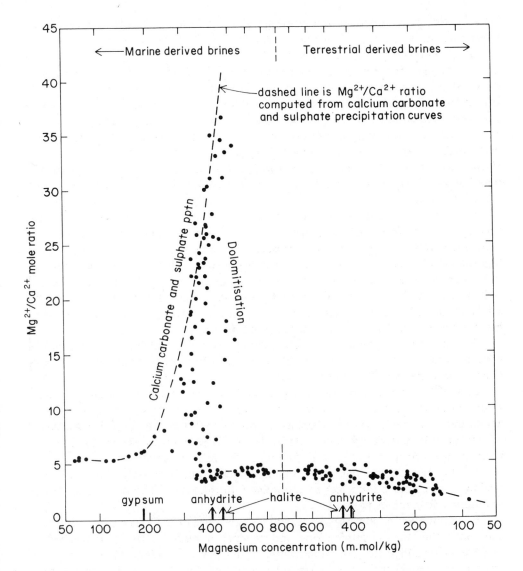

Figure 13.4 Variations in the magnesium-to-calcium ratio, with magnesium concentration, across the sabkha. From the left, sea water evaporates as it comes onto the sabkha. Then precipitation of calcium carbonate and calcium sulfate drives the magnesium-to-calcite ratio in the water very high. Apparently, dolomitization of preexisting calcium carbonate brings the magnesium-to-calcium ratio down again, even as further evaporation of the water continues. From the right side of the diagram, continental ground waters enter the sabkha and evaporate to halite deposition without the production of a high magnesium-to-calcium ratio. (After Butler, 1969.)

replaced somehow. Replacement comes through (1) the landward flow of marine water into the water table beneath the sabkha and (2) the seaward flow of continental groundwater from the interior of Arabia.

Continuing evaporation and continuing replenishment of the water lead to a steady-state condition of pore-water chemistry, such as is portrayed in Figure 13.4. Inland from the subtidal lagoons bordering the Gulf, pore water becomes more and more concentrated by evaporation. At some point in this process, there is an abrupt change in the magnesium-calcium ratio of the pore fluid. This change reflects precipitation of calcium sulfate minerals at or near the water table. The calcium sulfate may be gypsum or anhydrite. For our purposes, the distinction is unimportant. Evaporation continues until the water is almost saturated with sodium chloride.

The lithologic consequences of this process are depicted in Figure 13.3. Traversing from the intertidal zone across the sabkha surface, we first encounter an extensive development of algal flat deposits. This environment is only slightly above high-tide level. The sediment surface is kept generally moist by close proximity to the water table and occasional tidal flooding. Dessication polygons are typically well developed. Farther inland, algal deposits give way to carbonate and quartz sands of the sabkha deflation surface. Crystals of both gypsum and anhydrite become volumetrically important in these sediments.

Calcium sulfate crystals commonly grow by displacive precipitation; that is, the mineral does not cement nearby sedimentary particles together but instead pushes them aside as the calcium sulfate crystal grows.

Two striking textures result from this displacive precipitation process. Calcium sulfate beds on the order of centimeters to tens of centimeters thick tend to grow above the water table. As the displacive growth continues, these horizontal layers of calcium sulfate undergo intense deformation. With continuous expansion of the layer by new precipitated material, folding of the calcium sulfate layer ensues. We see then the first texture: highly crenulated, laminated beds of anhydrite. Individual laminae range from millimeters to centimeters in thickness. The entire bed can be as much as 50 centimeters thick.

As the process continues, the second texture develops. Contortion and expansion can obliterate the laminations and produce a more or less massive bed of "chicken-wire anhydrite," in which large aggregates of calcium sulfate are separated from one another by thin irregular strips of the sediment that the calcium sulfate is displacing.

Dolomitization in the sabkha environment is a particularly elusive problem, although all of the elements are here for any possible model. Dolomitization is indeed locally important, but its occurrence is spotty. Therefore, we must be cautious about making generalizations on the origin and importance of dolomite formation in association with supratidal calcium sulfate precipitation.

Coring beneath the anhydrite deposits of the sabkha deflation surface yields a vivid picture of the dynamics of tidal-flat progradation following the post-Wisconsin rise in sea level (Figure 13.3). The underlying subaerial erosion surface is closely overlain by subtidal carbonate mud. The vertical transition from subtidal carbonates to algal stromatolites to anhydrite deposits of the sabkha flat is an excellent vertical record of precisely the horizontal facies variations that are observed today in going from the subtidal lagoon across the algal mats onto the sabkha deflation surface.

Shark Bay, Western Australia, is another area in which the subtidal-supratidal transition has been studied in some detail (Logan *et al.,* 1970). One aspect of stromatolite sedimentation in this area is particularly worthy of our attention.

In discussing the Bahamas and the Persian Gulf, we noted the accumulation and preservation of algal stromatolites only in the supratidal area. Evidently, algae living in the subtidal and intertidal environments are heavily browsed by marine gastropods. Furthermore, any algal structure that escapes browsing will likely be disrupted by burrowing organisms.

In Hamblen Pool, Shark Bay, high salinity precludes any browsing and burrowing fauna. In the absence of these organisms, algal stromatolite structures extend well into subtidal environments. Thus, when it comes time for us to make a general model for the position of algal stromatolites within a paleogeographic reconstruction, we must bear in mind that these structures should be interpreted as intertidal-supratidal only in situations where it can be demonstrated that browsing and burrowing faunas also existed in related subtidal environments.

A word about deep-water evaporite sedimentation. *Deep-water evaporite sedimentation* is a concept that has developed out of the study of Ancient rocks. Its proponents base their arguments on position within the paleogeography, petrographic criteria for crystal growth above the sediment-water interface, and large sediment thicknesses that stratigraphic controls suggest must have been deposited very rapidly. (Buzzalini *et al.,* 1969, provide extensive discussion of these arguments.)

According to this concept, when large bodies of deep water are sufficiently isolated from open-marine circulation so that the entire thick column of basin water becomes highly concentrated by evaporation, then the evaporite minerals are precipitated: first, calcium carbonate; then, calcium sulfate; then, halite; and finally, the bittern salts. So-called deep-water evaporite basins commonly have alternations of carbonate and anhydrite. Some basins accumulate significant amounts of the more evaporative halite phase.

Although the idea of deep-water evaporite sedimentation has received considerable attention, no good Recent analogies have yet been documented. We are tempted to argue that conditions for the formation of deep-water evaporites were prevalent at other times in earth history but are not prevalent today. Again we must be cautious, however, in embracing such an argument; some geologists were saying the same things about dolomite as recently as the early 1960's.

Alternatively, we can look to the Recent epoch and speculate that the "deep-water evaporite model" is simply a detailed statement of our total lack of understanding of how these deposits originated. Two avenues of approach seem appropriate: Either the "deep-water" is not really deep, or "evaporite minerals" can form by processes other than evaporation.

Work in the supratidal flats of Baja California (Shearman, 1971) has documented halite fabrics that bear a remarkable similarity to supposed deep-water halites. Although the precise origin of these halite fabrics is not well understood, we cannot argue with the fact that they are there. Such is the empirical nature of much progress in geology.

Continuing Development of a Basic Model for Carbonate and Evaporite Sedimentation

The following discussion concerning the generalities of intertidal and supratidal sedimentation runs parallel to similar discussions in Chapter 12. Note well that these remarks will be simply additions to models already under development in the previous chapter.

Aerial distribution of contemporaneous sedimentary facies. We have previously observed that sediments in the subtidal shelf-interior environment are strongly influenced by the conditions of seawater circulation and the climate. The transition from subtidal to supratidal is subject to even greater variation as a function of these variables. If carbonate sedimentation is marginal to a clastic land mass, such as in British Honduras, then the transition from subtidal to supratidal along the mainland coast will occur in coastal clastic facies rather than in carbonate facies. Without a large clastic input, the transition from intertidal to supratidal is primarily dependent on the presence or absence of fresh water. Sediment types anticipated in this geographic position range from freshwater swamps (the Everglades, southern Florida) to mildly evaporating, dolomitic, stromatolitic tidal flats (the Bahamas) to highly evaporatic sabkha environments with crenulated anhydrite, chicken-wire anhydrite, or halite (the Persian Gulf). The stromatolitic, lower supratidal environment would be expected to provide a good environmental datum, with the one qualification that the absence of a subtidal browsing and burrowing fauna may allow stromatolite development in subtidal environments.

Progradation in intertidal-supratidal carbonates and evaporites. Certainly the concept of progradation has a straightforward application in examples like the sabkha flats of the Persian Gulf. Clearly, relationships between subtidal lagoonal lime mud, high intertidal to supratidal stromatolites, and sabkha evaporite flats are quite analogous to offshore silts, beaches, and topset marsh deposits of the Chenier Plain, for example.

The transition from northern Florida Bay into the freshwater swamps of the Everglades offers us an interesting example of just how complicated a prograding biogenic sequence can become. Once again observe that these complications are the direct result of very local production of carbonate and bituminous sediments.

It is difficult to imagine that such complications could arise in clastic environments where sediment must be transported to the site of deposition.

As in our discussion of topset accumulations in coastal clastic environments, we note that there is a serious space problem that precludes the accumulation of thick topset deposits under conditions of constant sea level.

Response of supratidal environments to conditions of emergence. Supratidal environments associated with carbonate deposition are closely related to the water table, which is in turn closely related to sea level. As the water table falls during conditions of emergence, former supratidal deposits are left high and dry. Freshwater swamps may become dry land and thus cease to produce carbonate mud and peat. Sabkhas will very likely undergo eolian deflation.

It is interesting to speculate on this possible phase to sabkha sedimentation. However, no one has yet discussed the Pleistocene sabkha deposits of the Persian Gulf. In sharp contrast, Pleistocene subtidal carbonates are today well exposed, well preserved, and well studied in such places as south Florida and Barbados. Apparently, Pleistocene sabkha deposits were subjected to a pronounced destructive phase associated with the lower water table of glacio-eustatic low stands of the sea.

Response of supratidal environments to conditions of submergence. We note once more the parallel between supratidal carbonates and the topset beds of coastal clastic sedimentation. In both cases, submergence is required if there is to be sufficient room to accommodate a thick section of these facies.

The rate of sea-level rise will also affect the geometric relationships of supratidal facies and related sediment. In the coastal swamp environment, rising water may accommodate vertical accumulation of subtidal to intertidal marine-carbonate deposits, brackish-water swamp deposits, and freshwater swamp deposits, more or less as they appear in map view. On the sabkha flats, rising sea level means a rising water table (deflation surface) and therefore opportunity for vertical accumulation of supratidal deposits. As long as net deposition of eolian detritus and precipitated evaporite minerals can keep this area above sea level, vertical contacts will be generated among evaporite facies, algal stromatolite facies, and subtidal lagoonal carbonate-mud facies. If sediment accumulation on the sabkha flat cannot keep pace with rising water, transgression will occur. Algal stromatolite facies may migrate landward over evaporite facies in a continuous fashion, or the entire sabkha flat may develop subtidal conditions by undergoing discontinuous floods.

An Ancient Example: The Helderberg Group, Devonian of New York

Let us now attempt to apply our general model to an Ancient stratigraphic sequence. Does the model explain what we see in Ancient rocks? Does the model help us to understand better the precise earth history recorded by these rocks?

Regional setting and earlier stratigraphic work. The Helderberg Group of western New York State has long been recognized as containing some of the oldest Devonian fossils in North America. The Group has been designated as the type section for the Helderbergian stage, the lowermost Devonian stage of North American biostratigraphy. Because of the biostratigraphic significance attached to the Helderberg Group, an understanding of its sedimentology and precise depositional history is fundamental to the understanding of the Silurian-Devonian contact. For example, is there or is there not a major worldwide transgressive episode separating the Silurian from the Devonian? If a major transgressive event can be demonstrated, then transgressive basal Devonian deposits may be synchronous around the world. If, on the other hand, no major transgressive event exists, then we must face the possibility that the biostratigraphically defined transition from Silurian to Devonian may be diachronous on a worldwide scale.

For many years, geologists believed that the Helderberg Group was a layer-cake arrangement of lithologically defined formations (Figure 13.5). Each lithology was thought to represent a discrete "time." The Rondout and Manlius formations were considered Upper Silurian, and the Coeymans was the base of the Lower Devonian Helderbergian stage.

Figure 13.5 Stratigraphic diagram of different interpretations of Helderberg stratigraphy in New York. Whereas the Helderberg Group had previously been considered layer-cake stratigraphy in which each rock unit represented a different period of time, Rickard (1962) demonstrated intertonguing relationships among these units. (After Laporte, 1969.)

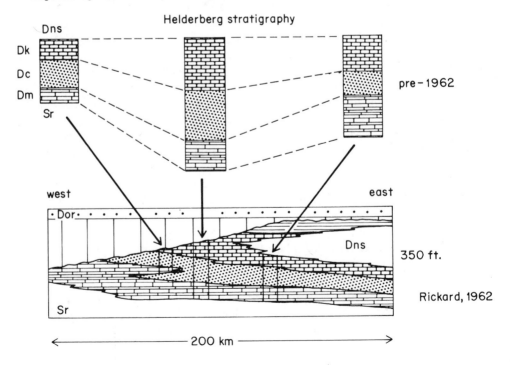

The carbonate sequences under consideration rest with an angular unconformity upon Middle Ordovician shales and sandstones. The Rondout and Manlius formations are poorly fossiliferous. The Manlius is primarily dark grey to black, fine-grained, carbonate rock, with occasional bituminous laminated horizons. The fine grain size, lack of fossils, and bituminous nature of the Manlius led early workers to think that this environment had been relatively deep (below wave base) stagnant water.

Coeymans rocks, with their diverse fauna and high abundance of skeletal sands, were taken to indicate "a time" of shallower water and better circulation. Muddy Kalkberg rocks suggest a deepening of the water and the first significant influx of clastic detritus into this area from the east. In "New Scotland time," marine, fine-grained, clastic sedimentation dominated the area.

Ricard (1962) demonstrated intertonguing relationships between Manlius and Coeymans (Figure 13.5). This demonstration of diachronous relationships required that geologists rethink the layer-cake version of earth history. If the Manlius formation does not represent "a time of . . . ," then what does it represent?

Sedimentological observations within the Manlius Formation. Recognizing the major questions of sedimentological history that are posed by Ricard's stratigraphic observations, Laporte (1967, 1970) undertook detailed sedimentological and paleoecological studies within the Helderberg Group. In the following discussion, we shall emphasize his work in the Manlius formation, because understanding the intertonguing relationship between Manlius and Coeymans provides the key to understanding the history recorded by the Helderberg Group.

Laporte recognized three distinct facies within the Manlius formation. Their characteristics are summarized in Table 13.1, where the facies are labeled *A, B,* and *C.* Let us work along with Laporte in attempting to assess their paleogeographic significance.

Table 13.1. Lithologic and Paleontologic Attributes of Manlius Facies [1]

Facies	Lithology	Paleontology
A	Pelletal, carbonate mudstones and reefy biostromes. Medium to massively bedded; in places "reefoid."	Biota relatively abundant and diverse. Stromatoporoids, rugose corals, brachiopods, ostracods, snails, and codiacean algea.
B	Interbedded, pelletal, carbonate mudstone and skeletal calcarenite; a few limestone-pebble conglomerates and mud cracks.	Fossil types few but individuals abundant. Ostracods, tentaculitids, brachiopods, oncolites, bituminous laminations.
C	Dolomitic, laminated mudstone. Disruption of laminations common.	Fossils scarce; bituminous laminations, ostracods, and burrows.

(1) After Laporte, 1967, with modification.

Figure 13.6 (Above) Negative print of a peel of facies *C*, Manlius, showing the discontinuity of laminations. Observe the concave-up tendency in disrupted laminae at the upper right. (Peel courtesy of L. F. Laporte.)

Figure 13.7 (Below) Negative print of a peel of facies *C*, Manlius, showing the concave-down arrangement of laminae. Whereas Figure 13.6 depicts laminae that appear to have been broken, many of these laminae actually seem to have grown in a discontinuous concave-down configuration. (Peel courtesy of L. F. Laporte.)

Consider first facies *A*. The fauna contains a bit of everything. The lithology is generally carbonate mudstone with only rare occurrences of moderately clean sands. This lithologic and paleontologic description sounds reasonable for a subtidal platform-interior environment. The presence of green algae and oncolite structures suggests only moderate water depth, but the precise depth is undetermined. Perhaps we can learn something about that feature from the other facies.

Next consider facies *B* in comparison with facies *A*. Green algae, corals, and echinoderms, which are present in facies *A,* are all absent from facies *B*. Well-preserved laminations, lacking in facies *A,* are quite common in facies *B*. Furthermore, the abundance of snails is reduced and spar-cemented carbonate sands are common. Indeed, occasional beds within facies *B* are coquinas of mollusc and possibly brachiopod debris.

In facies *C,* the fauna is still further reduced; spar-cemented sediments are absent; dolomite, detrital quartz, and clay are more abundant than elsewhere; and bituminous laminations are common.

These bituminous laminations are the features that led early geologists to conclude that the Manlius environment had been deep, stagnant water. Closer examination proves interesting. Laminae and groups of laminae are commonly disrupted (Figures 13.6 and 13.7). Typically, disrupted segments of laminae are somewhat curled; on occasion, curled chips have apparently been overturned and transported. In still other cases, disrupted laminations appear to have formed in a convex configuration, whereas intervening laminations appear to have formed in a concave configuration.

Table 13.2. Lithologic and Paleontologic Attributes of Manlius Facies, with Interpretation [1]

Facies	Lithology	Paleontology	Interpretation
C	Dolomitic, laminated mudstone; mud cracks; "birds's-eye."	Fossils scarce; algal laminae, ostracods, and burrows.	Supratidal
B	Interbedded, pelletal, carbonate mudstone and skeletal calcarenite; a few limestone-pebble conglomerates and mud cracks.	Fossil types few but individuals abundant. Ostracods, tentaculitids, brachiopods, algal stromatolites, and oncolites.	Intertidal
A	Pelletal, carbonate mudstones and reefy biostromes. Medium to massively bedded; in places "reefoid."	Biota relatively abundant and diverse. Stromatoporoids, rugose corals, brachiopods, ostracods, snails, and codiacean algae.	Subtidal

(1) After Laporte, 1967, with modification.

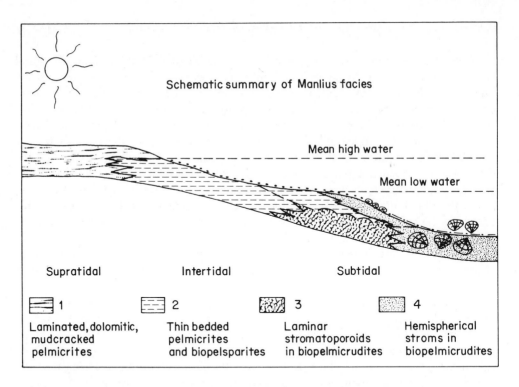

Figure 13.8 Schematic cross section showing interpreted relationships among four Manlius facies. Sediment types are as follows: (1) supratidal laminated, dolomitic, mudcracked pelmicrites; (2) intertidal thin-bedded pelmicrites and biopelsparites, with stromatolites, and mud cracks; (3) subtidal channel-floor deposits, predominantly biopelmicrudites with laminar stromatoporoids; (4) subtidal biopelmicrudites with hemispherical stromatoporoids and oncolites. In the intertidal to subtidal portion of the diagram, symbols below the dashed line represent tidal-channel deposits, whereas symbols above the dashed line represent tidal-flat and normal subtidal deposits. (After Laporte, 1967, with modification.)

Sedimentological interpretation. The interpretation placed on these observations by Laporte (1967) is summarized in Table 13.2 and Figure 13.8. Facies *A* is taken to represent moderately restricted, platform-interior sedimentation. Facies *B,* with restricted fauna and biopelsparites, is the intertidal environment. The bituminous laminations of facies *C* are interpreted as algal stromatolites, complete with dessication chips and mud cracks, as pictured in Figures 13.6 and 13.7. Facies *C* is identified as a supratidal environment.

These facies interpretations are exciting pieces of paleogeographic information. If we examine the Coeymans biosparites with diverse fauna, they look like shelf-margin facies. Kalkberg biomicrites become foreshelf facies. The fact that Kalkberg overlies Coeymans and Manlius to the southeast indicates that submergence occurred within a rather low-lying paleotopography. These views are put forward in detail by Laporte (1970).

Let us, therefore, acknowledge three major environments within the Manlius formation and recognize the paleogeographic relationships within the Helderberg Group. What can we then say about the history of the Helderberg transgression? Figure 13.9 summarizes facies relationships within the Manlius.

The Thachter member of the Manlius formation was apparently deposited on a broad, shallow shelf interior. Persistent occurrences of supratidal facies within predominantly subtidal Thachter facies confirm rather shallow-water conditions.

With continuing submergence, the shelf margin apparently succumbed to rising water and reestablished itself in the vicinity of localities (99) to (115). Again, the generally shallow character of the shelf interior during this time is confirmed by the occurrence of supratidal facies in among predominantly subtidal facies of the Olney member. Upward accumulation of sediment was in close balance with the rate of submergence, thus generating the relatively restricted zone of inter-

Figure 13.9 Environmental interpretation of the Lower Devonian Manlius formation with its five members. Lateral migration of facies depicted in Figure 13.8 resulted in the complex facies mosaic that we see in the Manlius today. (After Laporte, 1967, with modification.)

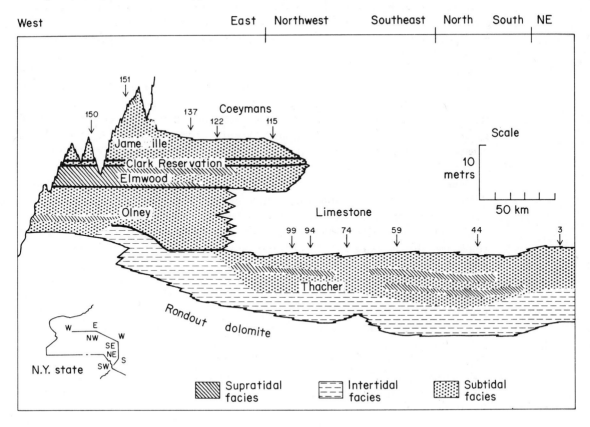

fingering between the Manlius and Coeymans lithologies throughout the deposition of more than 10 meters of sediment.

Note at locality (115); the occurrence of subtidal Manlius over Coeymans, a "regressive" relationship. Regression at this time is further confirmed by the supratidal Elmwood over the subtidal Olney member. But what kind of regression allows subtidal shelf-interior sediment to be deposited overlying shelf-margin biosparite facies?

According to our general model of carbonate sedimentation, this sequence of lithologies records progradation under conditions of net submergence. The Coeymans shelf-margin facies was capable of producing more sediment that was required to keep pace with rising water. Therefore, this facies prograded seaward. The Elmwood supratidal flats had no difficulty keeping up with rising water and so began to prograde out across the shallow subtidal environments. The subtidal Manlius in this interval was simply the shallow water that was left *behind* the prograding Coeymans shelf margin and *in front of* the prograding Elmwood supratidal flats.

Moving up the section, we see that the rate of submergence once more exceeded sedimentation rate. Jamesville subtidal Manlius overlies Elmwood supratidal Manlius, and Coeymans shelf-margin facies once again migrates westward.

Let us return to the problem of the Silurian-Devonian boundary, which we mentioned briefly in the introduction to the Helderberg Group. We are familiar with the age-old problem of where to place boundaries between systems and periods. If there is a major event, such as a worldwide transgression, we use it. Rocks above it can be separated from rocks below it, and the basal transgressive rocks are more or less synchronous around the world. Helderberg rocks are traditionally regarded as the lowest Devonian rocks of North America. Is the transgression that brought these rocks into existence likely to be of worldwide significance?

Manlius and Coeymans taken together never exceed 50 meters in thickness. Contrary to earlier beliefs that the Manlius represented relatively deep water (and therefore a major transgression), it now appears that the Manlius was deposited at or near sea level during conditions of submergence. Taking isostasy into account, we observe that a relative sea-level rise of less than 20 meters is probably sufficient to account for the measured thickness of Manlius and Coeymans sediments. Such a scale of sea-level fluctuation will hardly prove useful for worldwide correlation purposes. Thus, the Helderberg Group is probably just a little vignette in earth history. It does not record a worldwide event of profound importance; instead, another small piece of low-lying coastline happened to get a little water over it about 350 million years ago.

Thus, it would appear that the Silurian-Devonian boundary is best set by arbitrary agreement. Evidently, no major physical event took place to help out the biostratigrapher in this time range.

An Ancient Example: The Permian of West Texas

Permian strata of west Texas not only provide an excellent opportunity to test our carbonate sedimentation models, but they also exemplify the economic importance of carbonate rocks. In this particular example, the discovery of oil preceded systematic study of the carbonate rocks. However, let us review the facts so that this exploration province can serve as a model in which stratigraphic geology may lead to the discovery of petroleum, whether it be in a new field in the Permian of west Texas or the first discovery in some new and previously unexplored basin.

Regional setting and older stratigraphic work. The discovery of major petroleum accumulations in the Permian of west Texas occurred in the 1920's. The productive formations were at relatively shallow depths (700 to 1500 meters). As was the custom of the day, wildcatters punched down holes at random, wherever they could put together a large block of land and get sufficient backers to pay for the drilling of the hole. This process was rather unscientific, but it found a lot of oil.

As time went by, more and more data accumulated, and more scientific methods were applied to petroleum exploration in this region. Early drilling documented

Figure 13.10 General Permian paleogeography in west Texas and adjoining New Mexico.

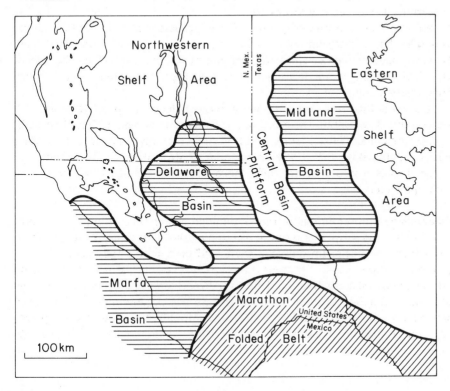

the existence of the Delaware basin and the Midland basin (Figure 13.10). Both of these basins are Permian topographic features that have no expression in the Cretaceous rocks now covering the area. Wells drilled in the eastern shelf area or on the central basin platform encountered Permian carbonate rocks at relatively shallow depth, and a significant number of such wells found oil. Wells drilled in the basin areas encountered thick sections of anhydrite and seldom found Permian oil.

As this paleogeography became apparent, geologists realized that rocks of equivalent age to those in the central basin platform and eastern shelf producing horizons cropped out in the Guadalupe Mountains. Clearly, detailed knowledge of the surface exposures of these economically important units could be valuable in further exploration of Permian strata in the subsurface of west Texas.

In the Guadalupe and Delaware mountains of west Texas and New Mexico, Permian paleogeography can be easily reconstructed. Fine-grained sandstones and occasional black limestones of the Delaware mountain group interfinger with limestones and dolomites of the Guadalupe Mountains (Figure 13.11). Furthermore, the high angle of dip and the abundant evidence of transportation led early workers to recognize the limestones of the eastern front of the Guadalupe Mountains as talus deposits marginal to shallow-water, carbonate-shelf deposits. Indeed, on the basis of field examination and stratigraphic relationships, it appeared quite reasonable to consider the massive gray to white limestone updip from these talus deposits as the "reef" from which the talus was derived. In this manner, the "Permian reefs" of western Texas became legendary.

Note well that from the very beginning of stratigraphic studies in the Guadalupe Mountains, the basic platform and basin paleogeography was recognized, and attempts were made to work with the stratigraphy in the context of the paleogeography. This starting point is tremendously elevated, compared to early work in the Helderberg Group, for example.

Yet, even with this early recognition of basic paleogeography, lithologic descriptions continued to emphasize gross aspects of the rock unit rather than sedimentology of the unit. Thus, Figure 13.11 is only partially analogous to studies of Recent sediment. With the stratigraphy under control, it was only a matter of time until geologists would want to know more about the details of sedimentation in the "Permian reefs."

Sedimentological observations in the Guadalupe Mountains. Figure 13.12 presents brief lithologic descriptions in modern sedimentological terms. Consider these descriptions from the basin, up onto the shelf, and back across the shelf.

The lower portion of the "forereef talus" is wackestone, which is carbonate sand with an abundant carbonate-mud matrix. Proceeding updip, we see that the carbonate sands contain less mud. Skeletal grainstones predominate at the top of the eastern part of the presentday Guadalupe Mountains. In terms of paleogeography, this is the area that separates flat-lying, shallow-water carbonates to

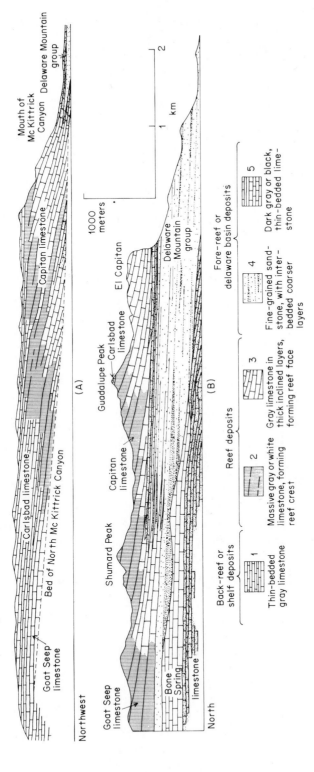

Figure 13.11 Two cross sections through the Permian rocks of the Guadalupe Mountains, west Texas. Section *A* is stratigraphically higher than section *B*. Rock types are as follows: (1) thin-bedded gray limestone, (2) massive gray or white limestone, (3) gray limestone in thick inclined layers, (4) fine-grained sandstone with interbedded coarser layers, (5) dark-gray or black thin-bedded limestone. (After King, 1948.)

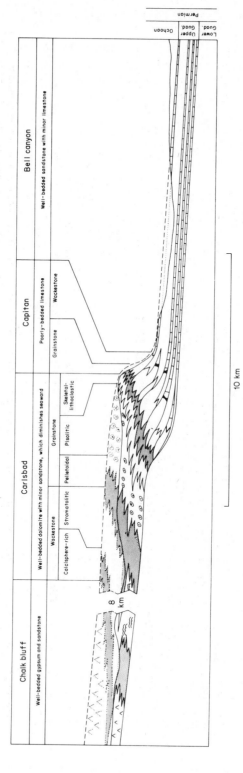

Figure 13.12 Sedimentary facies of Upper Guadalupian (Permian) carbonate rocks in the Guadalupe Mountains and surrounding area. (After Dunham, 1972.)

the west from gently dipping, "forereef" carbonates to the east. Interior from the skeletal grainstone facies is a pisolitic facies; interior to the pisolitic facies is pelletoidal grainstone; and so on further back onto the shelf, through stromatolitic wackestone, into interbedded evaporite and sandstone.

Sedimentological interpretation. After you examine Figures 13.11 and 13.12, it will come as no surprise that the Permian of west Texas has become a classic example of regressive shelf-margin facies with a restricted, shallow, subtidal to supratidal shelf interior. That the basinward progradation of the shelf margin occurred during a time of net submergence is amply confirmed by the thick stromatolitic and evaporite sequences of the shelf interior.

We shall want to return to these rocks for a more detailed look at cyclic sedimentation. However, the problems that will interest us go considerably beyond the scope of this chapter. For now, it is enough to note that no mention is made of "reef." Simply stated, massive *in situ* skeletal structures are not an important part of the shelf-margin complex in the Permian of west Texas. Terminology that seemed quite appropriate in the light of megascopic field examination (that is, massive gray or white limestone equals a reef crest) has simply not withstood detailed petrologic investigation.

Citations and Selected References

BORCHERT, H. 1969. Principles of oceanic salt deposition and metamorphism. Bull. Geol. Soc. Amer. 80: 821–864.
A summary paper in English by a deep-water evaporite specialist who usually publishes in German.

BRIGGS, L. I. 1958. Evaporite facies. J. Sed. Petrology 28: 46–56.

BUTLER, G. P. 1969. Modern evaporite deposition and geochemistry of coexisting brines, the sabkha, Trucial Coast, Arabian Gulf. J. Sed. Petrology 39: 70–89.
General stratigraphy and pore-water chemistry of dolomitization and evaporite deposition on supratidal flats.

———. 1970. Holocene gypsum and anhydrite of the Abu Dhabi Sabkha, Trucial Coast: an alternative explanation of origin, p. 120–152. *In* J. L. Rau and L. F. Dellwig (eds.), Third symposium on salt, Northern Ohio Geological Society.
An extensive and well-illustrated discussion.

BUZZALINI, A. D., F. J. ALDER, and R. L. JODRY (eds.). 1969. Evaporites and petroleum. Bull. Amer. Assoc. Petrol. Geol. 53: 775–1011.
Entire issue of the bulletin devoted to evaporite sedimentation.

DEFFEYES, K. S., F. J. LUCIA, and P. K. WEYL. 1965. Dolomitization of Recent and Plio-Pleistocene sediments by marine evaporite waters on Bonaire, Netherlands Antilles, p. 71–88. *In* L. C. Pray and R. C. Murray (eds.), Dolomitization and limestone diagenesis. Soc. Econ. Paleont. Mineral. Spec. Pub. 13.

DUNHAM, R. J. 1972. Capitan Reef, New Mexico and Texas: Facts and questions to aid interpretation and group discussion. Soc. Econ. Paleontologists and Mineralogists, Permian Basin Section (Midland, Texas) Publication 72–14, 213 p.

DeGROOT, K. 1969. The chemistry of submarine cement formation at Dohat, Hussain, in the Persian Gulf. Sedimentology 12: 63–68.

GEBELEIN, C. D. 1969. Distribution, morphology, and accretion rate of Recent subtidal algal stromatolites, Bermuda. J. Sed. Petrol. 39: 49–69.
A "breakthrough" study concerning the details of sediment entrapment by filamentous algae.

———, and P. HOFFMAN. 1971. Algal origin of dolomite in interlaminated limestone-dolomite sedimentary rocks, p. 319–326. *In* O. P. Bricker (ed.), Carbonate cements. Studies in Geology, 19. Johns Hopkins Press, Baltimore.

KENDALL, C. G. ST. C., and P. A. D. SKIPWITH. 1969. Geomorphology of a Recent shallow-water carbonate province: Khor Al Bazam, Trucial Coast, southwest Persian Gulf. Bull. Geol. Soc. Amer. 80: 865–892.
Especially noteworthy for its large foldout map (plate 6).

———. 1969. An environmental reinterpretation of the Permian evaporite-carbonate shelf sediments of the Guadalupe Mountains. Bull. Geol. Soc. Amer. 80: 2503–2526.

KIRKLAND, D. W., and R. Y. ANDERSON. 1970. Micro-folding in the Castile and Todilto evaporites, Texas and New Mexico. Bull. Geol. Soc. Amer. 81: 3259–3282.

LAPORTE, L. F. 1967. Carbonate deposition near mean sea level and resultant facies mosaic: Manlius formation (Lower Devonian) of New York State. Bull. Amer. Assoc. Petrol. Geol. 51: 73–101.
Modern reinterpretation of classical sequences that were grossly misunderstood by early workers.

———. 1969. Recognition of a transgressive carbonate sequence within an epeiric sea: Helderberg Group (Lower Devonian) of New York State, p. 98–118. *In* G. M. Friedman (ed.), Depositional environments in carbonate rocks. Soc. Econ. Paleont. Mineral. Spec. Pub. 14.

———. 1971. Paleozoic carbonate facies of the central Appalachian shelf. J. Sed. Petrology 41: 724–740.

LOGAN, B. W., R. REZAK, and R. N. GINSBURG. 1964. Classification and environmental significance of stromatolites. J. Geology 72: 68–83.

LUCIA, F. J. 1968. Recent sediments and diagenesis of south Bonaire, Netherlands Antilles. J. Sed. Petrology 38: 845–858.

MATTER, A. 1967. Tidal flat deposits in the Ordovician of western Maryland. J. Sed. Petrology 37: 601–609.

MURRAY, R. C., and F. J. LUCIA. 1967. Cause and control of dolomite distribution by rock selectivity. Bull. Geol. Soc. Amer. 78: 21–35.

Selective dolomitization was apparently controlled by the flow path of early diagenetic fluids.

————. 1969. Hydrology of south Bonaire, Netherlands Antilles: a rock selective dolomitization model. J. Sed. Petrology 39: 1007–1013.

PRAY, L. C., and R. C. MURRAY (eds.). 1965. Dolomitization and limestone diagenesis. Soc. Econ. Paleont. Mineral. Spec. Pub. 13. 180 p.

RAUPE, O. B. 1970. Brine mixing: an additional mechanism for formation of basin evaporites. Bull. Amer. Assoc. Petrol. Geol. 54: 2246–2259.

RICKARD, L. V. 1962. Late Cayugan (Upper Silurian) and Helderbergian (Lower Devonian) stratigraphy in New York. New York State Museum and Science Service Bull. 386. 157 p.

SCHOLL, D. W., and M. STUIVER. 1967. Recent submergence of southern Florida: a comparison with adjacent coasts and other eustatic data. Bull. Geol. Soc. Amer. 78: 437–454.

SHEARMAN, D. J., and J. G. FULLER. 1969. Anhydrite diagenesis, calcitization, and organic laminites, Winnipegosis formation, Middle Devonian, Saskatchewan. Bull. Canadian Petrol. Geol. 17: 496–525.
Argues a shallow-water depositional environment for basin-center evaporites.

SHINN, E. A. 1968. Practical significance of birdseye structures in carbonate rocks. J. Sed. Petrology 38: 215–233.

————, R. N. GINSBURG, and R. M. LLOYD. 1965. Recent supratidal dolomite from Andros Island, Bahamas, p. 112–123. *In* L. C. Pray and R. C. Murray (eds.), Dolomitization and limestone diagenesis. Soc. Econ. Paleont. Mineral. Spec. Pub. 13.

THOMAS, G. E., and R. P. GLAISTER. 1960. Facies and porosity relationships in some Mississippian carbonate cycles of western Canada basin. Bull. Amer. Assoc. Petrol. Geol. 44: 569–588.

WILLIAMS, R. E. 1970. Ground water flow systems and accumulation of the evaporite minerals. Bull. Amer. Assoc. Petrol. Geol. 54: 1290–1295.

14

Shelf-to-Basin Transition

at Continental Margins

Gravity is a predominant driving force in sedimentary processes. Alluvial environments are characterized by water and sediment moving downhill. At the shoreline, gravity-driven processes are masked by tidal processes and by the action of wind-driven waves and currents. At some distance from the shore, however, in perhaps 50 to 100 meters of water, these shallow-water processes yield once again to the dominance of gravity.

Persistent ocean currents may play a role in the ultimate distribution and fabric in deeper-water environments, but the sediment is supplied chiefly by gravitational mechanisms. Clay particles suspended in water near the bottom may make the water sufficiently dense so that it flows slowly basinward as a low-velocity density current. Newly sedimented silt and clay may undergo plastic deformation and creep basinward. As creep becomes severe, slumping may occur. As slump blocks move faster, the sediment may become resuspended and flow basinward as a high-velocity turbidity current.

Recent Sedimentation along Continental Margins

The generalizations of shelf-to-basin sediment transport just outlined seem quite plausible; indeed, for many years, they have constituted *ad hoc* models for sedimentary sequences observed in the stratigraphic record. In recent years, however, scuba-diving equipment, research submarines, and improved oceanographic techniques for subbottom profiling, bottom photography, and sampling have allowed geologists to investigate these processes as they occur today along the continental margins. The following two examples are particularly instructive.

Submarine canyons and basins off southern California. As we noted earlier (Figure 4.14), continuing tectonic activity has produced pronounced topographic relief in southern California. On land, this tectonic topography takes the form of valleys and mountain ranges. In the offshore area, tectonism has resulted in relatively deep basins adjacent to the modern coastline. Obviously, we would expect to find that sediments derived from the continent may have been deposited in this relatively deep water.

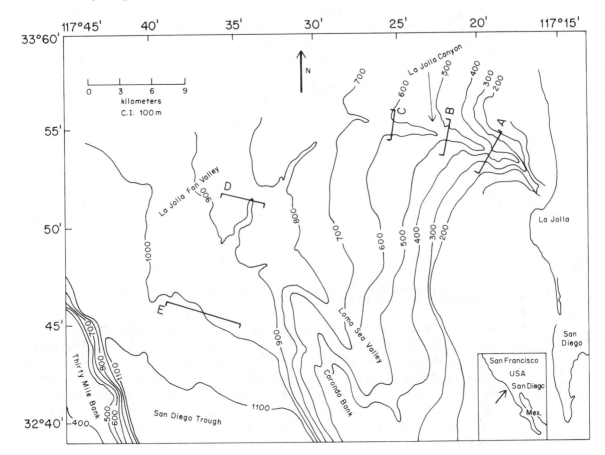

Figure 14.1 Bathymetry of La Jolla canyon, La Jolla fan, and San Diego trough.

Scientists have made detailed studies of the shore-to-basin transition in the vicinity of the La Jolla canyon and San Diego trough. The general bathymetry of the area is indicated in Figure 14.1. Observe the extremely narrow continental shelf here; the 100-meter contour lies within 4 kilometers of the shoreline. Outward from the 100-meter contour, the bottom slopes seaward at greater than 50 meters/kilometer. Below 800 meters, the topography begins to flatten out and

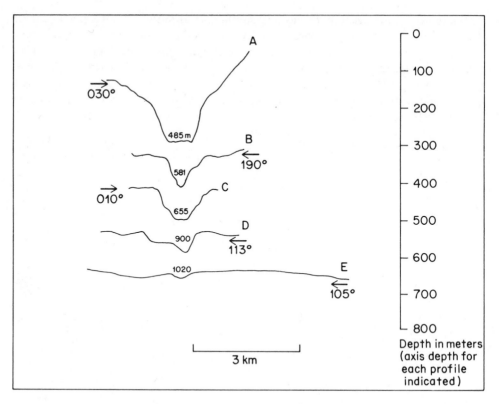

Figure 14.2 Bathymetric profiles of La Jolla canyon and La Jolla fan. (From the data of Shepard *et al.*, 1969.)

the general slope of the sediment surface turns southward into the San Diego trough. Note that the La Jolla canyon and its seaward extension remain essentially perpendicular to the regional strike of the sea floor from the shelf to the San Diego trough. Bathymetric profiles across the La Jolla canyon and its seaward extension are indicated in Figure 14.2. The shallow end of the canyon is deeply incised into the shelf and slope topography, whereas the seaward extension of the canyon is simply a pronounced low-relief channel. We can see from its topography alone that the La Jolla canyon must be an important element in regional sedimentation.

The story of clastic sedimentation in this area begins with the longshore transport of sediment that the rivers have delivered to the coast. Sorting of the sediment accompanies longshore transport. Sands continue to be moved southward along the shore, whereas silt and clay tend to be winnowed out and moved seaward (Figure 14.3).

Bottom observations indicate that at least some of the clay-sized material is transported seaward by slow-moving density currents. Turbid layers are commonly observed to occupy the first meter or so of the water column above the shelf sediments. It is presumed that the clay minerals in these turbid suspensions provide sufficient density so that the turbid water tends to flow slowly downslope. Although part of the silt and clay is deposited at the shelf margin and slope, some clay in the slow-moving density currents is also eventually channeled into the deep basins by submarine canyons. Apparently, the layers of turbid water move obliquely across the continental shelf until they are either funneled into a submarine canyon or spill over the continental slope. When he made seismic studies of unconsolidated fill in outer basins of the southern California continental borderline, Moore (1970) demonstrated striking differences in the amount of fill in adjacent basins. He found that these differences are related to the way the system of basins is connected to submarine canyons. Where an outer basin is ultimately

Figure 14.3 Directions of sand and mud transport in the La Jolla canyon and San Diego trough. Sand tends to be transported southward along the coast until it is funneled down La Jolla canyon. In contrast, mud travels in low-velocity turbid flows near the bottom across much of the shelf. [After Von Rad, "Comparison of Sedimentation in the Bavarian Flysch (Cretaceons) and Recent San Diego Trough (California)," *Jour. Sed. Petrology*, **38**, 1120–1154 (1968).] Courtesy of the Society of Economic Palentologists and Mineralogists.

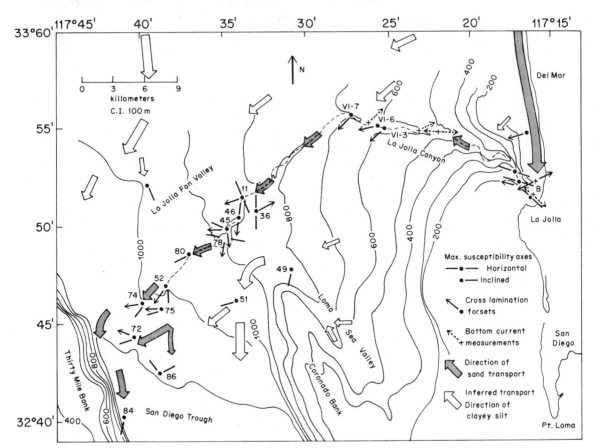

connected to basins with active submarine canyons, the basin receives abundant clay fill. An adjacent basin, topographically isolated from landward basins, receives very little clay fill. Prior to these observations, it had been assumed that the clay fill of the outer basins settled out of the entire water column. Moore's observations strongly suggest that low-velocity density currents, consisting of turbid water such as is observed on the outer continental shelf, are the transporting mechanism for outer-basin clay sedimentation. The observations further emphasize the importance of submarine canyons as funnels for these fine-grained sediments.

Let us return to the shore and pick up the story of sand transportation and deposition. Longshore transport of the sand-sized fraction ends abruptly at the La Jolla headland. Here the coastline presents a topography that is unfavorable to continued longshore transport. Sand-sized sediment, therefore, accumulates at the head of La Jolla canyon. This sediment will ultimately find its way by gravitational processes to the La Jolla fan, 20 to 40 kilometers seaward and beneath 800 to 1200 meters of water.

The precise mechanism of sand transport down the La Jolla canyon has been the subject of considerable study and discussion. Clearly, some high-energy currents must come down the valley from time to time; the canyon in places dissects rocks of Cretaceous and Eocene age, leaving vertical and overhanging walls. Boulders and cobbles of this material, as well as mud lumps of semiconsolidated shelf and slope sediment, accumulate in the central channel of the sediment fan that forms in front of the erosional portion of the La Jolla canyon. However, the strong currents that these sediment accumulations imply have not been observed firsthand.

Studies carried out by scuba divers have demonstrated a slow, downslope, sediment creep in the head of the canyon. It has also been noted that the sediment surface around the head of the canyon may undergo periodic deepening on the order of several meters. Presumably, such deepening is related to rather large-scale slumping, but details of the process are not known.

Thus, the study of La Jolla canyon reveals certain inconsistencies concerning the mode of sand transport from near shore into deep water. The sand does get transported; the La Jolla fan and San Diego trough contain large amounts of sand of demonstrable shallow-water origin. The morphology of the erosional portion of La Jolla canyon and the size of cobbles and boulders in the channel of the La Jolla fan certainly imply high-energy current systems. Yet observations of sediment transport today indicate primarily slow movement of sediment by creep or slump. Indeed, much of the floor of La Jolla canyon, La Jolla fan, and San Diego trough are now covered with a layer of silt and clay, as though the processes responsible for the movement of sand, cobbles, and boulders are dormant.

Primary structures revealed in box cores from the La Jolla fan and the San Diego trough (von Rad, 1968; Shepard *et al.,* 1969) provide a further complication concerning the generalities of sand sedimentation in this area. The primary structures are typical of traction transportation. Structureless graded beds, which experimentation and theory suggest should be typical of high-velocity turbidity-current deposition, are uncommon in the La Jolla fan. As shown in Figure 14.3, the data indicate that traction transport is southward down the axis of the San Diego trough. Whether this transport was the result of persistent marine currents or the result of currents associated with the La Jolla canyon is not certain. In either case, the process does not appear to be active today, because the entire region is covered with a layer of silt and clay.

Thus, the area from the head of La Jolla canyon, across La Jolla fan, and down the San Diego trough presents an interesting dichotomy. At some time in the near past, high-energy currents accomplished significant erosion within the walls of La Jolla canyon and delivered large volumes of sand, cobbles, and even boulders out onto the La Jolla fan. Bottom currents further reworked this material into traction-transported sand deposits. Yet, present-day large-scale sand transport in the La Jolla canyon is restricted to creep and slumping at the head of the canyon. Indeed, the lower reaches of the canyon, the La Jolla fan, and the San Diego trough are now covered with a thin accumulation of silty clay. Shepard *et al.* (1969) suggest that this contrast of modern conditions with more energetic conditions in the past may be associated with the possibility of a greater sediment supply from the land at a time when higher rainfall brought more sand to coastal areas. On the other hand, studies by Hand and Emery (1964) and Moore (1970) note that the eustatic lowering of sea level during the Pleistocene would have brought longshore sediment transport into the steeper reaches of existing sub-marine canyons, thus producing a gravitational instability more pronounced than exists today.

Gorsline and Emery (1959) propose another way that southern California basins may have been filled more easily at a time of eustatic sea-level lowering. Figure 14.4 depicts the development of deep-water sand fans where sediment supply comes from submarine canyons, a generalization more or less compatible with La Jolla fan sedimentation described earlier. This figure indicates the possi-bility of deriving basin fill when silt and clay slumps from the outer-shelf and slope environment. Whether the outer-shelf, clayey silt creeps, slumps, or flows into the adjacent basin may be considered a point of secondary importance; the important point is that outer-slope sediments deposited during a eustatic high stand provide a convenient source of relatively shallow-water, fine-grained sediment, which may be remobilized for deposition in the basin. We would expect this process to occur especially with the increasing energy conditions that are attendant to the eustatic lowering of sea level.

Estimates of sedimentation rate in the deep-water basins are generally in the range of 5 to 15 centimeters/1000 years. All of these estimates are based on the carbon-14 dating of sediments 10,000 years old and younger. As previously out-lined, there is good reason to suspect that modern sedimentation is occurring at a slower rate than may have prevailed during Pleistocene eustatic low stands of the sea or under wetter climatic conditions.

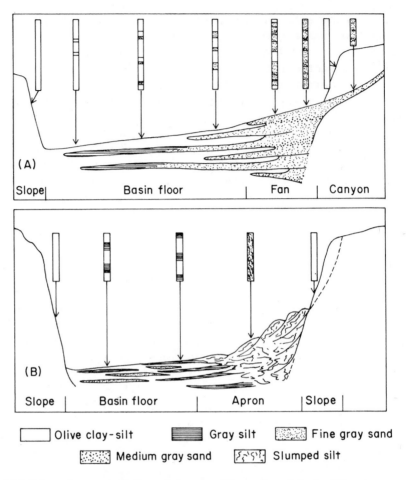

Figure 14.4 Schematic representation of two possible modes of basin filling in southern California. Where submarine canyons enter the basin, (*A*), sand transport down the canyon would be anticipated. Away from the canyon or in basins that do not attach to canyons, (*B*), slumping would occur as an important filling process. (After Gorsline and Emery, 1959.)

Shelf-to-basin transition, east coast of the United States and Canada. The transition from the eastern coastline of the United States to the abyssal plain of northwestern Atlantic provides an important contrast to sedimentation in the continental borderland basins of southern California. First and foremost, the east coast of the United States is not a tectonically active area. Whereas tectonism of southern California chops up the area into small units that must be considered individually, the lack of tectonism on the east coast allows us to apply broad generalizations to larger areas. The east coast, therefore, displays contrast in terms of both tectonic setting and scale.

298

Figure 14.5 indicates the general bathymetry of a portion of the western Atlantic. Consider the area from latitude 34° northward. There is (1) a broad continental shelf, 60 to 300 kilometers in width, which exhibits slopes on the order of .6 to 3 meters/kilometer; (2) a continental slope, usually ranging in the depth from 200 to 2000 meters and in width from 20 to 60 kilometers, with slopes averaging 40 meters/kilometer; (3) a continental rise, generally between the 2000- to 4500-meter bathymetric contour, 200 to 300 kilometers wide, and having an average slope of 10 meters/kilometer; and (4) an abyssal plain that is very nearly flat. Observe the sharp contrast between this bathymetry and scale and that of southern California (Figures 4.14 and 14.1). Here the continental shelf is extremely broad. Any basins that may have existed are now smoothed over by

Figure 14.5 Bathymetry of a portion of the northwest Atlantic.

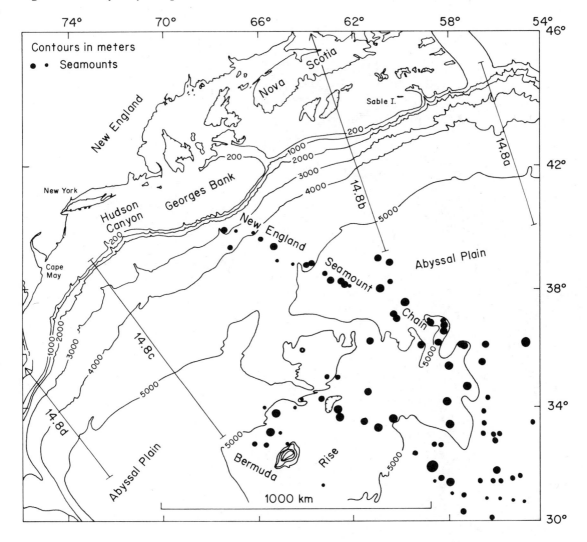

sedimentation. The continental slope and the continental rise have exceedingly similar profiles for more than 1000 kilometers along strike. If we consider the distance from continental shelf to abyssal plain as the measure of the shelf-to-basin transition, the transition zone here is hundreds of kilometers wide.

It is convenient to begin our discussion of the distribution of sediment types in the northwest Atlantic by referring to the characteristic of their reflections in echo-sounding profiles. Figure 14.6 presents a summary of such data. In general, a strongly reflecting bottom indicates the presence of firm substrate composed largely of sand. A weakly reflecting bottom indicates the predominance of clay

Figure 14.6 Physiography and sediment reflection characteristics of a portion of the northwest Atlantic. (After Emery *et al.,* 1970, with modification.)

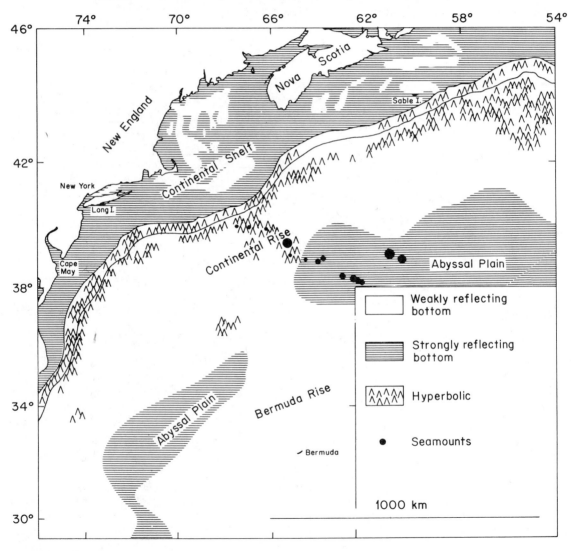

and silt-sized material. Hyperbolic reflections occur where local topographic relief allows high points to act as point-source reflectors as the survey ship passes by. Note that the northwest Atlantic can be divided into four physiographic-sedimentologic provinces on the basis of these data. The continental shelf is typically an area of strongly reflecting bottom; the continental slope, an area of weakly reflecting bottom with abundant hyperbolic reflections; the continental rise, a smooth area of weakly reflecting bottom; and the abyssal plain, a region with a stronger reflecting bottom. Sedimentologic studies allow us to translate these observations directly to sediment type and sedimentation processes.

Sediments of the continental shelf are generally relict sand deposits or coarse glacial sediment accumulated during Pleistocene eustatic low stands of the sea (Emery, 1968). The sediment load of rivers discharging into this area has been insufficient to keep pace with Holocene sea-level rise. River valleys along this coast are drowned estuaries. Modern deltas exist only at the heads of bays. Even the sand of modern beaches is derived from the relict deposits of the continental shelf.

Sedimentation on the continental rise runs strongly parallel to "hemipelagic" sedimentation in the outer basins of the southern California borderland. By relating variations in sedimentation rate in the outer basins to the presence or absence of a landward connection to submarine canyons, Moore (1970) inferred that "hemipelagic sediment" is not pelagic at all. Rather than settling down through the water column from the surface, these clays are deposited by low-velocity, turbid-density currents that follow bottom topography.

Similarly turbid bottom water exists above the continental rise off the eastern United States. However, the vast scale of this low-relief topography has allowed a very different current system to develop.

Ewing and Thorndike (1965) report on the observation of slightly turbid bottom water over the western Atlantic continental rise. *In situ* measurements made by optical-photometric methods indicate the lower 200 to 900 meters of the water column to be somewhat cloudy. Processing of a large volume of water samples suggests that bottom water contains on the order of 2 milligrams/liter of clay minerals not observed elsewhere in the water column.

Heezen *et al.* (1966) account for continental-rise sedimentation by means of a southward-flowing western boundary undercurrent, which follows the contours of the continental rise from Labrador to the Bahamas. The western boundary undercurrent is a geostrophic current that has its origin in the deep thermohaline circulation of the Atlantic. Near the western margin of the North Atlantic, the Coriolis force (to the right) is balanced by the gravitational force associated with the general slope of the continental rise (to the left). As the net result of the balance between these two forces, the current is guided along the contours of the continental rise for extremely long distances.

Heezen *et al.* (1966) suggest that this current transports the cloudy bottom water observed by Ewing and Thorndike (1965) and calculate that deposition of

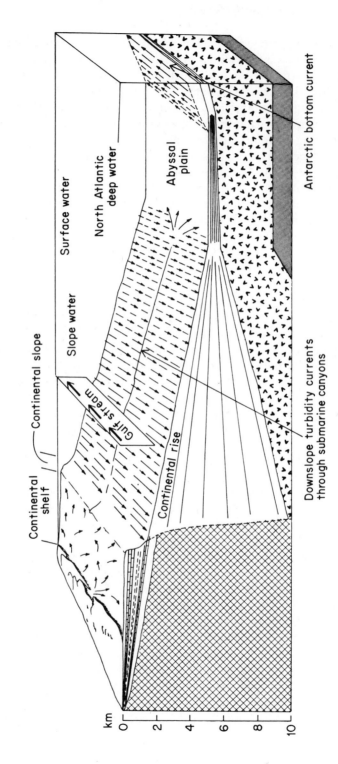

Figure 14.7 Diagram showing how the continental rise is shaped by geostrophic contour currents. [After B. C. Heezen *et al.*, "Shaping of the Continental Rise by Deep Geostrophic Contour Currents," *Science*, **152**, 502–508 (22 April 1966).] Copyright 1966 by the American Association for the Advancement of Science.

10% of the total sediment load would account for the total thickness of Recent continental-rise clay accumulation observed in piston cores.

The generalizations of Heezen *et al.* (1966) concerning the genetic relationships between continental shelf, continental rise, and abyssal plain are summarized in Figure 14.7. According to their generalizations, the continental shelf owes its origin largely to shallow-marine sedimentation processes, like those discussed in Chapters 11 to 13. The continental rise originated largely from clay sedimentation left by southward-flowing contour currents. They view the continental slope as the zone of discontinuity between continental-shelf processes and continental-rise processes. Clastic transport to the abyssal plain is limited to downslope turbidity currents that follow submarine canyons across the continental rise.

Studies of the continental slope in front of the Hudson canyon tend to bear out the generalities of Figure 14.7 concerning the genetic relationship between canyons and deep-sea sands (Ericson *et al.*, 1952, for example). Sediment cores taken in the topographic extension of the Hudson canyon 50 to 100 kilometers basinward from the continental slope and in 3000 meters of water contain graded beds of muddy gravel with pebbles of igneous, metamorphic, and sedimentary rocks up to 2 centimeters in diameter. Cores taken a few tens of kilometers to either side of this locality contain only the clay typical of the general continental-rise province.

Sediments of the abyssal plain are quartz sand and silt interbedded with deep-sea clays and pelagic ooze. Fauna contained in these sands indicate that the shallow water of the continental margin is the immediate source of the quartz sand and silt. Abrupt lower contact and a generally graded nature of many of these sand layers are considered evidence of deposition from turbidity currents. The occurrence of such layers as much as 1000 kilometers from any apparent source area seems to indicate that the turbidity currents were of considerable size and generated considerable momentum.

With this synoptic sketch of sediment types from the continental shelf, the continental rise, and the abyssal plain, it is important that we return to Figure 14.6 and consider once again the immense scale of these sedimentological provinces.

The continuous seismic profiling of Emery *et al.* (1970) provides additional insight on sedimentation and contemporaneous deformation on the continental rise. Figure 14.8 presents four profiles from the continental shelf to the abyssal plain. In each profile, horizon *A* is a prominent reflecting horizon, taken to be Middle Eocene to Upper Cretaceous in age; horizon *B* is the contact between sediment and oceanic basement. Observe carefully the vertical and horizontal scale on these diagrams. We are discussing truly large-scale sedimentation features. Of particular interest to us are the numerous evidences of large-scale slump structure. Note especially the upper kilometer of sediment in Figure 14.8(C) between 200 and 500 kilometers off the continental shelf. An enormous slice of continental-rise sediment appears to have slid along planes more or less parallel to bedding. This situation is truly gravity tectonics on a grand scale.

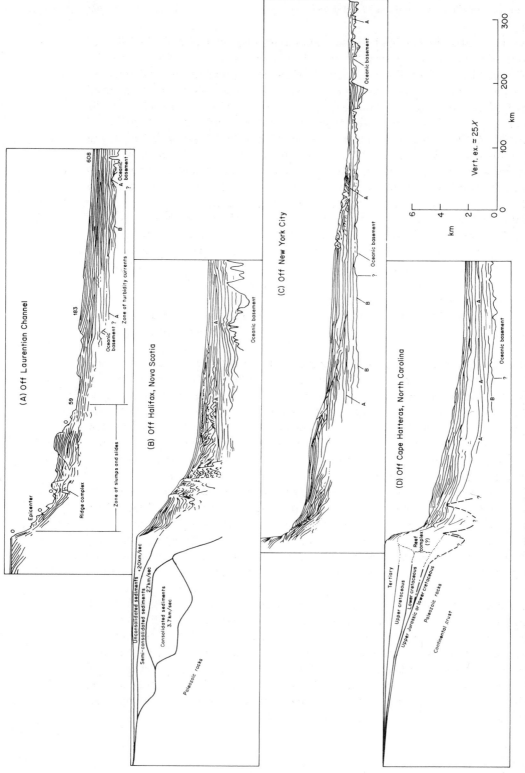

Figure 14.8 Continuous seismic profiles of the continental rise of southeastern Canada and eastern United States. Horizon *A* is a prominent reflecting horizon believed to be of Middle Eocene to Upper Cretaceous age. Horizon *B* is the reflection from the oceanic basement. (After Emery *et al.*, 1970.)

A similar example of large-scale sediment transport from continental slope to continental rise is apparent in Figure 14.8(A). Here we have the additional benefit of data from the 1929 trans-Atlantic cable break to add vivid color to these large-scale sediment-transport processes. During 1929, an earthquake occurred near the Grand Banks. The approximate position of the epicenter is indicated in Figure 14.8(A). Following the earthquake, trans-Atlantic cables began to give way. A number of cables failed at the time of the earthquake; others broke a considerable time after the earthquake. The positions and times of some cable failures are given in Figure 14.8(A). Figure 14.9 gives a summary of time-distance relationships for cable failures.

Figure 14.9 Distance-time plot for cable breaks following the 1929 Grand Banks earthquake. (After Emery *et al.*, 1970.)

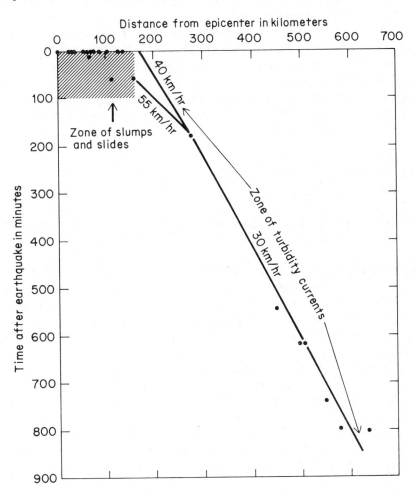

The failure of cables at or near the time of the earthquake appears to have been caused by either the shock of the earthquake or local slumping of continental-slope sediment attendant to the earthquake. As geologists, we would not consider these failures spectacular or surprising. In retrospect, it was just a bit of poor geological engineering. If you lay a cable along the side of a slope, you are simply asking for small movements on the slope to stretch your cable and break it. However, about 10 hours later, something rather spectacular did begin to happen. As indicated in Figure 14.9, seven cables, located 400 to 600 kilometers away from the epicenter, failed in a very systematic fashion. The first to go were the ones closest to the epicenter, and the last to go were the ones farthest away.

Inasmuch as these cable breaks occurred half a century ago and in 6000 meters of water, we shall probably never know what happened. However, a plausible and widely accepted explanation for the delayed breaks is that slumping and sliding on the continental slope and upper continental rise generated turbidity currents that flowed along the bottom at velocities in the range of 30 to 50 kilometers/hour. This interpretation would suggest that turbidity currents still contain considerable energy even after traveling 500 to perhaps 1000 kilometers outward from their source. In addition, it should be noted that the outer cable breaks occurred over an area several hundred kilometers in width, measured along the strike of the continental rise. Thus, the proposed turbidity currents were not only fast and far-reaching, they were also of wide lateral extent.

A Basic Model for Shelf-to-Basin Transition at the Continental Margin

The preceding discussions have been area-oriented. Let us now begin to extract the generalities and combine them into a flexible working model.

Sand sedimentation. In both examples just cited, the most striking generality concerning sand sedimentation is the intimate relationship between deep-water sands and submarine canyons. Shallow-marine processes, such as discussed in Chapter 11, deliver sand to the heads of submarine canyons. As the sand accumulates, gravitational instability ensues and the stage is set for some kind of down-slope transport.

Modern observations confirm that creep and slump move sand down submarine canyons. However, the topography of canyons and the existence of pebbles, cobbles, and boulders of soft sediment derived from canyon walls strongly suggest that any rapid movement of sand down the canyons occurs either periodically or under conditions different from those that may have existed for the last 5,000 years. Such currents may have been genuine turbidity currents, in which the greater density of the turbid water mass provided the impetus for rapid downhill transport. Alternatively, high-velocity water currents generated by tides and storms may have moved large volumes of sediment down the canyons by traction transport. The choice between these two mechanisms remains a matter for speculation; neither process has been observed to operate in the Recent epoch on the scale that is implied by deep-water sand deposits. In any case, we must realize that the erosion accompanying this large-scale transportation is extremely important in maintaining submarine canyons.

Silt and clay sedimentation. Whereas material of sand size and larger is most commonly transported by traction, clay and, to some extent, silt is transported in suspension by low-velocity turbid flows. The existence of widespread turbid bottom water on the continental shelf, slope, and rise is well documented. This turbid water may undergo gravity flow because of the slight density contrast provided by the suspended mineral matter, or the turbid water may be moved along by persistent bottom currents of other origin.

Lack of relationship between sand and clay sedimentation in the Recent epoch. As indicated in Chapter 11, shallow-marine processes tend to separate sand and clay. If poorly sorted material undergoes significant longshore transport, we anticipate that the mud fraction will be winnowed out of the sand. The sand will continue its movement along the shoreline, and the mud will be carried seaward.

In the previously cited examples of continental-margin sedimentation, deep-water sand and mud have been shown to accumulate by more or less independent processes. Therefore, a sedimentation model based strictly upon Recent studies must actually consist of two independent models: one for sand sedimentation and another for clay sedimentation.

To a large extent, the separation of sand and clay sedimentation in the Recent epoch is the direct result of the existence of continental shelves. Such a relatively broad, gently sloping platform allows sand to accumulate in a position of relative gravitational stability. Mud, however, tends to be carried to the edge of the shelf or off the shelf into the basin. Clearly, if there were no continental shelf, then shallow-water sand would accumulate in a position of much greater gravitational instability. Thus, a "no-shelf" model for basin-margin sedimentation would bring sand sedimentation and mud sedimentation into much closer relationship than is observed today.

Relationship of basin-margin sedimentation to sea-level fluctuations. In Chapter 11, we suggested that fluctuating sea level tends to modulate sediment supply to coastal environments. When the sea level falls, stream gradients usually increase, resulting in downcutting and more abundant sediment supply to the coastline. When the sea level rises, the sediment supply is trapped as alluvial valleys build upward to the new equilibrium profile and as deltaic sedimentation builds outward into newly formed estuaries. Obviously, these same considerations apply to the modulation of clastic sediment supply from the land to deep water.

Falling sea level has the additional effect of bringing the shoreline closer to the shelf margin. Sediment transported by longshore currents will then find its way to submarine canyons at a point where canyon topography is more pronounced. In Figure 14.1, for example, note that the present head of La Jolla canyon is a rather small area. In sharp contrast, a sea level 100 meters lower will intersect La Jolla canyon in a position where the canyon is 4 kilometers wide. Furthermore, there are numerous canyons that cut the outer shelf and slope but do not extend to the modern shoreline. With lowered sea level, longshore transport will tend to intersect the heads of many small canyons that are today bypassed by the coarse fraction as it moves along the present high-stand shoreline.

Finally, falling sea level undoubtedly causes higher-energy conditions to impinge upon the silt and clay deposits of the outer shelf and slope. Presumably, these sediments will be remobilized under these new conditions, be it by increased rate of creep, slumping, or turbid flow.

Thus, numerous lines of reasoning suggest that shelf-to-basin transport of clastic sediments would be more rapid during low stands of the sea than during high stands of the sea. Furthermore, the relatively broad width of continental shelves at times of high sea-level stand would afford greater opportunity for clean separation of sand-sedimentation processes from mud-sedimentation processes. During low stands of the sea, sediment is introduced at a shoreline much closer to gravitational instability. Therefore, we would suspect that there is less opportunity for separation of sand from mud prior to downslope transport.

Relationship to tectonic setting. Tectonic activity tends to create local basins and highs that may behave as almost independent sedimentation systems during their early history. Juvenile basins that are fed by large submarine canyons usually fill rapidly, whereas nearby basins may receive little or no sediment. As the juvenile basins fill to sill depth, a graded profile develops. Downslope transportation bypasses the filled basin and begins to build the next basin downslope (see, for example, Gorsline and Emery, 1959; and Moore, 1970). Given tectonic quiescence and continuing sedimentation, the basic sedimentation model takes on larger and larger proportions.

With reference to Recent examples cited earlier, this model is the transition from a San Diego trough model to a northwestern Atlantic model. Indeed, Emery *et al.* (1970) suggest that this transition occurred along the eastern coast of North America throughout the Mesozoic and into the early Tertiary. Seismic evidence on the continental shelves and slopes indicates local basin-filling throughout the Mesozoic with little contribution of turbidies to the abyssal plain. During the mid-Eocene, a relatively smooth continental shelf and slope grew up. Then the continental rise developed, and turbidity currents began depositing material on the abyssal plain.

An Ancient Example: The Taconic Sequence, Cambro-Ordovician, Eastern New York State

It has long been recognized that mountain chains often occupy the position formerly taken by continental margins. The western margin of the Americas provides a Mesozoic-Cenozoic example of this association; the Appalachians of the eastern United States and Canada are an equally spectacular Paleozoic example.

Yet for many years, geologists did not rely heavily upon the analogy between modern continental margins and the sedimentary sequences that preceded mountain building. We referred to those Ancient continental margins as *geosynclines,* and we were not certain that there was anything like them on the face of the earth

Figure 14.10 Geologic sketch map of the Taconic region. Major lithologic terranes of the area are as follows: (1) billion-year-old Grenville basement rocks; (2) Cambro-Ordovician sequence beginning with sandstones and giving way upward to carbonate rocks; (3) Cambro-Ordovician sequence predominantly of shale; (4) Cambro-Ordovician shale, sandstone, and volcanics now metamorphosed; (5) shales and sandstones deposited during major regional tectonism; (6) post-tectonic sediment, including the Helderberg and younger sequences of eastern New York and the Triassic red beds of the Connecticut Valley.

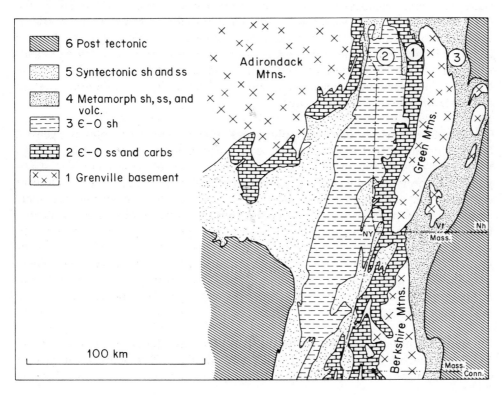

today. We were not sure why geosynclines subsided and filled with sediment, and we were not sure why they came up again to form mountain ranges.

With the rapid development of "the new global tectonics" during the 1960's, geologists began to unravel many of the apparent discrepancies between geosynclines and continental margins. As a broad generalization, for instance, Appalachian tectonism records the mid-Paleozoic closing of the Atlantic. Perhaps, therefore, early Paleozoic sediments of the Appalachians are a better analogy to the modern east coast continental margin than we had previously believed. With this possibility in mind, let us examine briefly one of the classical areas of Appalachian geology.

General lithostratigraphy of eastern New York and Vermont. Figure 14.10 is a generalized geological sketch map of the Taconic region. Billion-year-old Grenville rocks constitute the basement of the area and outcrop in the Adirondacks, Berkshires, and Green Mountains. The basement is overlain by Cambrian orthoquartzite sandstone, which gives way upsection to Cambro-Ordovician carbonate rocks. This section is exposed along the eastern margin of the Adirondacks and along the western margin of the Green Mountains. The area between the Adirondacks and the Green Mountains is largely occupied by Cambro-Ordovician shales, slates, and minor sandstones. This region is the Taconic facies. To the east of the Green Mountains, shale, sandstones, and volcanic rocks have undergone considerable metamorphism. Here chlorite schist predominates. To the west of the Taconic facies and southeast of the Adirondacks, there is a large area of shales and sandstones, deposited at the time of, and in intimate association with, major regional tectonism. Posttectonic sediments of Siluro-Devonian age occur in the southwestern and eastern portion of the map area.

Cambro-Ordovician biostratigraphic correlation. Figure 14.11 presents general facies information and biostratigraphic correlation among lithologies (2) through (4) of the map in Figure 14.10. The general location of these composite stratigraphic sections is indicated in the upper portion of Figure 14.10. Observe especially that lithostratigraphic units (2), (3), and (4) comprise the same biostratigraphic interval, namely Cambrian through Lower Ordovician. Geologists have long realized that sediments of the Taconic sequence must have accumulated somewhere to the east of their present position and must have been thrust westward prior to the metamorphism of their equivalents in eastern Vermont. These Taconic thrust sheets comprise an area some 20 kilometers wide and 200 kilometers long and appear to involve westward thrusting at least on the order of 30 to 50 kilometers.

Structure and stratigraphy within the Taconic thrust sheets. Figures 14.12 and 14.13 summarize the results of detailed structural and stratigraphic analysis of the Taconic thrust sheets. The physical limits of each thrust sheet are worked out

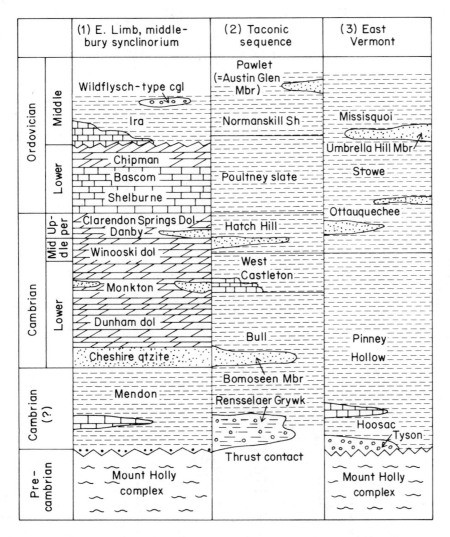

Figure 14.11 Biostratigraphic correlation of principal sequences within the Taconic region. See Figure 14.10 for location of composite sections. (After Zen, 1968.)

on the basis of field relationships. Within each thrust sheet, the biostratigraphic sequence has been determined. It is interesting to note that the first thrust sheets to be emplaced contain the most complete representation of Cambrian-Lower Ordovician stratigraphic units and that the youngest thrust sheets contain only lowermost Cambrian-Precambrian rocks. The emplacement of these thrust sheets is interpreted as a gravity-tectonic unroofing of the source area to the east.

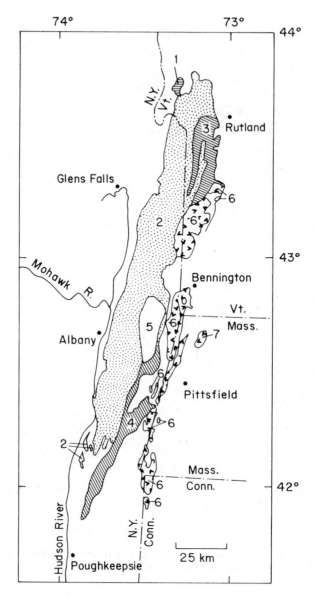

Figure 14.12 Sketch map depicting the extent of major thrust sheets within the Taconics. The thrust sheets are numbered more or less in order of their tectonic emplacement. See Figure 14.13 for additional details. [After E-An Zen, "Time and Space Relationships of the Taconic Allochthon and Authochthon," *Geol. Soc. American Spec.* Paper No. 97, 107 p. (1967).]

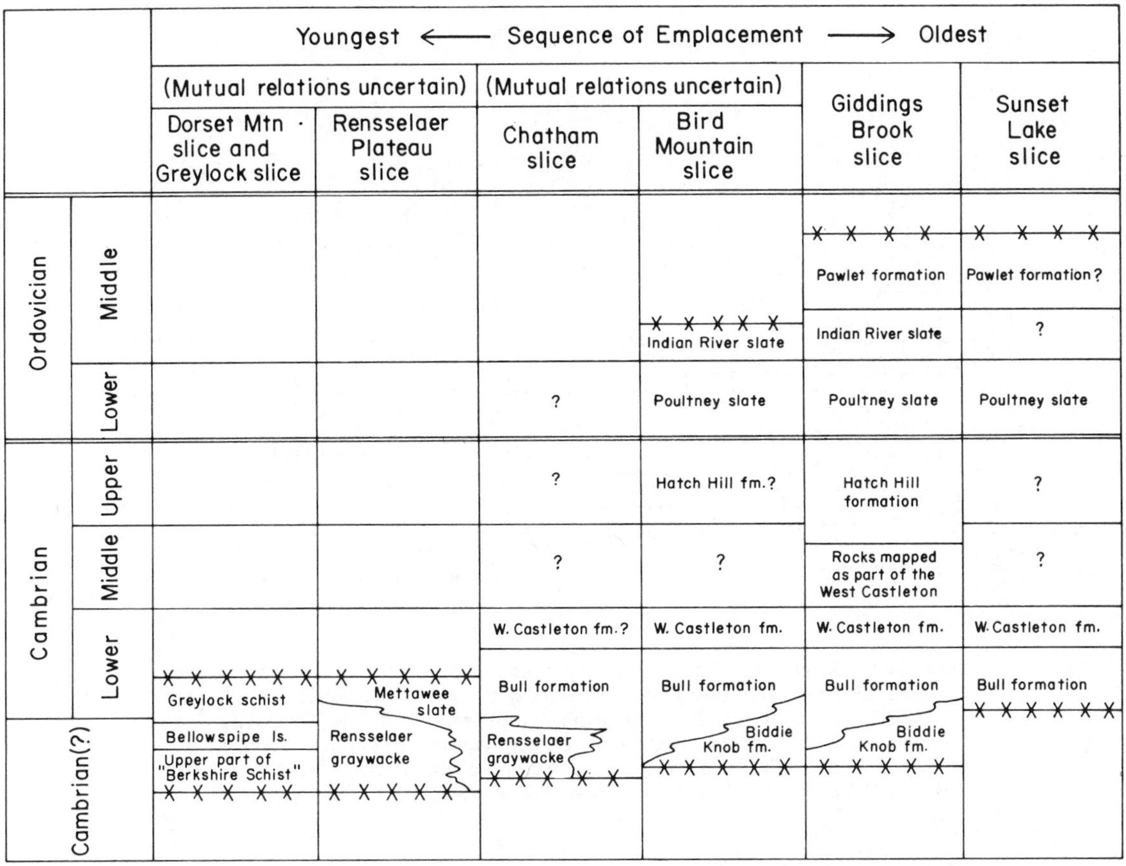

		Youngest ← Sequence of Emplacement → Oldest					
		(Mutual relations uncertain)		(Mutual relations uncertain)		Giddings Brook slice	Sunset Lake slice
		Dorset Mtn · slice and Greylock slice	Rensselaer Plateau slice	Chatham slice	Bird Mountain slice		
Ordovician	Middle					× × × × Pawlet formation	× × × × Pawlet formation?
					× × × × Indian River slate	Indian River slate	?
	Lower			?	Poultney slate	Poultney slate	Poultney slate
Cambrian	Upper			?	Hatch Hill fm.?	Hatch Hill formation	?
	Middle			?	?	Rocks mapped as part of the West Castleton	?
	Lower			W. Castleton fm.?	W. Castleton fm.	W. Castleton fm.	W. Castleton fm.
		× × × × × × Greylock schist	× × × × Mettawee slate	Bull formation	Bull formation	Bull formation	Bull formation × × × × × ×
Cambrian(?)		Bellowspipe ls. „Upper part of "Berkshire Schist" × × × ×	Rensselaer graywacke × × × ×	Rensselaer graywacke × × |× ×	Biddie Knob fm. × × × × × ×	Biddie Knob fm. × × × × ×	

Figure 14.13 Stratigraphic ranges of rocks within the various thrust slices of the Taconics. Lines with crosses indicate the beginnings and ends of sections within these slices. [After E-An Zen, "Time and Space Relationships of the Taconic Allochthon and Authochthon," *Geol. Soc. American Spec.* Paper No. 97, 107 p. (1967).]

Time-space synthesis. In Figure 14.14, we have a time-space schematic diagram of the Taconic region. Cambrian-Lower Ordovician sandstones and carbonates [lithology (2) of Figure 14.10] appear analogous to the modern continental shelf. Cambro-Ordovician fine-grained sediments of lithologies (3) and (4) of Figure 14.10 are interpreted as analogous to the modern continental-rise deposits. With the closing of the Atlantic, commencing in Middle Ordovician time, the continental-rise sedimentary prism underwent collapse and subsequent metamorphism as the Atlantic plate slipped beneath the North American plate. Gravity tectonics emplaced the Taconic thrust sheets early in this process. Continued compression formed the large uplift that shed clastic debris to the west to form the Catskill clastic wedge (Figure 5.4).

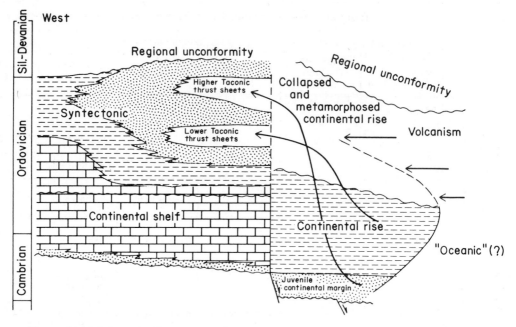

Figure 14.14 Schematic diagram showing time-space relationships in the Taconic region. During Cambrian-Early Ordovician time, the deposition of continental-shelf and continental-rise sequences took place. Middle Ordovician tectonism resulted from the collapse of the continental rise prism concurrent with consumption of the Atlantic plate beneath the North American plate. (After Bird and Dewey, 1970, with extreme simplification.)

Citations and Selected References

BIRD, J. M., and J. F. DEWEY. 1970. Lithosphere plate-continental margin tectonics and the evolution of the Appalachian orogen. Bull. Geol. Soc. Amer. 81: 1031–1060.

BOUMA, A. H., and A. BROUWER. 1964. Turbidites. Elsevier, Amsterdam. 264 p. A collection of papers. Extensive bibliography.

BRIGGS, G., and L. M. KLINE. 1967. Paleocurrents and source areas of late Paleozoic sediments of the Ouachita Mountains, southeastern Oklahoma. J. Sed. Petrology 37: 985–1000.
Turbidity current deposition is dominated by current movement down the axis of the geosyncline throughout the deposition of 4000 meters of sediment.

BUFFINGTON, E. C., D. G. MOORE, R. F. DILL, and J. W. VERNON. 1967. From shore to abyss: near-shore transport, slope deposition and erosion, canyon transport, and deep-basin sedimentation [Abstract]. Bull. Amer. Assoc. Petrol. Geol. 51: 456–457. One of the most informative abstracts ever written.

CUMMINS, W. A. 1962. The greywacke problem. Liverpool and Manchester Geol. J. 3: 51–72.
Possible diagenetic origin of clay matrix in these sands.

DAVIES, D. K. 1968. Carbonate turbidites, Gulf of Mexico. J. Sed. Petrology 38: 1100–1109.
The Campeche shelf supplies carbonate sediment to the abyssal plain.

DIETZ, R. S. 1963. Collapsing continental rises: an actualistic concept of geosynclines and mountain building. J. Geology 71: 314–333.

———, and J. C. HOLDEN. 1966. Miogeoclines in space and time. J. Geology 74: 566–583.

DOTT, R. H., JR. 1963. Dynamics of subaqueous gravity depositional processes. Bull. Amer. Assoc. Petrol. Geol. 47: 104–128.
A readable summary of the various transport processes that deliver clastic sediments to deep-water environments.

EMERY, K. O. 1968. Relict sediments on continental shelves of the world. Bull. Amer. Assoc. Petrol. Geol. 52: 445–464.

———, E. UCHUPI, J. D. PHILLIPS, C. O. BOWIN, E. T. BUNCE, and S. T. KNOTT. 1970. Continental rise off eastern North America. Bull. Amer. Assoc. Petrol. Geol. 54: 44–108.
Reports extensive geophysical surveys concerning structure, sedimentation, and history of the northwestern Atlantic. This entire issue of the bulletin is devoted to the geology of continental margins.

EMILIANI, C., and J. D. MILLIMAN. 1966. Deep-sea sediments and their geologic record. Earth Science Rev. 1: 105–132.
A good starting point for those interested in deep-sea sediments. Extensive bibliography.

ENOS, P. 1969. Anatomy of a flysch. J. Sed. Petrology 39: 680–723.

ERICSON, D. B., M. EWING, and B. C. HEEZEN. 1952. Turbidity currents and sediments in the north Atlantic. Bull. Amer. Assoc. Petrol. Geol. 36: 489–511.

EWING, M., and E. M. THORNDIKE. 1965. Suspended matter in deep ocean water. Science 147: 1291–1294.

GORSLINE, E. S., and K. O. EMERY. 1959. Turbidity current deposits in San Pedro and Santa Monica basins of southern California. Bull. Amer. Assoc. Petrol. Geol. 70: 279–290.

HAND, B. M., and K. O. EMERY. 1964. Turbidites and topography of the north end of San Diego trough, California. J. Geology 72: 526–542.

HANER, B. 1971. Morphology and sediment of Redondo submarine fan, southern California. Bull. Geol. Soc. Amer. 82: 2413–2432.

HAYES, D. E., and M. EWING. 1970. North Brazilian ridge and adjacent continental margin. Bull. Geol. Soc. Amer. 54: 2120–2150.

HEEZEN, B. C. 1963. Turbidity currents, p. 742–775. *In* M. N. Hill (ed.), The sea: ideas and observations on progress in the study of the seas. Interscience, New York.

———, and C. HOLLISTER. 1964. Deep sea current evidence of abyssal sediments. Marine Geol. 1: 141–174.
Bottom photographs provide widespread evidence of current transport in the deep sea.

———, and C. L. DRAKE. 1964. Grand banks slump. Bull. Amer. Assoc. Petrol. Geol. 48: 221–224.
Slump and turbidity currents presumably triggered by an earthquake.

———, C. D. HOLLISTER, and W. F. RUDDIMAN. 1966. Shaping of the continental rise by deep geostrophic contour currents. Science 152: 502–508.

HERSEY, J. B., and M. EWING. 1949. Seismic reflections from beneath the ocean floor. Amer. Geophys. Union, Trans. 30: 5–14.

HUANG, T. C., and H. D. GOODELL. 1970. Sediments and sedimentary processes of eastern Mississippi cone, Gulf of Mexico. Bull. Amer. Assoc. Petrol. Geol. 54: 2070–2100.

HUBERT, J. F. 1964. Textural evidence for deposition of many western North Atlantic deep-sea sands by ocean-bottom currents rather than turbidity currents. J. Geology 72: 747–785.

JACKA, A. D., C. M. THOMAS, R. H. BECK, K. W. WILLIAMS, and S. C. HARRISON. 1969. Guadalupian depositional cycles of the Delaware basin and Northwest shelf, p. 152–197. *In* J. G. Elam and S. Chuber (eds.), Cyclic sedimentation in the Permian basin. West Texas Geological Soc. Pub. 69–56.
Considers basinal sandstones to be of deep-water origin and analogous to modern sedimentation of the continental borderland basins of southern California.

KLEIN, G. DEV. 1966. Dispersal and petrology of sandstones of Stanley-Jackfork, boundary, Ouachita foldbelt, Arkansas and Oklahoma. Bull. Amer. Assoc. Petrol. Geol. 50: 308–326.

KUENEN, P. H. 1967. Implacement of flysch-type sand beds. Sedimentology 9: 203–243.
A veteran geologist takes strong exception to those who question the turbidity current hypothesis for implacement of deep-water sands.

LAJOIE, J. (ed.). 1970. Flysch sedimentology in North America. Geol. Assoc. Canada Spec. Paper 7. 272 p.

MCBRIDE, E. F. 1962. Flysch and associated beds of the Martinsburg formation (Ordovician), Central Appalachians. J. Sed. Petrology 32: 39–91.
An exhaustive study of a formation containing numerous turbidites.

MOORE, D. G. 1970. Reflection profiling studies of the California continental border-land: structure and Quaternary turbidite basins. Geol. Soc. Amer. Spec. Paper 107. 142 p.

NORMARK, W. R. 1970. Growth patterns of deep-sea fans. Bull. Amer. Assoc. Petrol. Geol. 54: 2170–2195.

RODGERS, J. 1968. The eastern edge of the North American continent during the Cambrian and early Ordovician, p. 141–150. *In* E. Zen *et al.* (eds.), Studies of Appalachian geology: northern and maritime. Wiley-Interscience, New York.

———. 1970. The Taconics of the Appalachians. Wiley-Interscience, New York. 271 p.

SANDERS, J. E. 1965. Primary sedimentary structures formed by turbidity currents and related resedimentation mechanisms, p. 192–219. *In* G. V. Middleton (ed.), Primary sedimentary structures and their hydrodynamic interpretation. Soc. Econ. Paleont. Mineral. Spec. Pub. 12.
Proposes that many so-called "turbidites" are not deposited from turbulent suspensions and suggests criteria for distinction among various deep-water resedimentation mechanisms.

SHEPARD, F. P., and R. F. DILL. 1966. Submarine canyons and other sea valleys. Rand McNally, Chicago. 381 p.

———, R. F. DILL, and U. VON RAD. 1969. Physiography and sedimentary processes of La Jolla submarine fan and fan-valley, California. Bull. Amer. Assoc. Petrol. Geol. 53: 390–420.
Excellent summary of physiography, sediment types, and sedimentary processes.

STAUFFER, P. H. 1967. Grain-flow deposits and their implications, Santa Ynez Mountains, California. J. Sed. Petrology 37: 487–508.
Discusses 4000 meters of conformable lower Tertiary sequence from deep-water turbidites through nonmarine beds.

THOMSON, A. F., and M. R. THOMASSON. 1969. Shallow to deep water facies development in the dimple limestone (lower Pennsylvania), Marathon region, Texas, p. 57–78. *In* G. M. Friedman (ed.), Depositional environments in carbonate rocks. Soc. Econ. Paleont. Mineral. Spec. Pub. 14.

VAN DER LINGER, G. J. 1969. The turbidite problem. New Zealand J. Geol. Geophys. 12: 7–50.
Do we really know what they look like and are they really all that abundant?

VON RAD, U. 1968. Comparison of sedimentation in the Bavarian flysch (Cretaceous) and recent San Diego trough (California). J. Sed. Petrology 38: 1120–1154.
A convenient and well-illustrated discussion concerning the choice between turbidity current or normal bottom-current origin for deep-water sediment accumulation. Extensive bibliography.

WALKER, R. G. 1966. Shale grit and grindslow shales: transition from turbidite to shallow water sediments in the upper Carboniferous of northern England. J. Sed. Petrology 36: 90–114.
Ancient analog of La Jolla valley fan and continental shelf sedimentation.

———. 1967. Turbidite sedimentary structures and their relationships to proximal and distal depositional environments. J. Sed. Petrology 37: 25–43.

———. 1967. Upper flow regime bed forms in turbidites of the Hatch formation, Devonian of New York State. J. Sed. Petrology 37: 1052–1058.

———, and R. G. SUTTON. 1967. Quantitative analysis of turbidites in the upper Devonian Sonyea Group, New York. J. Sed. Petrology 37: 1012–1022.
Criteria for recognizing proximal and distal turbidites. Discusses application of the scheme to the interpretation of the basin-filling process.

ZEN, E. 1967. Time and space relationships of the Taconic allochthon and authochthon. Geol. Soc. Amer. Spec. Paper 97. 107 p.

———, W. S. WHITE, J. B. HADLEY, and J. B. THOMPSON, JR. 1968. Studies of Appalachian geology: northern and maritime. Wiley-Interscience, New York. 475 p.

———. 1968. Nature of the Ordovician orogeny in the Taconic area, p. 129–140. *In* Zen *et al.* (eds.), Studies of Appalachian geology: northern and maritime. Wiley-Interscience, New York.

Section *IV*

Cyclicity

in the Stratigraphic Record

In Section III, we emphasized the interaction between sedimentary environments and single events of submergence or emergence. We went no further than to conclude in each discussion that a certain small episode of earth history could be understood in terms of a sedimentation model based on the study of Recent sedimentary environments.

In many sequences of sedimentary rocks, lithologies are repeated over and over again. There is obviously cyclic repetition of environmental conditions. To understand this cyclicity, we must still build on previous discussions. So let us now take all of the points that we have made about alluvial clastics, coastal clastics, shelf carbonates, and so on, and combine them into larger, more inclusive models.

First we shall review cyclic sedimentation in the Pleistocene. If our geological tools cannot help us properly understand Pleistocene cyclicity, how can we really expect to understand cyclicity in Ancient sedimentary sequences? Secondly, we shall examine the classic cyclicity of Paleozoic epeiric sea sedimentation of the continental interior and the more complicated cyclic sedimentation that occurred in areas of greater topographic relief.

15

The Quaternary

as the Key to the Past

Although geologists have often claimed that the present is the key to the past, they have also said that Pleistocene glaciation makes the whole Quaternary period an extremely anomalous time in earth history. This statement may or may not be true, but it is decidedly difficult to prove.

In the following chapters, we shall take a different approach. We shall entertain the possibility that the Pleistocene epoch *appears* to be highly anomalous only because it is so close to the Recent epoch that we can really get a good look at it. If, for example, we had lived several hundred million years ago and had been able to investigate Lower Paleozoic sediments before they became covered by younger sediments, altered by diagenesis and low-grade metamorphism, and mangled by orogeny, we might have found that they were just as complicated as the Pleistocene. In short, let us assume that probably age tends to obscure complexity. Therefore, we should be suspicious concerning the simplicity of *any* geologic sequence and learn to recognize evidences of complexity throughout the geologic record.

Consideration of the Quaternary as the key to the past has a further basis in the fundamental processes of logic. Because Quaternary sediments are young and accessible, we not only know a great deal about them but also can learn a great deal more rather easily. Because of high-resolution radiometric dating techniques, we have far more precise control over Quaternary dynamics than we have over the dynamics of any other time in earth history. This model is the only one that nature allows us to know with any certainty. If it more or less fits our data concerning older rocks, we should not dream up *ad hoc,* untestable models. Only when hard data contradicts the data from the Quaternary are we logically allowed to consider seriously any other *ad hoc* hypothesis. No matter how reasonable that hypothesis may appear, it must be based on data that are *not* compatible with the Quaternary model or it cannot be accepted.

We have previously discussed the dynamics of sea-level fluctuations and tectonism as we know them from the Quaternary. In this chapter, we shall examine the cyclicity of the Pleistocene as the most obvious one for us to use when we approach problems of cyclicity elsewhere in the stratigraphic record. We shall be interested in the dynamics within the various phases of the cycle and in any fundamental frequencies that may exist.

If we can identify certain frequencies and infer that they recur in Ancient cyclic sequences, we can make important conclusions about the interaction among sea-level fluctuation, tectonism, and sedimentation rates. Fundamental frequencies may be used as a tuning fork of geologic time, against which rates can be measured regardless of the absolute age of the cyclic sequence. In such a manner, we can acquire a very precise understanding of time relations within a sequence, without knowing exactly when that sequence of events occurred in the absolute sense of time.

Early Interest in Pleistocene Geology

For many years, scientific studies of Pleistocene geology centered around the spectacular sediments left on the continents by glaciers that once covered much of the populated area of North America and western Europe. We find it difficult to realize that not too long ago several kilometers of ice lay where man's tallest buildings stand today.

At first, the study of continental Pleistocene geology ran into insurmountable difficulties because adequate criteria for time-stratigraphic subdivision was lacking. Numerous advances of the ice sheets obviously had occurred. All of the glacial sequences were clearly discontinuous; each series of glacial deposits were separated by soil zones. Yet there were no adequate means of relating local historical reconstruction to that of another area. Carbon-14, the only radiometric dating method of widespread applicability, places the late Wisconsin ice advance at 20,000 to 25,000 years B. P. All older continental glacial deposits are beyond the range of carbon-14.

Pleistocene History of the Ocean Basis

Oceanic sediments and shallow-water carbonate deposits offer a much more promising possibility for the reconstruction of precise Pleistocene history, primarily because these deposits can be dated by longer-lived radiometric techniques and by magnetic stratigraphy. Furthermore, oceanic sediments reveal a continuous record of Pleistocene history. For these two reasons—the possibility of developing an absolute radiometric time scale and the continuity of the deep-ocean record—the following discussion of Pleistocene cyclicity will emphasize Pleistocene history of the oceans rather than of continental areas.

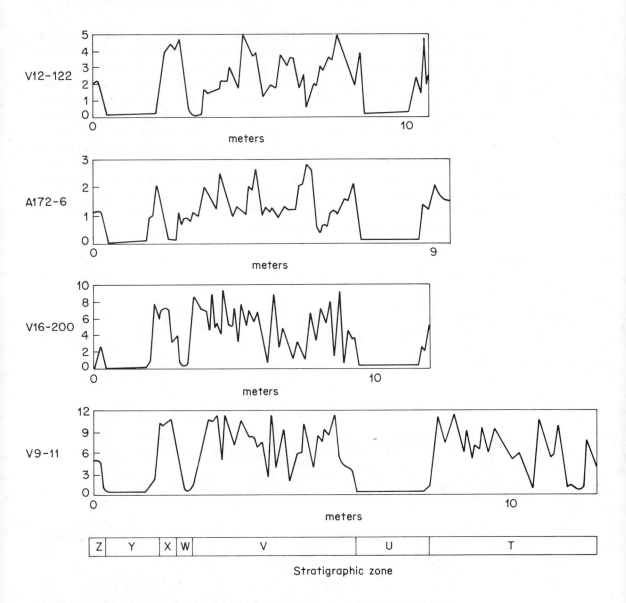

Figure 15.1 Percentage abundance of the warm-water pelagic foraminifera *Globorotalia menardii* in four cores from the Caribbean (V112-122, A71-6) and tropical Atlantic (V16-200, V9-11). Although the two areas are more than 2000 kilometers apart, the warm-cool stratigraphy of *G. menardii* abundance affords good correlation. (After D. B. Ericson and G. Wollin, "Pleistocene Climates and Chronology in Deep Sea Sediments," *Science,* **162** 1227–1234, 13 December 1968). Copyright 1968 by the American Association for the Advancement of Science.

Early recognition of warm-cool stratigraphy in Pleistocene deep-sea cores. Early scientists who studied the foraminifera of Pleistocene deep-sea cores recognized, among other things, that the tropical species, *Globorotalia menardii,* which is moderately abundant in modern tropical and subtropical waters, was alternately absent and present at intervals down Pleistocene cores from the Caribbean and tropical Atlantic. The presence of this species was assumed to indicate that conditions had been similar to the present time. Its absence was taken to represent cooler conditions at the ocean's surface. Carbon-14 dates on the upper portion of such cores allowed estimation of sedimentation rate; then, if a constant sedimentation rate was assumed, the dates provided a crude absolute time stratigraphy for the rest of the core. Later, longer half-line radiometric methods and magnetic stratigraphy were used in correlating absolute time among the cores. The results of such efforts are summarized in Figures 15.1 and 15.2.

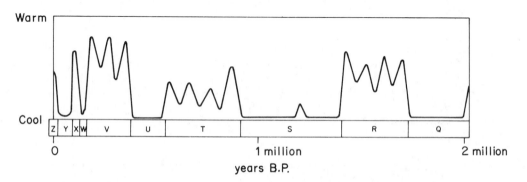

Figure 15.2 Generalized warm-cool history of the Atlantic and Caribbean, as recorded by *G. menardii* abundance. Direct radiometric dating of cores and magnetic stratigraphy allow the estimation of an absolute time scale. (After D. B. Ericson and G. Wollin, "Pleistocene Climates and Chronology in Deep Sea Sediments," *Science,* **162** 1227–1234, (13 December 1968). Copyright 1968 by the American Association for the Advancement of Science.

Ericson assigned letter designations to his zones of the deep-sea Pleistocene. Z stood for the Recent zone, a warm one; then, alternating back in time, Y was cool; X, the next warm; W, the next cool; V, the next warm; U, the next cool; and so on. Radiocarbon dates on the Y zone agree well with radiocarbon dates on the major advance of the Wisconsin glaciation. Cool surface conditions in the Caribbean and tropical Atlantic, therefore, corresponded to glacial advances on the continental areas. The X-zone of the deep-sea cores is dated at around 80,000 to 120,000 years B. P. Presumably, this time was the Sangamon interglacial epoch that separated the Wisconsin from the Illinoisan glacial epochs of continental stratigraphy of North America.

Thus, for the first time, scientists studying the Pleistocene had strong evidence for an absolute chronology of Pleistocene history beyond the range of radiocarbon dating. Unfortunately, this early work recognized a long warm *V*-zone, a point that has been seriously challenged by other geologists. The resulting disagreements have led to a profuse and contradictory literature and have caused most workers to shy away from any attempts to relate deep-sea Pleistocene chronology to continental history beyond the Sangamon.

Pleistocene history of the oceans as recorded by the oxygen-18 content of pelagic foraminifera. The oxygen-18 content of pelagic foraminifera also records ocean history as it is related to the buildup of continental ice caps. Two effects are important.

First, the amount of O-18 in the oceans is increased in proportion to the amount of water tied up in continental glaciers. This fact results from the fractionation of O-18 throughout the hydrologic cycle. As water evaporates from the surface of the oceans, heavy water tends to remain behind. As water vapor condenses to form droplets, the heavy water also tends to condense first and rain out of the atmosphere over the oceans or coastal areas. Thus, it is light water that usually accumulates in continental and mountain glaciers.

Secondly, temperature affects the amount of O-18 taken up by carbonate-secreting organisms. The cooler the water, the more O-18 will be taken into the carbonate skeleton. There is considerable controversy concerning which of these effects, ice volume or temperature, is the most important in determining O-18 values on pelagic foraminifera down a Pleistocene core. These arguments need not concern us here; according to most of the information, the two effects seem to reinforce each other.

The essential generality of Oxygen-18 history in deep-sea cores is presented in Figure 15.3. The O-18 data agree with the *G. menardii* data back to the *V*-zone. The *V*-zone, considered by early paleontological studies as generally warm (Figure 15.2), is shown to be quite cyclic by the O-18 data. Whereas the early foraminifera data suggested that glacial and interglacial conditions alternated on a somewhat random pattern, O-18 data indicate a pronounced first-order rhythm to alternation between glacial and interglacial conditions. Note further that the cycle is consistently asymmetric. Translated into melting ice and therefore sea-level fluctuations, the asymmetry suggests rapid eustatic submergence followed by gradual eustatic emergence.

When variations begin to look systematic, geologists become very excited. Oxygen-18 data, such as presented in Figure 15.3, has become the basis for the development of a more detailed understanding of Pleistocene cyclicity.

Pleistocene sea-level history as recorded by uplifted coral reefs. As indicated in Chapter 12, consideration of the various rates involved leads to the conclusion that tectonically uplifted Pleistocene reef tracts record high stands of the sea. Tectonic uplift has simply served to raise each older terrace up out of the way of each new high stand of the sea, leading to the preservation of each discrete sea-level

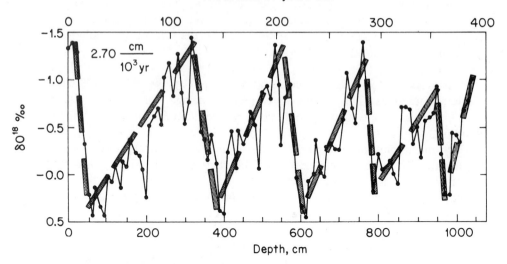

Figure 15.3 A plot of variations in oxygen-18 content of foraminifera tests versus depth in two Pleistocene deep-sea cores. The dots are the data points, and the heavy dashed line calls attention to the primary sawed-tooth cycle. The time scale is that of Broecker and Van Donk (1970). Other geologists have argued for a 20% shorter time scale. Throughout our discussion, we shall adhere to the views of the Broecker studies. [After W. S. Broecker and J. Van Donk, "Isolation Changes, Ice Volume, and the O18 Record in Deep-Sea Cores," *Review of Geophysics and Space Physics,* **8,** 169–198 (1970).]

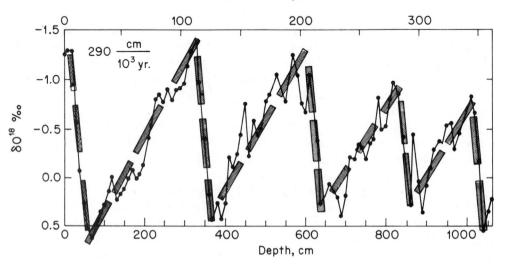

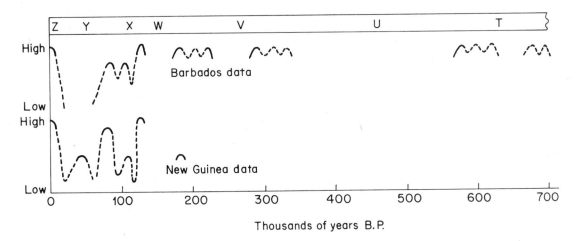

Figure 15.4 Absolute chronology of Pleistocene coral-reef terraces of Barbados, in the West Indies, and New Guinea. Data indicated by solid lines are based on radiometric dates; data indicated by dotted lines are inferred by other methods. (From data of Mesolella *et al.,* 1969, and Veeh and Chappel, 1970.)

event rather than to a compilation of sea-level events one on top of the other in an unrecognizable collage. Furthermore, aragonite corals can be easily dated by the thorium-230 growth method and by the helium growth method. These methods involve a minimum of assumptions, so they provide the best available age determinations for the time range between that dated by Carbon-14 and the Brunhes-Matuyama boundary of magnetic stratigraphy. Thus, uplifted coral-reef tracts provide an easily datable record of Pleistocene sea-level stands.

The Pleistocene coral-reef terraces of Barbados, the West Indies, and of New Guinea, in the south Pacific, have been mapped and dated by radiometric methods. The resultant record of high sea-level stands is presented in Figure 15.4.

The high-stand data are in good agreement with other lines of evidence back to the *V*-zone. Within the *V*-zone, the presence of several distinct terraces tends to confirm the cyclicity of that zone as suggested by O-18 deep-sea data. On the other hand, the absence of terraces in the range of 350,000 to 500,000 years B. P. seems to emphasize that the *U*-zone indeed represents a time set apart from the cool intervals within the *V*-zone.

Quantitative estimates of Pleistocene temperature and salinity based on pelagic foraminifera data. The controversy generated by apparent discrepancies between early foraminifera data and O-18 data led Imbrie and co-workers (Imbrie and Kipp, 1971, for example) to apply quantitative statistical techniques to foram-abundance data for the 20 common pelagic species present in Recent and Pliestocene foraminiferal deep-sea sediments. Factor analysis indicates that a large percentage of the variability within the data matrix can be explained as various combinations of four faunas (that is, the factors). A regression equation can be generated for any parameter of the modern ocean as a function of Z-zone foramini-

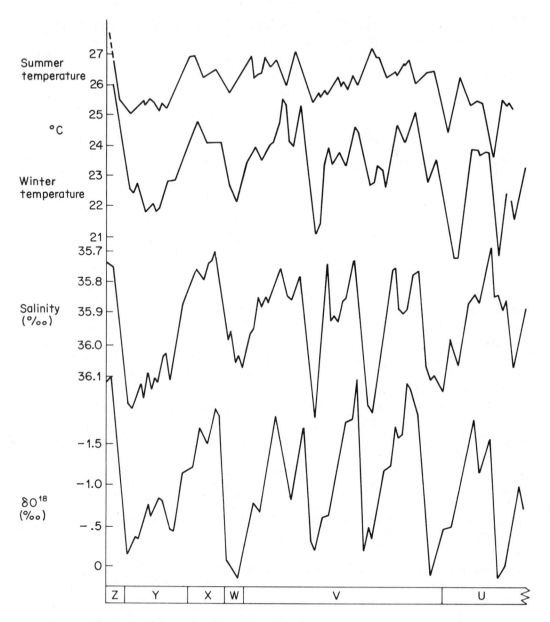

Figure 15.5 Quantitative estimates of Pleistocene temperature and salinity record of a single core based on pelagic foraminifera content. Oxygen-18 determinations are included for comparison. (After Imbrie and Kipp, 1971.)

fera populations from sediments of many widespread localities. These regression equations may then be used to estimate the values of that parameter for Pleistocene oceans at the spot where any individual core was taken.

The results of such an analysis of a Caribbean core are indicated in Figure 15.5. Note that the technique allows independent estimation of summer temperature, winter temperature, average salinity, or whatever parameter of the modern ocean in which we may choose to write a regression equation. The potential of this technique is readily apparent. For example, for the first time we have a hard-data estimate that winter temperature fluctuations are greater than summer temperature fluctuations. Moreover, summer and winter temperature estimates appear to be generally in phase; whereas salinity estimates near the *U-V* boundary are out of phase with temperature estimates, yet still in phase with the O-18 data. The fact that O-18 data correlate well with salinity and are out of phase with temperature suggests that the O-18 data reflect primarily ice volume rather than water temperature. A warming trend in the Caribbean had begun, yet continental ice caps continued to grow for several thousand years. Thus, we are beginning to make some rather precise statements concerning dynamics of the Pleistocene cycles.

The Milankovitch Hypothesis Concerning Pleistocene Climatic Fluctuations

As we continue to study the Quaternary period as a model for cyclicity in the stratigraphic record, we face a choice. We may simply take the empirical data and work with it, or we may try to use a unifying theory.

Most theories of Pleistocene glaciation are of little potential interest to the stratigrapher. Such theories relate only to the Pleistocene; they call on some "special" configuration of the present earth; and they make no statement concerning chronology or repeatability of glacial events.

One theory, however, proposes an external control that must be continuous in time and predictable in its periodicity. This idea is the astronomical theory of climatic change, the Milankovitch hypothesis.

The hypothesis. Stated most generally, the Milankovitch hypothesis holds that variations in the tilt of the earth's axis, precession of the equinoxes, and variation in the eccentricity of the earth's orbit all add together to produce a systematic variation in the distribution of solar radiation between summer and winter. Thus, major climatic change is produced. If, indeed, it can be demonstrated that perturbations of the earth's orbit have played a significant role in Pleistocene climatic change, then the stratigrapher is faced with a fascinating prospect. These orbital changes have been a part of earth history since time began. Their effects on climate may be recorded in the stratigraphic record as sea-level fluctuations or as faunal or floral alternations. In short, the perturbations of the earth's orbit are potentially a tuning fork for geologic time.

The tilt of the earth's axis (that is, the obliquity of the ecliptic) varies slightly with a periodicity of about 40,000 years. The higher the angle of tilt, the more pronounced will become the difference between summer and winter insolation in high latitudes. This particular effect is rather simple to understand. The other two effects—the precession of the equinoxes and variations in eccentricity of the earth's orbit—require a more elaborate explanation.

Figure 15.6 summarizes the combined effect of eccentricity and precession. The eccentricity of the earth's orbit varies with a periodicity of about 90,000 years. The periodicity of the precession of the equinoxes is about 20,000 years. At a time of high orbital eccentricity, the years when the summer solstice occurs at or near the perihelion will be times of hot summers.

Although other variations of the hypothesis are possible, one common version of the astronomical theory of climatic change is that the earth today has a climatic regime that generally favors the accumulation of continental ice caps. Only during

Figure 15.6 Diagram showing the effect of variations in orbital eccentricity upon the insolation intensity of precession maxima and minima. High orbital eccentricity combined with precession maximum result in maximum northern-hemisphere summer insolation and minimum northern-hemisphere winter insolation. High orbital eccentricity combined with precession minimum result in minimum northern-hemisphere summer insolation and maximum northern-hemisphere winter insolation. Low orbital eccentricity tends to modulate these effects. (After Mesolella *et al.,* 1969.)

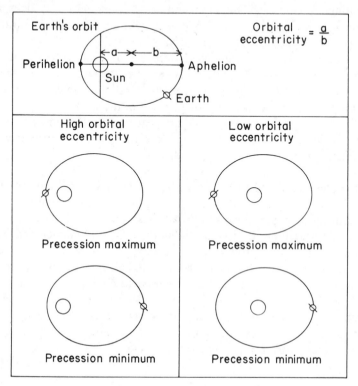

times of unusually hot summers will there be sufficient melting to cause a general retreat of glaciers and therefore an accompanying sea-level rise.

Test of the hypothesis by the Pleistocene empirical data. Precise chronology of Pleistocene periodicity provides a test for the astronomical theory of climatic change. The perturbations of the earth's orbit are sufficiently well known that calculations of northern-hemisphere summer insolation can be accurately carried back at least into the *U*-zone.

Figure 15.7 compares an absolute chronology of calculated northern-hemisphere summer insolation and orbital eccentricity with stratigraphic data. Obviously, the stratigraphic implications of the Milankovitch hypothesis deserve consideration. There is a significant possibility, therefore, that we can use changes in the earth's orbit as the tuning fork of geologic time.

Summary. In ocean sediments, the most clearly demonstrated generality concerning Pleistocene cyclicity is the existence of a fundamental cycle of approximately a 90,000-year frequency. This cycle is well displayed in the oxygen-18 data (Figure 15.3), in the quantitative paleoecological data (Figure 15.5), and in the absolute chronology of uplifted Pleistocene reef tracts. The O-18 data further suggest that the cycle is asymmetrical; that is, rapid sea-level rise was followed by gradual sea-level lowering. Discrepancies between quantitative paleotemperature estimates and the O-18 data indicate that the asymmetry of the cycle is the result of a kickback mechanism rather than a one-to-one response to temperature. Below some threshold condition, continental glaciers are stable. Above some threshold, continental glaciers are unstable and wane rapidly, only to build again.

A second- or higher-order cycle of approximately a 20,000-year frequency is suggested by the data concerning uplifted Pleistocene reef tracts (Figure 15.4) and, to a certain extent, by quantitative paleoecological data (Figure 15.5) and the O-18 data (Figure 15.3).

Both the 90,000-year cycle and the 20,000-year cycle are consistent with the Milankovitch hypothesis. There is a strong implication that the perturbations of the earth's orbit may provide fundamental frequencies in cyclic processes and may thereby afford the stratigrapher a tuning fork for geologic time. We shall explore this possibility in Chapter 17.

Within the available data, we can recognize a vague suggestion of a longer wavelength cyclicity. In the *G. menardii* data, for example (Figure 15.2), *V*-, *T*-, and *R*-zones seem to be clusters of warm oscillations (*G. menardii* often present) separated from one another by long intervals of cool cycles (*G. menardii* generally absent). This same frequency appears in the broad *Y-U* half-cycle of Figure 15.5 paleotemperature estimates. However, the *G. menardii* data are somewhat unsophisticated, and the quantitative paleoecological data do not cover a sufficient time span. So it is difficult to state with certainty that this frequency is real. If it is real, it has a frequency of about 500,000 years. Theoretical explanation of such a frequency is lacking.

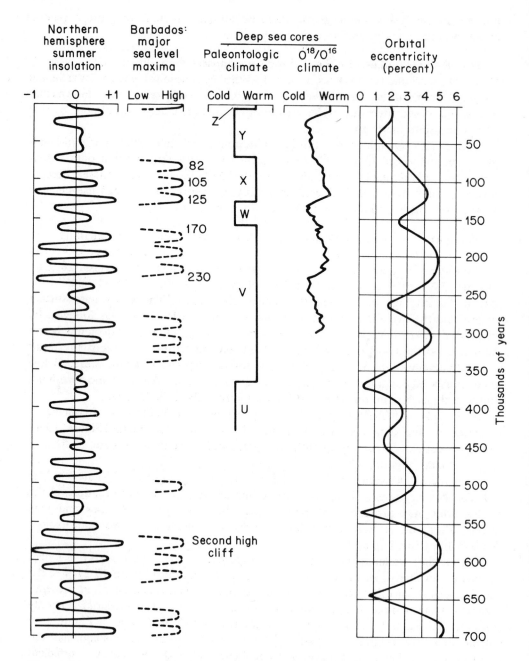

Figure 15.7 Comparison of the various types of oceanic Pleistocene climatic data with the northern-hemisphere summer insolation curve and orbital eccentricity curve of the Milankovitch hypothesis. (After Mesolella *et al.,* 1969.)

Citations and Selected References

BROECKER, W. S., and J. VAN DONK. 1970. Insolation changes, ice volume, and the O^{18} record in deep-sea cores. Rev. Geophys. Space Phys. 8: 169–198.
Easy reading concerning the O^{18} record in Pleistocene deep-sea cores and the general problems of establishing the precise chronology of these cores.

EMILIANI, C., and J. GEISS. 1959. On glaciations and their causes. Geologische Rundschau 46: 576–601.
A thoughtful discussion of possible mechanisms of glacial cyclicity.

———. 1966a. Isotopic paleotemperatures. Science 154: 851–857.

———. 1966b. Paleotemperature analysis of the Caribbean cores, P6304-8 and P6304-9, and a generalized temperature curve for the past 425,000 years. J. Geology 74: 109–126.

ERICSON, D. B., and G. WOLLIN. 1964. The deep and the past. Alfred A. Knopf, Inc., New York. 292 p.
Describes attempts to establish Pleistocene history from deep-sea cores. Nontechnical.

———, and G. WOLLIN. 1968. Pleistocene climates and chronology in deep sea sediments. Science 162: 1227–1234.

IMBRIE, J., and N. G. KIPP. 1971. A new micropaleontological method for quantitative paleoclimatology: application to a late Pleistocene Caribbean core, p. 71–182. *In* K. Turekian (ed.), The Late Cenozoic glacial ages. Yale Univ. Press, New Haven.
Application of statistical techniques to paleotemperatures and to analyses of cyclicity in one late Pleistocene Caribbean core.

KONISHI, K., S. O. SCHLANGER, and A. OMURA. 1970. Neotectonic rates in the central Ryukyu islands derived from thorium-230 coral ages. Marine Geol. 9: 225–240.

KUKLA, J. 1969. The cause of the Holocene climate change. Geol. en Mijnbouw 48: 307–334.
A well-thought-out model based on small fluctuations in seasonal solar input to the northern hemisphere.

MESOLELLA, K. J., R. K. MATTHEWS, W. S. BROECKER, and D. L. THURBER. 1969. The astronomical theory of climatic change: Barbados data. J. Geology 77: 250–274.
Summarizes radiometric dating of the Pleistocene coral-reef terraces of Barbados and relates these data to the Milankovitch hypothesis.

VEEH, H. H., and J. CHAPPELL. 1970. Astronomical theory of climate change: support from New Guinea. Science 167: 862–865.

WRIGHT, H. E., JR., and D. G. FREY. (eds.). 1965. The Quaternary of the United States. Princeton Univ. Press, Princeton, N. J. 922 p.
An excellent review volume for most aspects of geology and history. Some material is becoming slightly dated.

16

Cyclic Sedimentation
in Paleozoic Epeiric Seas
of Central North America

The decidedly cyclic nature of many stratigraphic sequences has intrigued geologists for many years. Examples are numerous, and general discussions of cyclicity fill volumes. The following pages are intended only as a short introduction to cyclicity in Ancient sequences and to the application of Quaternary dynamics to those Ancient sequences.

Well-documented examples of cyclic sedimentation are the cyclothems of the Pennsylvanian in the continental interior of North America. Because of the excellent exposure of these rocks and because of the vast economic importance of the coal beds contained within them, this section has probably received more detailed study than any other section on the face of the earth. (See especially the works of Moore, Wanless, and Weller; all represented in Merriam, 1964.)

The general paleogeography of these deposits consists of a more or less permanent seaway on the west, up through the central part of western United States, with emerging mountain belts in the Appalachians to the east and in the Ouachitas to the south (see Figure 5.1). Examples of sedimentary cycles in western Kansas contain predominantly marine rocks. In eastern Kansas, the section is perhaps two-thirds marine sediments and one-third nonmarine sediments. Passing further eastward into the Illinois basin, we see that the cyclic sediments are about one-half marine and one-half nonmarine. In the Appalachian fold belt from Pennsylvania to Tennessee, Pennsylvanian cycles consist of nonmarine sandstones alternating with coal beds and only occasional marine beds.

Cyclothems of Eastern Kansas

We shall emphasize the section of eastern Kansas because it involves good representations of both marine and nonmarine sediments.

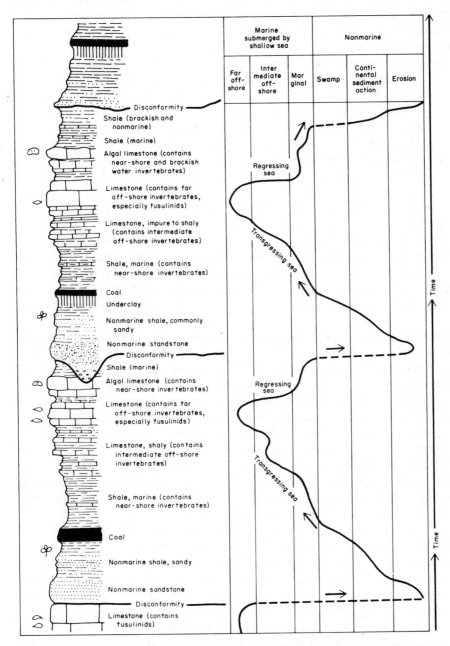

Figure 16.1 Diagrammatic section of successive cyclothems, showing typical lithologic and paleontologic attributes. A typical cycle is 30 to 60 meters thick. The diagram to the right indicates the classical interpretation of the earth history recorded by these sedimentary sequences. (After Moore, 1964.)

335

The Kansas cycle. The generalized concept of a Kansas cyclothem is depicted in Figure 16.1. The cycle is based on lithologic terminology. The general interpretation of earth history recorded by the cycle, however, was estimated simply on the basis of common sense.

According to stratigraphic tradition, descriptions of lithologic sequences begin and end at the most obvious discontinuity surfaces. In the cyclothems of eastern Kansas, the most obvious discontinuity surfaces are those cut by the migration of meandering-stream channels. Thus, the Kansas cyclothem begins with channel-lag gravel, continues upward into nonmarine cross-bedded sands, then passes upward into nonmarine shale. According to the traditional interpretation of the cyclothem, the fining of grain size from channel-lag gravel up to the underclay beneath the coal is assumed to indicate the arrival of a transgressing sea. A coal bed typically separates nonmarine shales below from marine shales above. Because of this fact, the coal has been regarded as a paralic swamp marginal to the transgressing sea.

Marine shales pass upward into marine limestones, which usually contain their greatest diversity of invertebrate fauna near the middle of the limestone unit. The upper portion of the limestone unit generally contains abundant algae and a greatly reduced invertebrate fauna. At the top, marine to brackish-water shale indicates the regression of coastal clastical environments back across the area. The next cycle begins with erosion into these units or perhaps down as deep as the limestone unit.

The place of the Kansas cyclothem in the context of "the Quaternary as the key to the past". We experience a certain amount of difficulty when we try to find examples from Recent sedimentation studies that may strengthen the conceptual model presented within Figure 16.1. The dynamics of the model seem particularly incompatible with the Quaternary model. Whereas the Quaternary model (Figures 4.17, 15.3, and 15.4, for example) indicates rapid submergence followed by gradual emergence, Figure 16.1 interprets the sequence to represent prolonged transgression followed by rapid regression. We are immediately faced with two choices: Either this sequence is subject to reinterpretation in the light of modern sedimentation models, or the epeiric seas of the late Paleozoic in the continental interior were so different from modern depositional environments that we cannot possibly hope to understand them in terms of modern sedimentation.

To test the latter hypothesis, we shall next turn our attention to a sequence not involving clastic influx. In the cyclothems of eastern Kansas, we are clearly dealing with interaction between an event of submergence and the general supply of clastic sediments to the depositional site. If we can move to a study site that does not involve clastic influx, we can remove one exceedingly complicated variable from the model. By understanding the interaction between submergence and carbonate facies in an area of no clastic influx, perhaps then we can return to eastern Kansas with our simple carbonate model and begin to complicate it further with the addition of clastic influx.

A Model For Clear-Water Sedimentation In Epeiric Seas

Irwin (1965) and Shaw (1964) present a theoretical model for carbonate and evaporite sedimentation in an epeiric sea. The development of their model begins with the premise that these shallow seas are rather unlike anything we have studied in the Recent epoch. In particular, they propose that exceedingly restricted conditions can develop within the epeiric seas without the necessity of physical barriers. Just the sheer expansiveness of the shallow-marine environment can itself serve to restrict circulation and thereby lead to hypersaline conditions. Their basic model is summarized in Figure 16.2, and the lithologic sequences of the model are indicated in Figure 16.3.

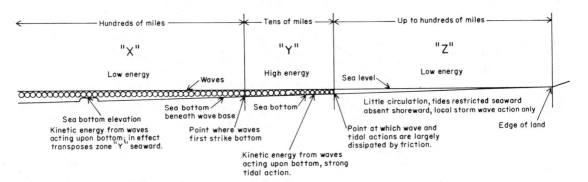

Figure 16.2 (Above) Schematic section showing a theoretical energy zone that may develop in epeiric seas. Not to scale. (After Irwin, 1965.)

Figure 16.3 (Below) Theoretical sediment types to be expected for the various energy zones set forth in Figure 16.2. (After Irwin, 1965.)

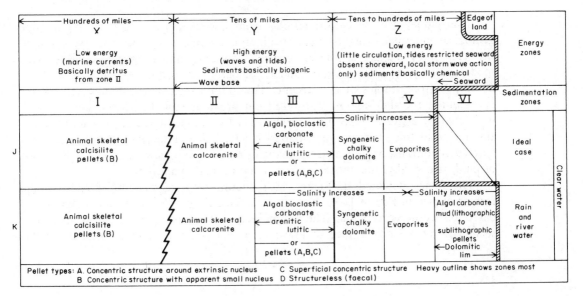

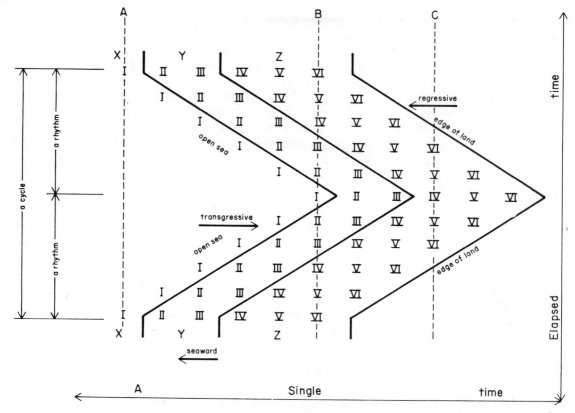

Figure 16.4 A simple model for stratigraphic sections generated by transgression and regression within the clear-water epeiric sea sedimentation model. (After Irwin, 1965.)

This model was developed strictly from common sense, that is, by *ad hoc* geological inference from Ancient stratigraphic sequences. With the exception of the scale and the possible subtidal origin of dolomite and evaporite, however, this model is not very different from models for carbonate sedimentation that we have developed from studies of the Recent. The animal-skeletal calcarenite (facies II) is essentially the shelf-margin portion of our previous models. Here "shelf margin" is defined by effective wave base rather than by dominant preexisting topography. Interior from the shelf margin, animal-skeletal calcarenite gives way to more restricted conditions, as indicated by changing fauna, frequency of pellets, abundance of mud, and finally, the appearance of dolomite and evaporites.

The facies depicted in Figure 16.3 should respond to submergence in the fashion shown in Figure 16.4. Note that the best time lines within this diagram are based on event correlation. The submergence event is recorded in stratigraphic section *B* by the appearance of low-energy, open-marine, animal-skeletal calcilutites. In section *C,* the same submergence event is recorded by the shift from evaporite sedimentation to dolomite sedimentation. In actual practice, stratigraphic studies suggest that the shelf-margin facies II is capable of maintaining its position by vertical accumulation of sediment during submergence. The consequences of this situation, in terms of event correlation, are indicated in Figure 16.5.

Event correlation based on lithologic response to submergence is presented in Figure 16.6 for three idealized sections. Observe the decided asymmetry to lithologic variations in this sequence. Systematic upward transition to more and more restricted facies is abruptly terminated by a sudden freshening of the area (that is, by submergence).

Figure 16.5 Illustration of alternate responses to conditions of submergence, such as may be noted in individual stratigraphic sections. The normal transgressive sequence will simply progress in stepwise fashion from the most restricted at the bottom to the most open-marine at the top. Should a high-energy, animal-skeletal reef be able to keep pace with rising water, it may hold its position and accumulate vertically to significant thickness. Finally, in the backreef position, we may expect to see a very sudden jump from moderately restricted conditions clear out to low-energy, open-marine conditions at such time as the reef environment finally succumbed to rising water. (After Irwin, 1965.)

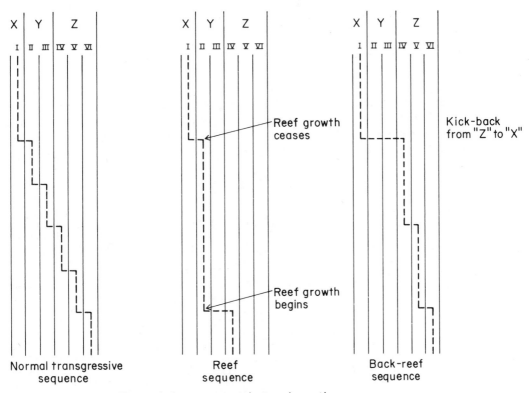

Elapsed times are equal at each section

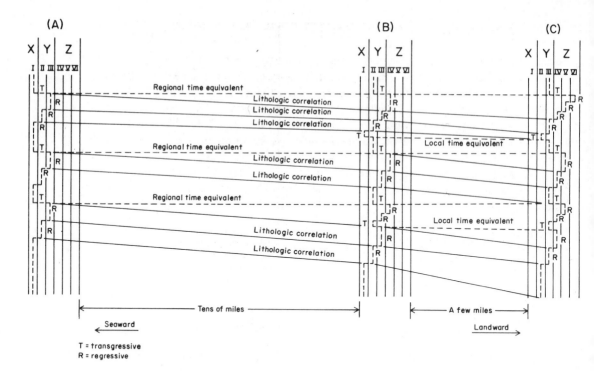

Seaward ←

Landward →

T = transgressive
R = regressive

Figure 16.6 Correlation of *kickbacks* (that is, rapid transgression) to establish time correlation within the clear-water epeiric sea sedimentation model. (After Irwin, 1965.)

It is interesting to note that this abrupt pulse of submergence is precisely the style of Pleistocene sea-level fluctuation discussed in previous chapters. Rapid submergence is followed by progressive restriction of the environment as (1) sediment accumulates vertically, or (2) sea level gradually falls, or (3) both processes occur.

Figure 16.7 (Right) Earth history recorded by eastern Kansas cyclothems, as may be interpreted by variations of the clear-water epeiric sea sedimentation model. The two columns up the center of the interpretation diagram are essentially a compressed version of the original clear-water epeiric sea sedimentation model. Because clastic sediments are important in eastern Kansas, we must add marine clastics and marginal-marine swamps to the left portion of our diagram and progradational clastics and topset backswamp deposits to the right portion of our diagram. Sedimentation from *A* to *B* probably occurs under genuinely transgressive conditions. Sedimentation from *B* to *C* is essentially a basin-filling process that need not be related to eustatic sea-level lowering or any other emergence event. Sedimentation from *C* to *D* records progradation of coastal clastic deposits or meandering-stream deposits, or both, out across an area filled in to sea level by carbonate sedimentation in restricted environments. According to this model, coal deposits may be largely freshwater backswamp rather than the paralic swamps shown in Figure 16.1.

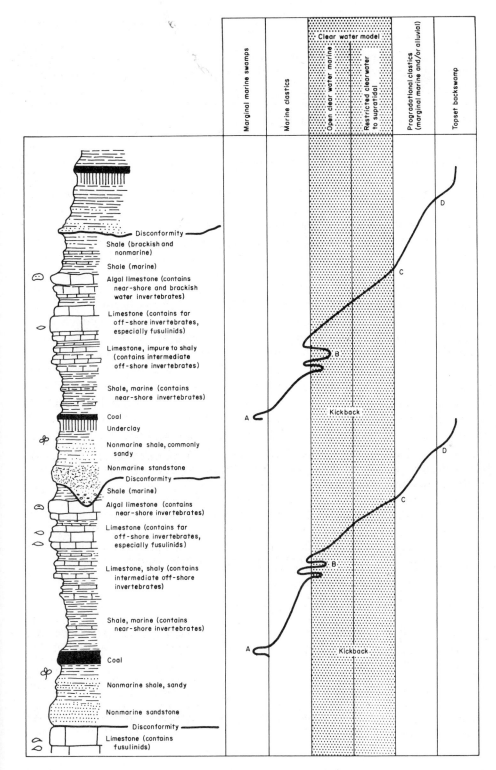

Thus, the lithologies making up this clear-water epeiric sedimentation model are relatively familiar to us from our studies of Recent carbonate environments, and the style in which the lithologies fit together to make cycles is suggestive of Quaternary cyclicity. Made confident by these generalizations, we shall now re-examine the classical cyclothem of eastern Kansas.

An Alternate View of the Kansas Cyclothem

Let us develop a model for the Kansas cyclothem (Figure 16.1) that combines the Irwin-Shaw epeiric clear-water sedimentation model with some modern views of clastic sedimentation. First, let us consider where the cycle starts. Presumably, the main event we want to trace here is the rapid submergence that we have come to expect from the Quaternary and the clear-water sedimentation models. We begin a cycle, therefore, with the first marine rocks in the sequence (Figure 16.7). In the absence of clastic influx, marine-carbonate sediments overlie the most restricted facies present. With the presence of clastic influx, a variable amount of marine shale precedes carbonate sedimentation. Animal-skeletal carbonate sediments pass upward into algal limestone; presumably the transition II–III of the clear-water sedimentation model. Marine to brackish-water shale overlying the algal limestones records the regression of coastal clastic sedimentation, whether it was associated with an emergent sea-level event or whether it was simply progradation associated with basin filling. Finally, meandering streams have deposited a fining upward unit of nonmarine sandstone and shale, which is more or less unrelated to either submergence or emergence. The area is so big and so flat that whether sea level is a few tens of meters lower probably makes little difference. An extremely low-lying land area is covered with upward-fining cycles of meandering-stream facies and the backswamps (coal) associated with them.

The purpose of this discussion has been not to demonstrate that Figure 16.7 is a better model than Figure 16.1 but rather to demonstrate that a model compatible with Recent sedimentation and Quaternary dynamics can be offered for Paleozoic epeiric sea cyclicity. We cannot accept an *ad hoc* model until it has been demonstrated that hard data are incompatible with a Quaternary model.

Citations and Selected References

BEERBOWER, J. R. 1961. Origin of cyclothems of the Dunkard Group (Upper Pennsylvanian-Lower Permian) in Pennsylvania, West Virginia, and Ohio. Bull. Geol. Soc. Amer. 72: 1029–1050.

DAPPLES, E. C., and M. E. HOPKINS (eds.). 1969. Environments of coal deposition. Geol. Soc. Amer. Spec. Paper 114. 204 p.
Six papers concerning accumulation of organic sediments in Recent and Ancient sedimentary environments.

DeRaaf, J. F. M., H. G. Reading, and R. G. Walker. 1965. Cyclic sedimentation in the lower Westphalian of North Devon, England. Sedimentology 4: 1–52.
Clastic cycles of turbidite through deltaic depositional environment.

Fischer, A. G. 1964. The Loafer cyclothems of the Alpine Triassic, p. 107–150. *In* D. F. Merriam (ed.), Symposium on cyclic sedimentation. State Geol. Survey of Kansas, Bull. 169.
Recognizes approximately 200 cycles in subtidal to supratidal carbonates. A grand cycle is often composed of 5 to 8 cycles.

Irwin, M. L. 1965. General theory of epeiric clear water sedimentation. Bull. Amer. Assoc. Petrol. Geol. 49: 445–459.
Kickback correlation on the basis of a theoretical model.

Merriam, D. F. (ed.). 1964. Symposium on cyclic sedimentation. State Geol. Survey of Kansas, Bull. 169. 636 p.
Numerous papers by authors representing many divergent views.

Momper, J. A. 1966. Stratigraphic principles applied to the study of the Permian and Pennsylvania systems in the Denver basin, p. 87–90r. Wyoming Geological Association twentieth annual conference.
Deduces by independent lines of evidence that there is a 100,000-year and a 500,000-year cycle in these sequences.

Moore, R. C. 1964. Paleoecological aspects of Kansas, Pennsylvania, and Permian cyclothems, p. 287–380. *In* D. F. Merriam (ed.), Symposium on cyclic sedimentation. State Geol. Survey of Kansas, Bull. 169.

School, D. W. 1969. Modern coastal mangrove swamp stratigraphy and the ideal cyclothem, p. 37–62. *In* E. C. Dapples and M. E. Hopkins (eds.), Environments of coal deposition. Geol. Soc. Amer. Spec. Pub. 14.
A thoughtful comparison of Recent south Florida swamp stratigraphy and the cyclothem of Pennsylvania strata of the mid-continent.

Shaw, A. B. 1964. Time in stratigraphy. McGraw-Hill, New York. 365 p.

Van Houten, F. B. 1964. Cyclic lacustrine sedimentation, Upper Triassic Lockatong formation, Central New Jersey and adjacent Pennsylvania, p. 497–532. *In* D. F. Merriam (ed.), Symposium on cyclic sedimentation. State Geol. Survey of Kansas, Bull. 169.
Recognizes three scales of cyclicity, presumably relating to climatic fluctuations. Short cycles, 14 to 20 feet in thickness, are related to the 21,000-year precession cycle. Intermediate cycles of 70 to 90 feet and long cycles of 325 to 350 feet are also recognized.

Wanless, H. R. 1964. Local and regional factors in Pennsylvanian cyclic sedimentation, p. 593–607. *In* D. F. Merriam (ed.), Symposium on cyclic sedimentation. State Geol. Survey of Kansas, Bull. 169.

———, and J. R. Cannon. 1966. Late Paleozoic glaciation. Earth Science Rev. 1: 247–286.

17
The Multimodel Approach
to Cyclic Sedimentation

In one of the most challenging approaches to cyclic sedimentation, models with entirely different facies alternate in time over the same paleogeography. For example, high-stand deposits alternate with low-stand deposits, all woven together in a complicated stratigraphy. This concept has produced a major breakthrough in our understanding of the environmental stratigraphy of sequences, which previously had defied explanation by any single sedimentation model. The simplest form of the concept is known as reciprocal sedimentation. Reciprocal sedimentation simply proposes that a complicated stratigraphy is explained not by a single sedimentation model but rather by two sedimentation models alternately occupying the same paleogeography.

The concept of reciprocal sedimentation was popularized by Wilson (1967) in the Upper Pennsylvanian strata of the Sacramento Mountains and the Oro Grande basin of New Mexico. However, we shall use the Middle Permian stratigraphic sequence of west Texas for our example of the multimodel approach to cyclic sedimentation.

Stratigraphic Relationships of the Shelf-to-Basin
Transition, Guadalupian (Middle Permian) of
West Texas

Figure 13.11 provides a cross section from shelf to basin through the Guadalupe Mountains of west Texas. Figure 17.1 indicates stratigraphic correlation within these rocks. The distribution of clastic sediments within this correlation chart has long puzzled sedimentologists and stratigraphers interested in precisely how these physical relationships came about. Sediments well back onto the shelf consist of interbedded evaporites and sandstones. These rocks grade laterally into subtidal

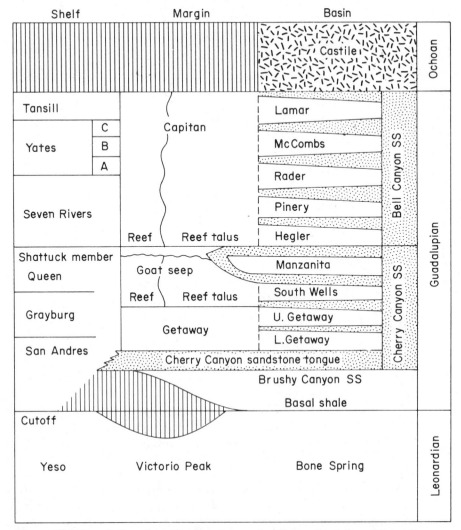

Shelf		Margin	Basin	
			Castile	Ochoan
Tansill		Capitan	Lamar	
Yates	C		McCombs	
	B			
	A		Rader	
Seven Rivers			Pinery	Bell Canyon SS
		Reef \ Reef talus	Hegler	
Shattuck member		Goat seep	Manzanita	
Queen		Reef \ Reef talus	South Wells	Cherry Canyon SS
Grayburg		Getaway	U. Getaway	
			L. Getaway	
San Andres		Cherry Canyon sandstone tongue		
			Brushy Canyon SS	Guadalupian
			Basal shale	
Cutoff				
Yeso		Victorio Peak	Bone Spring	Leonardian

Figure 17.1 Correlation chart for Permian rocks of the Guadalupe Mountains, Texas and New Mexico. (After Newell *et al.*, 1953, with modification.)

lagoonal and shelf-margin carbonate sediments that contain very little detrital sand. As we proceed basinward, we see abundant interlocking relationships that demon-state the basic contemporaneity of reef talus, basinal limestones, and detrital clastics.

The problem posed by these relationships is rather apparent. How did clastic sediments get across the outer shelf and shelf margin for contemporaneous deposition in the basin? The question is further complicated by the contention of some geologists that sands of the Delaware Group contain primary structures and flora indicative of shallow-water sedimentation. How can these sediments be shallow-water when a contemporaneous, subtidal, carbonate, bank-margin facies stands more than 300 meters above them in paleographic reconstruction?

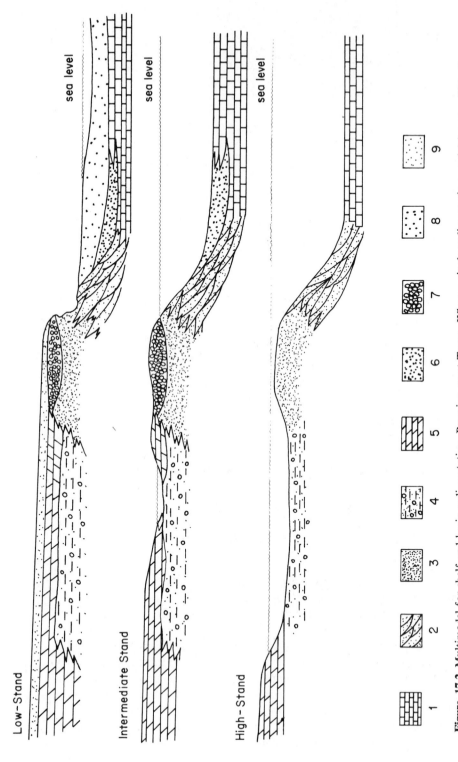

Figure 17.2 Multimodel for shelf and basin sedimentation, Permian, west Texas. When a single sedimentation model is not sufficient to explain the stratigraphic complexity encountered, a cyclic multimodel should be considered. Schematic implications concerning scale or bed thickness are not intended here. (See text for discussion.) Sediment types are as follows: (1) dark micritic limestone with pelagic fauna; (2) forereef detritus, predominantly carbonate; (3) shelf-margin facies, possibly reefs, skeletal sands, oolites, or eolian sands; (4) subtidal shelf-interior, degree of restriction variable and unspecified; (5) intertidal-supratidal carbonates, dolomites, evaporites, and locally coastal terrigenous clastics; (6) submarine debris flows, possibly involving shelf carbonates or clastics; (7) bank-margin, subaerial carbonate facies, eolianite, or such subaerial alteration products as pisolites and caliche; (8) low-stand basin clastics; (9) nonmarine clastics, eolian or alluvial.

346

A Multimodel Approach to the Middle Permian of West Texas

A multimodel such as portrayed in Figure 17.2 offers an explanation for the first-order complications of Middle Permian stratigraphy of west Texas. At one time, the high stand of the sea may have put subtidal conditions over the preexisting shelf topography. During such a time, it will be very difficult for clastic sediments to reach the deep basin. Any suspended matter carried from the shelf out into the basin is likely to be carbonate mud, and any debris flows or turbidites that come off the shelf into the basin are likely to be carbonate sands rather than clastic sands. The net result, therefore, is the accumulation of basin sediment consisting primarily of fine-grained carbonate with a sparse pelagic fauna and occasional debris tongues of displaced carbonate-shelf, skeletal detritus.

An intermediate model may result from a high stand that was somewhat lower or simply from restriction brought about by continued shelf sedimentation during a high stand. In such a situation, development of islands may characterize the shelf margin. The restricted and constricted lagoon is now a less formidable barrier to terrigenous clastic transport across the shelf. For example, eolian sand dunes may migrate across sabkha flats to deliver clastic sand to the deep basin. Thus, terrigenous clastics begin to show up in the basin sequence, whereas they are not recorded on the shelf at this stage in the cycle.

Finally, with a low stand of the sea, the entire shelf may stand as a high plateau above the low-stand sea level in the basin. Bioclastic sedimentation has ceased. Mass wasting of the shelf margin under subaerial conditions may produce carbonate debris tongues out into basin sediments, which are otherwise dominated by terrigenous deposition. Note that the emergent shelf platform may or may not accumulate sediment during this stage of the cycle. Terrigenous clastics of eolian or meandering-stream facies may be expected. Alternatively, the subaerially exposed limestone platform may simply provide the surface transportation of sand to the basin without any net accumulation of detrital sediments on the carbonate platform.

The Scale of Cyclicity in the Middle Permian of West Texas

On what scale do we apply the preceding multimodel? Does a single cycle explain the accumulation of a few meters of sediment, or does it explain several hundred meters of sediment? The best answer is probably a noncommittal, "yes." Our previous experience with sedimentation models gives us some feeling for the variability that may occur from on cycle to another. Moreover, we have seen in Pleistocene cyclicity how short-term cyclicity may accompany longer-term periodicity.

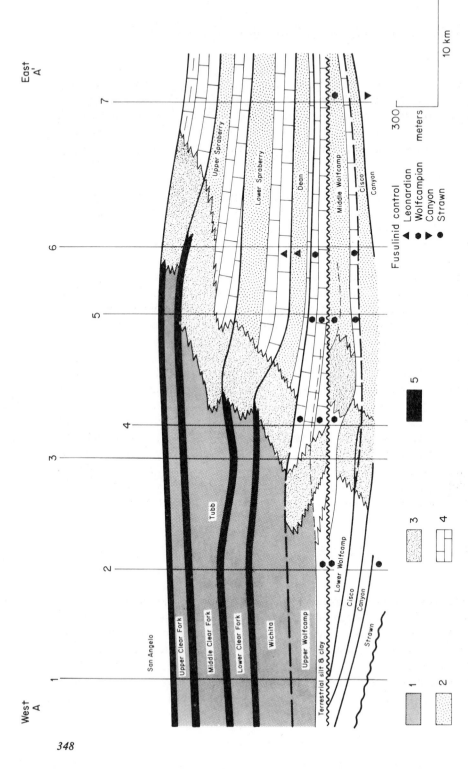

Figure 17.3 Cross section depicting shelf to basin transition, northwest margin of the Midland basin. The basic stratigraphic generalities are explained by a two-stage multimodel. Facies (1), (3), and (4) are associated with high stands of the sea; facies (2) and (5) are associated with the intervening lowstands of the sea. Sediment types are as follows: (1) shelf evaporites and carbonates; (2) basin clastics; (3) shelf-margin carbonates; (4) basin carbonates; (5) shelf clastics. (After Silver and Todd, 1969.)

First-order cyclicity. Figure 17.3 is a cross section from shelf to basin on the northwest flank of the Midland basin. It presents an interpretation of the first-order cyclicity of Middle Permian strata according to Silver and Todd (1969). Note that the cycles are best defined on the shelf and in the basin; correlation across the shelf margin is more or less pushed through. On the shelf, sequences of 150 to 300 meters of shelf evaporite and carbonate sediments are periodically interrupted by sheets of sandstone and shale 15 to 80 meters thick. The basinal cycle consists of a couplet of carbonate overlain by detrital clastics. The carbonate is typically dark, laminated micrite. The terrigenous detritus consists of quartzitic siltstones and claystones.

Each cycle begins with submergence. This fact places carbonate and evaporite sedimentation on the shelf and at the same time restricts the supply of terrigenous clastics to the basin. Consequently, the micritic, carbonate basin sediment is allowed to accumulate. Each cycle ends with the progradation of continental clastic sediments across the shelf, thus renewing the supply of clastic sediments to the basin.

Silver and Todd argue convincingly that this cyclicity is primarily the result of eustatic sea-level fluctuations. Their argument is based primarily on the uniformity of sediment thickness within each cycle over large areas of west Texas. If the cycles were the result of local tectonic activity, unusually thick sections should occur in areas of greater tectonic subsidence and unusually thin sections should occur in areas of less rapid tectonic subsidence. The absence of such thickness variations implies a widespread controlling mechanism, presumably eustatic by default.

Second-order cyclicity. Reeves Field is located in the same general northwestern portion of the Midland basin as is depicted in Figure 17.3. Oil production comes from a portion of the San Andres formation, the next major cycle above the Upper Clear Fork of Figure 17.3. From detailed petrographic study of the producing horizon in Reeves field, Chuber and Pusey (1969) recognized four distinct cycles within less than 30 meters of shelf-margin section. Their typical shelf-margin cycle is shown in Figure 17.4.

The lithologic marker that separates cycles is a brown, burrow-mottled, organic shale, typically 70 centimeters thick, and having sharp lower contact and gradational upper contact. Chuber and Pusey consider that the cycle both begins and ends within this shale. The shale presumably represents a soil zone that becomes reworked by the transgressing sea as submergence begins.

Note that the cycle is decidedly asymmetrical. The lowest unit is dark-gray lime mudstone, which becomes lighter in color and increasingly fossiliferous up the section, changing gradationally into abundantly fossiliferous light-gray wackestone. The wackestone passes gradationally upward into gray to brown fossiliferous packstone. The packstone is capped with a thin unit of oolitic grainstone, which may represent either shallow, subtidal, oolite deposition or development of a thin, caliche crust under subaerial conditions.

The cycle is interpreted as representing rapid submergence followed by shallowing conditions. Such a situation may result from sediment accumulation or from gradual emergence, or from both. Judgments about the scale of each submergent event depend to a considerable extent on the assumptions concerning paleogeography at the time of submergence. The occurrence of stromatolitic deposits overlying skeletal wackestones in cycles of the shelfward portion of Reeves Field led Chuber and Pusey to suggest that the submergence events were on the scale of 10 meters or less.

It is interesting to consider just how many of these small cycles may be present in any one of the first-order cycles discussed earlier. The entire San Andres formation in Reeves Field is approximately 300 meters thick. Inasmuch as Chuber and Pusey have demonstrated that 30 meters of section contain four cycles, we may assume, therefore, that the entire San Andres formation in this area may contain as many as 40 cycles.

Figure 17.4 The basic sedimentation cycle of the producing zone, Reeves Field, Midland basin west Texas. Whereas the gross aspects of Permian basin lithology can be subdivided into cycles involving hundreds of meters of section (such as depicted in Figure 17.3), detailed petrographic investigation of shelf-margin carbonate facies reveals cyclicity on a much smaller scale. (After Chuber and Pusey, 1969.)

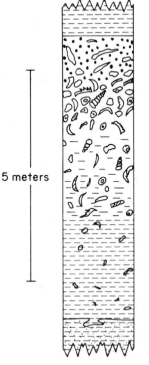

5 meters

Sharp contact
   grey or brown oolitic and fossiliferous packstone

Gradational contact
   grey or brown fossiliferous packstone

Gradational contact

   grey fossiliferous wackestone

Gradational contact

   dark grey mudstone, becoming lighter and more fossiliferous upwards

Gradational contact
   brown, burrow-mottled organic shale

Many geologists, however, consider that the San Andres formation must surely represent two of the first-order cycles, such as indicated in Figure 17.3. (Indeed, this complication is probably the diagram from which Figure 17.3 was adapted.) Thus, if we attempt to place the second-order cyclicity of Chuber and Pusey within the first-order cyclicity of Silver and Todd, we can expect that each first-order cycle contains approximately 20 second-order cycles.

Comparison with Quaternary cyclicity and Recent sedimentation rates. Recent sedimentation rates, the basic asymmetry of Pleistocene cycles, and the periodicities of Pleistocene cycles provide the basic ingredients for a multimodel. If a Quaternary model does not work, then we are stymied; we must put together our own *ad hoc* model and just say it is the best we can do. On the other hand, if we can demonstrate that a model based on the Quaternary is a reasonable model, we may be a step closer to understanding that "tuning fork of geologic time" discussed in Chapter 15.

Clearly, the pronounced asymmetry of second-order cycles encourages us to think of the analogy with Pleistocene sea-level fluctuations. As discussed in Chapter 4, eustatic sea-level fluctuations seem to be about the only mechanism that can produce submergence rates exceeding shelf-margin carbonate-sediment production rates. The asymmetry of second-order cyclicity must be taken as strong evidence for glacio-eustatic control of Middle Permian cyclic sedimentation. We mentioned earlier that Silver and Todd reached the same conclusion by an independent line of reasoning.

Thinking further on cycles within the Pleistocene, we logically must consider the last 130,000 years of earth history. Rapid, major eustatic sea-level rise occurred around 125,000 years ago and was followed by alternating low stands and high stands. High stands of a lower degree occurred at 105,000, 82,000, and perhaps around 60,000 and 30,000 B. P. The present high stand was another major rapid eustatic sea-level rise comparable to the 125,000 B. P. event. Can this grand cycle from 125,000 B. P. to the Recent be the basic model for the first-order cyclicity of Permian sedimentation?

Probably not. First and foremost, such a model would imply extremely rapid sedimentation rates on the Permian carbonate shelves. Permian cycles typically involve as much as 300 meters of carbonate-shelf sediment. With high stands occupying no more than 50% of the time represented by the grand cycle, carbonate-shelf sedimentation rates of 6 meters/1000 years are indicated. Such a sedimentation rate is extremely rapid but perhaps not completely unreasonable.

The proposition is further damaged by our estimates concerning the number of second-order fluctuations for each first-order cycle. The last 120,000 years of Pleistocene history lead us to suspect no more than five second-order cycles per large cycle. Again, this rate is perhaps not totally incompatible with the Permian data, but it is certainly not very consistent.

Seeking a still lower frequency for first-order cyclicity within the Pleistocene, we have to consider the similarities among the *Y, U, S,* and *Q* zones (Figure 15.2). Could these similarities represent the primary frequency of Permian cyclic sedimentation? Such a first-order cycle would represent approximately 500,000 years.

Considering one-half of this time to be higher order high stands and one-half of the time to be higher order low stands, we estimate a maximum shelf-sedimentation rate of approximately 1.2 meters/1000 years.This figure is certainly reasonable in comparison with Recent sedimentation rates. Furthermore, the sequence from Pleistocene *Y* zone to *U* zone appears to contain 14 high stands of one magnitude or another. (See Figure 15.7; the warm peaks are counted as high stands.) This fact agrees with our estimate of 20 higher-order cycles per first-order cycle in the Permian of west Texas.

As a final test of how well this model fits the Permian data, let us calculate the rate of tectonic-isostatic subsidence implied by the time scale that we have assumed. In Figure 17.3, observe that shelf sedimentation begins and ends with approximately the same shallow-water facies over the shelf areas. Thus, the thickness of sediment accumulated is a proper measure of the net submergence of the shelf by some combination of net eustatic sea-level rise, tectonic downwarping, and isostatic downwarping under sediment load. For the sake of argument, let us assume no net eustatic sea-level rise. Thus, approximately 1000 meters of tectonic-isostatic downwarp would have occurred within approximately 2 million years. One-half meter/1000 years is a reasonable rate of tectonic-isostatic subsidence by analogy with Quaternary examples.

In summary, the attempt to fit a theory of Pleistocene cyclicity to Permian data is at least encouraging. If the models fit at all, the higher-order cyclicity documented by Chuber and Pusey (1969) is not second-order but rather third-order. If the Milankovitch theory works at all, a second-order cyclicity should exist, and it would regulate the position and thickness of the third-order cycles documented by Chuber and Pusey.

Thus, an attempt to test the Quaternary model in Permian strata has led to a hypothesis concerning the detailed stratigraphy of higher-order cyclicity. The stratigraphic and perhaps economic significance of such a hypothesis is intriguing.

Citations and Selected References

CHUBER, S., and W. C. PUSEY. 1969. Cyclic San Andres facies and their relationship to diagenesis, porosity, and permeability in the Reeves oil field, Yoakum County, Texas, p. 136–151. *In* J. G. Elam and S. Chuber (eds.), Cyclic sedimentation in the Permian basin. West Texas Geol. Soc. Pub. 69–56.

DUNHAM, R. J. 1969. Vadose pisolite in the Capitan Reef (Permian), New Mexico and Texas, p. 182–191. *In* G. M. Friedman (ed.), Depositional environments in carbonate rocks.
Soc. Econ. Paleontologists and Mineralogists Spec. Pub. 14, 209 p.

————. 1970. Stratigraphic reefs vs. ecologic reefs. Bull. Amer. Assoc. Petrol. Geol. 54: 1931–1932.

HAMILTON, W., and D. KRINSLEY. 1967. Upper Paleozoic glacial deposits of South Africa and southern Australia. Bull. Geol. Soc. Amer. 78: 783–800.

NEWELL, N. D., J. K. RIGBY, A. G. FISCHER, A. J. WHITEMAN, J. E. HICKOX, and J. S. BRADLEY. 1953. The Permian reef complex of the Guadalupe Mountains region, Texas and New Mexico. W. H. Freeman and Co., San Francisco. 236 p.

SILVER, B. A., and R. G. TODD. 1969. Permian cyclic strata, northern Midland and Delaware basins, west Texas and southeastern New Mexico. Bull. Amer. Assoc. Petrol. Geol. 53: 2223–2251.

WILSON, J. L. 1967. Cyclic and reciprocal sedimentation in Virgilian strata of southern New Mexico. Bull. Geol. Soc. Amer. 78: 805–818.

Glossary

This glossary is intended only as a convenient supplement to the text. Explanations and definitions are brief and are limited to the meanings that are relevant to the context in which the words or phrases are used in this book. For more complete definitions of these and other geological terms, consult the *Glossary of Geology,* published by the American Geological Institute.

Aquifer Permeable rocks that will produce fresh water from wells drilled into them.

Asthenosphere That portion of the earth's mantle immediately interior to the lithosphere. The asthenosphere is much less rigid than the lithosphere. Transition from lithosphere to asthenosphere occurs generally around a depth of 100 kilometers.

Bar (1) A unit of pressure more or less equal to atmospheric pressure near sea level. (2) A positive topographic feature formed by sand or gravel deposits in rivers, near beaches, etc.

Bay A general term for an indentation in a shoreline. Smaller than a gulf, larger than a cove.

Benthonic Bottom-dwelling.

Bentonite A clay layer presumably formed by alteration of a volcanic ash.

Bioclastic sedimentary particles Broken fragments of skeletal remains, such as molluscs, corals, and coralline algae. Common components in limestones.

Bird's-eye fabric A common pattern in supratidal carbonates in which former gas bubbles become preserved as open or calcite-filled cavities. Cavities are typically 2 to 5 millimeters in diameter and may constitute 50% of the rock.

Bottom stability   The degree to which the sediment-water interface is immobile. Many marine organisms have life styles adapted to stability or lack of stability of the substrate.

Bryozoa   A phylum of attached and incrusting marine invertebrates. Locally important contributors to bioclastic limestones.

Burrow-mottled fabric   The burrow of an organism typically leaves features discernible in the sediment. When a number of burrows are superimposed, the identity of the individual burrow may be lost but the characteristic mottled appearance is easily recognizable on outcrop, in slab, or in thin section.

Caliche   Calcium carbonate precipitated by evaporative processes in or near a soil zone.

Chicken-wire anhydrite   A common recrystallization fabric in which impurities are concentrated along surfaces bounding large aggregates of white anhydrite. In slab, the zones of impurities produce a resemblance of wire mesh.

Comminution   Particle-size reduction by physical processes.

Contemporaneous   Existing or occurring at the same time.

Coquina   A clean sediment composed largely of shells of marine invertebrates, usually molluscs.

Creep   Slow, more or less continuous, downslope movement.

Cumulative probability scale   A scale on a graph so devised that a normal distribution plots as a straight line.

Curie point   The temperature at which a cooling mineral acquires permanent magnetic properties that record the surrounding magnetic-field orientation and strength at the time of cooling. Above the Curie point, the magnetic properties of a mineral change as the surrounding field changes. Below the Curie point, magnetic properties of the mineral do not change.

Cut-and-fill structures   The sedimentary record of local alternation between erosion and sedimentation. Applied particularly to sandbars with complicated internal structure.

Cyclothem   A series of beds deposited during a single sedimentary cycle of the type that prevailed during the Pennsylvanian period.

Deflation surface   A layer of coarse sediment derived from the sediment below after wind has transported the fine fraction.

Density current   Any current that flows downslope because it is more dense than the fluid occupying the lower elevation.

Diachronous   Differing in geologic age. Thus, a diachronous sedimentary unit is one that crosses time lines, be they real or hypothetical.

Diagenesis   Those physical and chemical changes that occur in a sediment after initial deposition but before metamorphism.

Ebb tide   The outgoing portion of the tidal cycle.

Ecology   The study of the mutual relationships between organisms and their environments.

Effective wave base The depth at which movement of sediment by wave action ceases to be a significant geologic process. Generally on the order of 30 to 100 meters, it varies with wave intensity and grain size of the sediment.

Emergence The upward movement of a point within the sediment relative to sea level. Sediment may be lifted (tectonic emergence) or sea level may be lowered (eustatic emergence) to produce the same relative motion.

Environmental datum Any lithology that must have formed under specific conditions and that in turn provide an indication of depositional conditions of associated lithologies.

Estuary A drowned river valley now under the influence of coastal tidal action.

Eustatic Pertaining to simultaneous, worldwide changes in sea level. Eustatic sea-level changes commonly result from a change in the volume of continental glaciers.

Evaporites Sediments precipitated from aqueous solution as a result of the evaporation of the water. Anhydrite ($CaSO_4$) and halite ($NaCl$) are common evaporite minerals.

Facies The sedimentological record of a depositional environment.

Factor analysis A data analysis technique that identifies natural groupings of parameters within a data matrix, thereby reducing the number of terms needed to describe variation among the samples.

Fauna The animals of any given environment or geologic age.

Flexural rigidity A measure of the stiffness of an elastic plate under the influence of a bending couple.

Flexure zone In Gulf Coast geology, that line downdip from which a formation thickens rapidly.

Flood tide The incoming portion of the tidal cycle.

Flora The plants of any given environment or geologic age.

Flume A laboratory apparatus constituting a long, straight channel in which sediment transport studies are carried out.

Fluvial Produced by rivers.

Gene pool Genes are the fundamental units governing hereditary characters. The gene pool is the sum of all genes within an interbreeding population.

Glauconite A green clay mineral commonly found in marine sandstones. It usually appears to be authigenic and so is potentially useful for radiometric dating, paleosalinity estimations, etc.

Hemipelagic Deep sea sediments that contain abundant clay derived from the continents.

Horizon A surface or distinctive layer within layered sedimentary rocks.

Hypersaline Having a salinity substantially greater than that of sea water.

In situ In the situation in which it was originally formed or deposited, as opposed to being transported from one situation to its final resting place. The term is usually italicized.

Intertidal   The benthic zone near sea level that lies between normal high and low tides.

Intraclast   A large, composite sedimentary particle formed contemporaneously in nearby sedimentary environments, as in mud chips on a supratidal flat.

Isopach map   A contour map depicting the thickness of a sedimentary unit.

Lagoon   The relatively quiet-water environment behind a barrier beach or reef.

Lithosphere   The relatively rigid outer shell of the earth comprising the crust and upper mantle. It averages approximately 100 kilometers in thickness.

Longshore current   A persistent nearshore current moving essentially parallel to the coast. Usually generated by waves breaking obliquely to the shoreline.

Mantle   The layer of the earth between crust and core.

Maxwell solid   A material that behaves elastically under the stress of "short" duration and plastically under the stress of "long" duration. "Silly putty" is a Maxwell solid.

Melobesia   Delicate red algae that live on the surface of marine grasses. A common source of high-magnesium calcite mud in the Recent epoch.

Net sand map   A contour map indicating the total thickness of sand units within a stratigraphic interval.

Niche   That subdivision of an environment occupied by a single species. Some attributes of the environment may be shared among several niches, but each niche is a unique combination of attributes.

Oncolites   Pebble to cobble-sized, laminated, carbonate bodies constructed by algae.

Oolites   Spherical, carbonate sand particles composed of concentric laminae of microscopic aragonite needles.

Paleogeography   A geography that existed at some specific time in the past. Commonly reconstructed as the physiographic context within which sedimentation of a specific unit is believed to have occurred.

Pellet   Silt or sand-sized aggregation of carbonate mud. Generally fecal in origin, it is a common grain type in nonskeletal carbonate sediments.

Penecontemporaneous   At nearly the same time.

Permeability   A measure of the ease with which a fluid can be passed through the pore space of a rock.

Physiography   The spatial arrangement of the upper surface of the lithosphere and the conditions existing upon it.

Pisolites   Pebble-sized, laminated, carbonate particles typically formed on subaerial exposure surfaces on carbonate sediments.

Plankton   Floating organisms.

Point bar   The sandbar deposited on the inside of a meander loop.

Porosity   The portion of the total volume of a rock that is not occupied by solid mineral matter.

Progradation   The seaward advance of the shoreline resulting from sediment deposition.

Provenance   The terrane or parent rock from which a sediment was derived.

Red Beds   Red sedimentary rocks; usually sand stones and shales of fluvial origin.

Reef   A wave-resistant structure constructed by sedentary carbonate-secreting organisms. Note that this definition has straightforward application in the Recent epoch but must be inferred from related sediments in Ancient sequences.

Reef apron   The shallow-water zone of reef-derived sand lagoonward of the reef flat.

Reef flat   The shallow-water zone of cobble to boulder-sized rubble immediately behind a livingcoral barrier reef.

Regression   A nonspecific term meaning that the shoreline moved toward the center of the basin. It may be caused by the prograding of sediments or by tectonic or eustatic emergence.

Relaxation time   A measure of the rate at which stress is released by plastic deformation of a material.

Relict sediment   Surface sediment representative of an environment that previously occupied the area under discussion. Not deposited under present conditions.

Sabkha   Supratidal, evaporative, deflation surfaces on the Arabian peninsula. Site of extensive evaporite sedimentation.

Sessile benthos   Bottom-dwelling aquatic organisms that live permanently attached to the bottom.

Shelf margin   The break in slope separating the continental shelf from the continental margin.

Slump   The downslope movement of a mass of unconsolidated sediment as a coherent unit. Typically involves major dislocation along a few preferred planes of weakness.

Soft sand   Carbonate mud formed into soft pellets. In the depositional environment, the soft particles react to wave and current action as though they were sand grains. In the lab, they are washed through the sand sieve like any other mud.

Sole mark   The marking recorded on the bottom of a sandstone bed where the sand has filled in depressions and tool marks in the underlying shale bed. They are best preserved by the sandstone because the shale tends to flake into small pieces on outcrop, whereas the sandstone commonly breaks out as large blocks.

Solstice   The points in the ecliptic at which the perpendicular to the sun makes the largest angle with the plane of the equator. As seen from the northern hemisphere, the summer solstice takes place when the sun is "as far north as it comes" and the winter solstice occurs when the sun is "as far south as it goes."

Stromatolites   Planar to head-like laminated structures constructed by algal entrapment of sediment and usually found in the high intertidal and low supratidal zones. Most commonly preserved in carbonate sediments.

Subaerial Formed or existing at or near a sediment surface significantly above sea level.

Submergence The downward movement of a point within the sediment relative to sea level. The opposite of emergence.

Substrate The sediment or rock upon which benthonic organisms reside. The distribution of rock, sand, and soft-mud substrates is a major factor controlling local distribution of marine benthonic organisms.

Subtidal The shallow-marine environment lying below normal low tide.

Supratidal The coastal environment lying above normal high tide.

Swale A marshy depression in otherwise generally level land.

Talus High-angle debris deposits at the base of a pronounced physiographic feature.

Tectonic Pertaining to structural deformation of the earth's lithosphere.

Terrigenous clastic sediment Sediment derived from the land by the weathering of preexisting rocks and transported by wind or water to the sedimentary environment in which we now find it. It contrasts with carbonates, evaporites, and some clays, all of which may form *in situ* by chemical or biochemical precipitation from aqueous solution.

Thallasia A shallow-marine grass common in areas of Recent carbonate sedimentation.

Time line A line in a stratigraphic cross section so constructed that all points represent the same instant in geologic time. A time line is the vertical tracing of a time surface, but this latter word is seldom used.

Time transgressive Said of a lithologic unit that is older in one area than it is in another area. For example, a prograding beach deposit is youngest toward the basin center and oldest toward the land. The lithologic unit crosses (that is, transgresses) time lines.

Tool mark The structure produced on a bedding-plane surface where transported objects or turbulent currents leave a mark indicating the direction of transport.

Transform fault A strike-slip fault bounded at each end by an area of crustal spreading that tends to be more or less perpendicular to the strike-slip fault.

Transgression A nonspecific term meaning that the shoreline moved away from the center of the basin. It may be caused by tectonic or eustatic submergence.

Turbidite A graded bed of sand and mud that presumably was deposited by a turbidity current.

Turbidity current A density current that owes its density to suspended mineral matter.

Index

A

Absolute time in stratigraphic record, 125–133
  geologic time scale, 130–132
  radioactive isotopes and, 125
  radiometric dating and, 125, 127–129, 132
Aggradation, 71–72
Allen, J. R. L., 185, 186
Andros Island, the Bahamas, carbonate sedimentation of, 226, 239
Antarctic Ice Cap, 6
Appalachians, general stratigraphy, 76–83
  Catskill "Delta" of, 80–81, 168–171
  Helderberg Group of, 277–284
  Reedsville and Oswego formations of, 204–208
  Shawangunk and Tuscarora formations of, 162–168
  Taconic Sequences of, 309–315

B

Bahamas, carbonate sedimentation of, 225, 226, 234–236, 245, 246, 255, 258

Baja California, subtidal-supratidal transition in, 269, 276
Ball, M. M., 251
Bank-margin carbonate sedimentation, 231–239
Barbados, West Indies, Pleistocene coral reefs of, 260–264
Bar-finger sands, 179–180
Barrier-bar sedimentation, 202, 204
Barrier island, Galveston, Texas, clastic sedimentation in, 187–189, 202
Basement, 91–92
Basin filling:
  isostasy and, 50–52
  by vertical accumulation of bottomset beds, 200
Basins:
  interior lowlands of United States, 76
  ocean, Pleistocene history of, 322–329
  See also Shelf-to-basin transition at continental margins
Bed load, 18
Bedding planes, 42–43
Belts, mobile, 76–82
Bernard, H. A., 137, 147, 148–151
Bioclastic debris, 24, 25
Biostratigraphy, 96–113

Biostratigraphy (cont.)
  evolution, overview of, 98–105
  fossils, 98
  niche and biogeography, complications of, 105–110
  nomenclature, 110–113
  rock units, 97–98
  time-rock units, 97–98
  time units, 97–98
Bioturbation, 39
Bird, J. M., 314
Bird, John A., 79
Blatt, H., 30
Bloom, A. L., 43, 56–58, 247
Bonaire, Netherlands Antilles, subtidal-supratidal transition in, 269
Bottomset sediment, 175
  basin filling by vertical accumulation of, 200
  in carbonate sedimentation, 252–254
  transition to meandering stream, 204–208
Braided stream sedimentation:
  Ancient example, 162–168
  composition of sediment, variation in, 156–157
  grain size, variation in, 156
  versus meandering-stream sedimentation, 159
  within mountainous region, 137–143
  beyond source area of sediments, 143–147
Brazos River, clastic sedimentation in, 147–151
British Honduras barrier reefs, 226, 230, 231–232, 234, 239–242, 247–251, 253
Buchanan, H., 234, 251
Burrow-mottling, 195
Butler, G. P., 272
Buzzalini, A. D., 275

C

Calcium sulfate precipitation, 272–274
California, southern:
  fault system in, 64

California (cont.)
  submarine canyons and basins of, 293–298
Campeche Bank of southeastern Mexico, 236–238
Canyons, submarine, shelf-to-basin transition at, 293–298, 306–307
Carbonate sedimentation:
  of intertidal and supratidal shelf interior, 268–291
    Ancient examples, 277–289
    basic model for, 276–277
    Recent, 268–276
    subtidal to supratidal transitions, 268–275
  of shelf margin and subtidal shelf interior, 224–267
    Ancient example, 259–264
    areal distribution of contemporaneous sedimentary facies, 251–252
    basic model for, 251–259
    bottomset, foreset, and topset beds in, 252–253
    progradation in, 254
    Recent, 226–251
    sea level and, 254–259
    topographic control of, 247–249
    vertical accumulation in, 253–254
Carbon-14 dating, 126, 127, 250
Catskill "Delta," Devonian of New York State, 168–171
Cement, 17–18
Chappell, J., 67, 71, 264, 267
Chenier Plain of southwestern Louisiana, clastic sedimentation in, 189–192, 202
Chert, 20, 22
Christensen, M. N., 67, 68
Chuber, S., 349, 350, 351
Clastic province, Texas-Louisiana Gulf Coast, 82–86
Clastic sedimentation:
  in coastal environments, 173–223
    Ancient examples, 204–220
    barrier island, Galveston, Texas, 187–189
    basic model for, 194–204

Clastic sedimentation (cont.)
  Chenier Plain of southwestern Louisiana, 189–192
  coarsening-upward cycle, 194–197
  continental shelf off Gulf of Mexico, 193
  foreset progradation, 199–200
  isostatic considerations, 200–201
  lagoons and topset deposits, 197–199
  Mississippi delta, 175–185, 202
  Niger delta, 175–185, 202
  Recent, 174–194
  sea level and, 201–204
  tidal-flats, 193–194
  in stream environments, 137–172
    Ancient examples, 162–169
    basic model for, 155–162
    Brazos River, 147–151
    Donjek River, 137–143
    graded stream response to dynamic events, 160–161
    grain size and sorting variation, 156
    isostatic considerations, 162
    Platte River, 143–147
    primary structures, 157–159
    progradation by meandering-stream deposits, 161–162
    Recent, 137–155
    Red River, 151–155
    sediment composition variation, 156–157
  time-stratigraphic correlation and, 122
Clay minerals, 18
Clear-water sedimentation, 171
  in epeiric seas, 337–342
Climatic fluctuations:
  Pleistocene, Milankovitch hypothesis concerning, 329–332
  time-stratigraphic correlation based on, 119–121
Clinton formation, 162, 163, 165, 168
Coarsening-upward cycle, 194–197
Coastal clastic wedge, 196–204
Coastal environments, clastic sedimentation in, 173–223
  Ancient examples, 204–220
  basic model for, 194–204

Coastal environments (cont.)
  coarsening-upward cycle, 194–197
  foreset progradation, 199–200
  isostatic considerations, 200–201
  lagoons and topset deposits, 197–199
  sea level and, 201–204
  Recent, 174–194
    barrier island, Galveston, Texas, 187–189
    Chenier Plain of southwestern Louisiana, 189–192
    continental shelf off Gulf of Mexico, 193
    Mississippi delta, 175–185, 202
    Niger delta, 185–187
    tidal-flats, 193–194
Coastal plain of continental margin, 82–88
Composite set, 33
Conglomerate, 18
Contacts, 91–92
Continental margin:
  coastal plain of North America, 82–88
  shelf-to-basin transition at, 292–318
    Ancient example, 309–314
    basic model for, 306–308
    east coast of United States and Canada, 298–306
    Recent, 292–306
    submarine canyons and basins off southern California, 293–298
Coral reefs, 226, 231–239, 251
  Pleistocene of Barbados, West Indies, 260–264
  Pleistocene sea-level history as recorded by, 325–327
  progradation, 254
Cox, A., 115, 117
Craton, stable, 74–76
Cross-stratification, 33–36, 195
Cyclic sedimentation:
  in Paleozoic epeiric seas of central North America, 334–343
  Irwin-Shaw Kickback model, 337–341
  Moore model, 335–336
  multimodel approach, 344–353
  Permian of West Texas, 344–353
  Pleistocene epoch, 321–323

Cyclic sedimentation (cont.)
  early interest in, 322
  Milankovitch hypothesis, 329–332
  ocean basins, 322–329
Cyclothems of eastern Kansas, 334–336

D

Deep-sea cores, 120, 322–329
Deep-sea trenches, 62–64, 78
Deep-water evaporite sedimentation, 275–276
Deltaic sedimentation, 174–187
Desiccation features, 39
Dewey, J. F., 79, 314
Diversity, physiological, 99–102
Doeglas, D. J., 143
Dolomite, 18
  supertidal, 268–275
Domes, 76
Donjek River, clastic sedimentation in, 137–143
Dunham, R. J., 24, 26, 27, 288
Dynamics in stratigraphic record, 48–73
  eustatic sea level fluctuations, 69–71
  global, of lithosphere, 60–69
  isostasy, 49–60
    basin filling and, 50–52
    flexural rigidity of lithosphere, 52–56
    relaxation times and viscosity of mantle, 52
    response to single eustatic sea-level rise, 58–59
    response to small loads, 56–58
  transgression and regression, 71–72

E

Earthquakes, 65
Eicher, D. L., 125
Electrical properties of sedimentary rocks, 43–44
Emergence, 71–72
Emery, K. O., 297, 298, 300, 301, 303, 304, 305, 308
Epeiric seas of central North America, Paleozoic:

Epeiric seas (cont.)
  cyclic sedimentation in, 334–343
Ericson, D. B., 303, 323, 324
Eugeosyncline sediments, 79–80, 82
Eustatic sea-level fluctuations, 69–71
Evaporite sedimentation, 275–277
Evolution, biostratigraphic overview of, 98–105
Ewing, M., 301
Exogeosynclinal sediments, 80–82

F

Faul, H., 125
Fault system, southern California, 64
Fining-upward unit, 158
Fisk, H. N., 180, 183
Florida carbonate sediments, 87–88, 226, 232, 233, 234, 239, 242–245, 247, 249, 250, 252, 254, 255
Florida Bay, Florida Everglades and, subtidal-supratidal transition, 269–270, 276
Flume studies, 31–32
Flysch sedimentation, 82
Folk, R. L., 24, 25, 30
Foreset sediments, 175, 199–200, 252–253
Formation, defined, 91–92
Fossil fuels, 6–7
Fossils, 98

G

Gallup sandstone, in Bisti oil field, New Mexico, 218–219
Gebelein, C. D., 271, 272
Geologic time scale, 130–132
Gilbert, C. M., 20, 22
Ginsburg, R. N., 88, 233, 242, 243, 245, 271
Glacial retreat, 56
Goreau, T. F., 227, 231, 250
Gorsline, E. S., 297, 298, 308
Gould, H. R., 175, 190, 191
Gradational contacts, 42
Graded bed, 30
Graded stream, 159–162

Grains, 17–18
Graywacke, 20, 22
Great Bahama Bank carbonate sediments, 225, 226, 234–236, 245, 255, 258
Great Barrier Reef carbonate sediments, 227
Gregory, J. L., 209–215
Group, defined, 91–92
Guadalupe Mountains:
  sedimentological observations in, 286–289
  shelf-to-basin transition in, 344–346
Gulf of Mexico clastic sediments, 175–185, 187–193, 208–215

H

Hand, B. M., 297
Harms, J. C., 151–155
Hays, J. D., 119
Heezen, B. C., 301–303
Helderberg Group, Devonian of New York, 277–284
Hemipelagic sediment, 301
Hodgson, W. D., 189
Hoffman, P., 272
Hoffmeister, J. E., 249
Horowitz, B. H., 204–208
Hurley, P. M., 125

I

Igneous rocks, age brackets on sedimentary sequences from radiometric dates on, 129, 132
Imbrie, J., 27, 28, 234, 246, 251, 327, 328
Intertidal shelf interior, carbonate and evaporite sedimentation on, 269–291
  Ancient examples, 277–279
  basic model for, 276–277
  Recent, 268–276
  subtidal to supratidal transitions, 268–275
Intraclasts, 24, 25
Irwin, M. L., 337–340
Isacks, B., 60

Isopach map, 213, 214
Isostasy:
  basin filling and, 50–52
  coastal clastic wedges and, 200–201
  flexural rigidity of lithosphere, 52–56
  relaxation times and viscosity of mantle, 52
  response to single eustatic sea-level rise, 58–59
  response to small loads, 56–58

J

Judson, S., 49

K

Kansas cyclothem, 334–336, 342
Kay, M., 79
King, Phillip B., 80, 81, 83
Kipp, N. G., 327, 328
Klein, G. deV., 168, 193, 194
Kolb, C. R., 181, 183
Konishi, A., 67, 71
Krumbein, W. C., 71
Krynine, P. D., 19, 20

L

La Jolla canyon, shore-to-basin transition at, 293–298
Lagoons, 197–199
Laminae, 30
Laminar flow, 31
Land, L., 250
Laporte, L. F., 278–284
LeBlanc, R. J., 189
Leet, L. D., 49
Lehner, P., 85
Limestones, 18, 224–291
  classification of, 24–27
  in clastic marine sequences, 122
  Portland Point, 171
  Tully, 171
Lithosphere:
  flexural rigidity of, 52–56
  global dynamics of, 60–69

Logan, B. W., 229, 237, 271, 275
Longitudinal bars, clastic sedimentation on, 139–147, 156
Longshore sedimentation, 187–189
Louisiana, clastic sedimentation in Chenier Plain of, 189–192
Lowman, S. W., 89

M

McAlester, A. L., 100
McBride, E. F., 22
McCave, I. N., 168–171
McConnell, R. K., Jr., 52, 53
McFarlan, E., Jr., 183, 190, 191
McKee, E. D., 30, 33
Magnetic stratigraphy, 114–119
Manlius Formation, 278–284
Mantle, viscosity of, 52, 54, 55
Martinsburg formation, 163, 165
Matrix, 17
Matsuda, T. K., 69
Matthews, D. H., 62
Matthews, R. K., 67, 242, 248, 256
Maxwell, W. G. H., 227
Meander loop, 148
Meandering stream sedimentation, 147–149, 161
    Ancient example, 162, 168–171
    bottomset transition to, 204–208
Member, defined, 91–92
Merriam, D. F., 334
Mesolella, K. J., 67, 71, 261, 262, 264, 327, 330, 332
Micrite, 24, 25
Middle Permian of west Texas, 344–352
Middleton, G. V., 36
Mid-ocean ridges, 62–63
Milankovitch hypothesis, 329–332
Miogeosynclinal sediments, 79–82
Mississippi delta clastic sedimentation, 175–185, 202
Mobile belts, 76–82
Molasse sedimentation, 82
Moore, D. G., 295–296, 301, 308
Moore, R. C., 334, 335

Multer, H. G., 249
Murray, R. C., 30

N

New Jersey, Shawangunk Conglomerate and Tuscarora Quartzite, Silurian of, 162–168
New Mexico, oil trap in coastal clastic sands of the subsurface Upper Cretaceous of, 214, 216–220
New York:
    Catskill "Delta," Devonian of, 168–171
    Helderberg Group, Devonian of, 277–284
    Shawangunk Conglomerate and Tuscarora Quartzite, Silurian of, 162–168
    Taconic Sequence, Cambro-Ordovician, 309–314
Newell, N. D., 345
Niger delta clastic sedimentation, 185–187
North America, central:
    coastal plain and continental margin, 82–88
    cyclic sedimentation in Paleozoic epeiric seas of, 334–343
    geologic framework of, 74–88
    mobile belts, 76–82
    stable craton, 74–76

O

Ocean basins, Pleistocene history of, 322–329
Ocean-floor spreading, 5–6
Oceanic sediments, magnetic stratigraphy and, 116–119
Oolite sands, 24, 25, 231–239, 251, 257
Oopelmicrite, 24, 25
Oopelmicrudites, 24, 25
Opdyke, N. D., 118, 119
Orthoquartzite, 20, 22
Overbank deposits, 150, 157, 170
Oxygen-18, 121, 325, 331

P

Paleogeography, recognition of, 92
Paleozoic epeiric seas of central North America, cyclic sedimentation in, 334–343
Pellets, 24, 25
Pelsparrudites, 24, 25
Pennsylvania:
    Reedsville shale, transition to Oswego sandstone, Upper Ordovician of, 204–208
    Shawangunk Conglomerate and Tuscarora Quartzite, Silurian of, 162–168
Permian of West Texas, 285–289
Persian Gulf:
    carbonate sedimentation in, 227, 229, 238–239, 245, 246
    subtidal to supratidal transition in, 272–274, 276
Pettijohn, F. J., 21, 22
Phillips, J. D., 119
Physical stratigraphy, 90–95
Plafker, G., 65, 66
Plate tectonics, mobile belts and, 79
Platte River clastic sedimentation, 143–147
Pleistocene epoch, 321–333
    early interest in, 322
    Milankovitch hypothesis, 329–332
    ocean basins, 322–329
Plumley, W. J., 156, 157
Point-bar sedimentation, 148–155, 157
Portland Point limestone, 171
Potassium-argon dates, 127, 128
Primary sedimentary structures, 30–39
Progradation:
    in carbonate sedimentation, 254
    defined, 71, 72
    foreset, 199–200
    in intertidal-supratidal carbonates and evaporites, 276–277
    by meandering-stream deposits, 161–162
Protactinium-231 dating, 126, 127
Purdy, E. G., 27, 28, 67, 234, 246, 247, 248
Pusey, W. C., 349, 350, 351

Q

Quartz, 18, 20, 22
Quaternary period, 321–333

R

Radioactive isotopes, 125
Radiometric dating, 125–133
    chief methods of, 126
    of contemporaneous volcanic sediment, 128–129
    direct, 127–128
    geologic time scale, 130–132
    of igneous rocks, age brackets on sedimentary sequences from, 129, 132
Red River clastic sedimentation, 151–155
Reedsville shale, transition to Oswego sandstone, Upper Ordovician of Central Pennsylvania, 204–208
Regression, 71–72, 161
Relaxation times, viscosity of mantle and, 52
Resistivity, 43–44
Reverse-graded bed, 30
Rezak, B. W., 271
Rhoads, D. C., 39
Rickard, L. V., 278, 279
Ridges, mid-ocean, 62–63
Rock units, 91–92, 97–98
Rudite, 24, 25
Rust, B. R., 137–142
Rutile, 18

S

Sabins, F. F., Jr., 216–220
St. Bernard delta, 182–183
San Diego trough, shore-to-basin transition at, 293–298
Sand dunes, eolian, 198
Sandbars, clastic sedimentation on:
    longitudinal, 139–147, 156
    point, 148–155, 157
    transverse, 143–147, 156

Sands:
  bar-finger, 179–180
  oolite, 24, 25, 231–239, 251, 257
Sandstones:
  classification of, 19–23
  composition of, 18
  cross-stratification in, 33, 34
  in hypothetical stratigraphic example,
    9–13
  textural maturity of, 18–19
Scholl, D. W., 270
Scruton, P. C., 178, 182
Sea-floor spreading, 62
Sea-level fluctuations:
  basin-margin sedimentation and, 307–
    308
  carbonate sedimentation response to,
    256–259, 263–264, 277
  coastal clastic wedges and, 201–204
  eustatic, 69–71
  isostatic response to, 56–60
  as selection pressure on shallow-marine
    benthonic invertebrates, 103–105
Sediment accumulation, geologic frame-
    work of, 74–89
  coastal plain and continental margin,
    82–88
  mobile belts, 76–82
  stable craton, 74–76
Sedimentation rates, 13–14
Sediments and sedimentary rocks:
  basic elements of classification of, 17–28
  dating of, *see* Radiometric dating
  electrical properties of, 43–44
  importance of, 1–8
  of mobile belts, 79–82
  physical characteristics of units of, 90–
    95
  primary structures, 30–37
  sequential relationships in, 39–43
  size distribution, 28–30
  stratification, 30
  transgressive-regressive events, 121–122
Shale, 18
  in hypothetical stratigraphic example,
    9–13

Shallow-marine benthonic communities:
  crossing geographic barriers, 108–109
  ecological limits of, 102
  sea-level fluctuation and, 103–105
  survival potential of, 103
  time correlation and, 109–110
  time-transgressive first appearance in re-
    sponse to changing climate, 108
  variability within, 100–102
  variation in depositional environments
    and, 105–108
Shark Bay, Western Australia:
  carbonate sedimentation in, 229
  subtidal-supratidal transition in, 269, 275
Shaw, A. B., 109–110, 337
Shawangunk Conglomerate, Silurian of
    Pennsylvania, New Jersey, and New
    York, 162–168
Shearman, D. J., 276
Shelf interior, carbonate sedimentation of:
  intertidal and supratidal, 268–291
    Ancient examples, 277–289
    basic model for, 276–277
    Recent, 268–276
    subtidal to supratidal transitions, 268–
      275
  subtidal:
    Ancient example, 259–264
    areal distribution of contemporaneous
      sedimentary facies, 251–252
    basic model for, 251–259
    bottomset, foreset, and topset beds,
      252–253
    progradation, 254
    Recent, 239–247
    sea level and, 254–259
    to supratidal transitions, 268–275
    vertical accumulation, 253–254
Shelf margin, carbonate sedimentation of:
  Ancient example, 259–264
  basic model, 251–259
    areal distribution of contemporaneous
      sedimentary facies, 251–252
    bottomset, foreset, and topset beds,
      252–253
    progradation, 254

Shelf margin (cont.)
    sea level and, 254–259
    vertical accumulation, 253–254
    Recent, 231–239
Shelf-to-basin transition at continental margins, 292–318, 344–346
    Ancient example, 309–314
    basic model for, 306–308
    Recent, 292–306
        east coast of United States and Canada, 298–306
        submarine canyons and basins off southern California, 293–298
Sheppard, F. P., 70, 294, 296
Shepps, V. C., 168
Shinn, E. A., 271
Siltstone, 18
Silver, B. A., 348, 349
Simons, D. B., 31
Skewness, 30
Sloss, L. L., 71
Smith, N. D., 137, 143–147, 162–168
Sparite, 24
Spontaneous potential, 43–44
Stable craton, 74–76
Stockman, K. W., 243, 250
Stream environments, clastic sedimentation in, 137–172
    Ancient examples, 162–169
    basic model for, 155–162
        graded stream response to dynamic events, 160–161
        grain size and sorting variation, 156
        isostatic considerations, 162
        primary structures, 157–159
        progradation by meandering-stream deposits, 161–162
    Recent, 137–155
        Brazos River, 147–151
        Donjek River, 137–143
        Platte River, 143–147
        Red River, 151–155
        sediment composition variation, 156–157
Strike-slip fault system, 64
Stuiver, M., 270

Submergence, defined, 71, 72
Subtidal shelf-interior, carbonate sedimentation of:
    Ancient example, 259–264
    basic model for, 251–259
        areal distribution of contemporaneous sedimentary facies, 251–252
        bottomset, foreset, and topset beds, 252–253
        progradation, 254
        sea level and, 254–259
        vertical accumulation, 253–254
    Recent, 239–247
    subtidal to supratidal transitions, 268–275
Supratidal shelf interior, carbonate and evaporite sedimentation on, 268–291
    Ancient examples, 277–289
    basic model for, 276–277
    Recent, 268–276
    subtidal to supratidal transitions, 268–275
Suspended load, 18
Swinchatt, J. P., 227

T

Taconic Sequence, Cambro-Ordovician, Eastern New York, 309–314
Tectonic basins and highs, 62–64
Tectonic deformation, rates of, 64–69
Texas:
    Galveston Island, clastic sedimentation of, 187–189, 202
    Permian strata of, 285–289
    southwest, coastal clastic sediments in subsurface Lower Oligocene of, 208–215
Texas-Louisiana Gulf Coast clastic province, 82–86
Thorium-230 dating, 126, 127
Thorndike, E. M., 301
Tidal-flat sedimentation, 193–196
Time, absolute, in stratigraphic record, 125–133

Time (cont.)
  geologic time scale, 130–132
  radioactive isotopes and, 125
  radiometric dating and, 125, 127–129,
    132
Time units, 97–98
Time-rock units, 97–98
  fossils and, 98
  nomenclature, 110–113
Time-stratigraphic correlation, based on
    physical events of short duration,
    114–124
  climatic fluctuations, 119–121
  magnetic stratigraphy, 114–119
  rapid depositional events, 123
  transgressive-regressive events, 121–122
Todd, R. G., 348, 349
Tool marks, 38–39
Topset beds, in carbonate sedimentation,
    252–253
Topset deposits, 197–199
Tourmaline, 18
Transform faults, 64
Transgression, 71–72
Trangressive-regressive events, time-strati-
    graphic correlation based on, 121–
    122
Transverse bars, clastic sedimentation on,
    143–147, 156
Trechmann, C. T., 261
Trenches, deep-sea, 62–64
Trough cross-stratification, 146, 154
Tully limestone, 171
Turbidites, 37
Turbulent flow, 31
Tuscarora Quartzite, Silurian of Pennsyl-
    vania, New Jersey, and New York,
    162–168

U

Unconformities, 40–41, 91–92
Uplifts, 76
Uranium series dating, 126, 127

V

Van Lopik, J. R., 181, 183
Van Straaten, L. M. J. U., 194
Veeh, H. H., 67, 71, 264, 327
Vicksburg formation, sedimentological ob-
    servations in, 209–215
Vine, F. G., 62
Visher, G. S., 36
Volcanic ashfalls, time-stratigraphic cor-
    relation and, 123
Volcanic rocks, magnetic stratigraphy and,
    115, 116, 118
Volcanic sediments, radiometric dating of,
    128–129
Von Rad, U., 295, 296

W

Walcott, R. I., 53, 54, 55, 58
Walker, R. G., 37
Wanless, H. R., 247
Warm-cool foraminifera stratigraphy of
    deep-sea cores, 120
Williams, P. F., 137–142
Wilson, J. L., 344
Winland, H. D., 245
Winnowing, 192
Woolin, G., 323, 324

Z

Zen, E., 311, 312, 313
Zircon, 18